*Christiane Gottschalk,
Judy Ann Libra, and
Adrian Saupe*
**Ozonation of Water and
Waste Water**

Related Titles

Wiesmann, U., Choi, I.S., Dombrowski, E.-M.
Biological Wastewater Treatment
Fundamentals, Microbiology, Industrial Process Integration

391 pages with 135 figures and 61 tables
2007
Hardcover
ISBN: 978-3-527-31219-1

Reemtsma, T., Jekel, M. (eds.)
Organic Pollutants in the Water Cycle
Properties, Occurrence, Analysis and Environmental Relevance of Polar Compounds

368 pages with 88 figures and 65 tables
2006
Hardcover
ISBN: 978-3-527-31297-9

Ganoulis, J.G.
Risk Analysis of Water Pollution

Second, revised and expanded edition
2009
ISBN: 978-3-527-32173-5

Dr. Catherine Gonzalez (Editor), Dr. Richard Greenwood (Editor), Prof. Philippe Quevauviller (Co-Editor)
Rapid Chemical and Biological Techniques for Water Monitoring

440 pages
June 2009
Hardcover
ISBN: 978-0-470-05811-4

Andrzej Benedykt Koltuniewicz, Enrico Drioli
Membranes in Clean Technologies: Theory and Practice, 2 Volume Set

909 pages
July 2008
Hardcover
ISBN: 978-3-527-32007-3

Christiane Gottschalk, Judy Ann Libra, and Adrian Saupe

Ozonation of Water and Waste Water

A Practical Guide to Understanding Ozone and its Applications

Second, completely revised and updated edition

WILEY-VCH Verlag GmbH & Co. KGaA

The Authors

Dr. Christiane Gottschalk
ASTEX GmbH
Wattstr. 11
13355 Berlin
Germany

Dr. Judy Ann Libra
acatech-
Deutsche Akademie der
Technikwissenschaften
Mauerstr. 79
10117 Berlin
Germany

Dr. Adrian Saupe
Projektträger Jülich
Zimmerstr. 26–27
10969 Berlin
Germany

Further Contributors

Martin Jekel, Professor of Water Quality Control at the Technical University of Berlin, has been involved in oxidation research since 1976, specializing on ozone/biological activated carbon, microflocculation mechanisms of ozone and advanced oxidation processes for water and wastewaters. He contributed the original manuscript for Chapter 3.3.

Anja Kornmüller, Dr.-Ing. in environmental engineering from the Technical University of Berlin. Currently she is the Head of Development Veolia Water Solutions & Technologies Germany, Austria and Switzerland and specialized in advanced oxidation processes in seawater and drinking water. She contributed Section 9.1.2 and 9.3.

Dieter Lompe holds a PhD from the Technical University of Berlin. He has worked many years in research and full scale projects with three phase advanced oxidation processes. As Professor at the University of Applied Sciences Bremerhaven (Germany) he is currently working on water technologies and solid waste treatment processes. He coauthored Section 9.1.2.

1. Edition 2000
2. completely revised and updated edition 2010

All books published by Wiley-VCH are carefully produced. Nevertheless, authors, editors, and publisher do not warrant the information contained in these books, including this book, to be free of errors. Readers are advised to keep in mind that statements, data, illustrations, procedural details or other items may inadvertently be inaccurate.

Library of Congress Card No.:
applied for

British Library Cataloguing-in-Publication Data
A catalogue record for this book is available from the British Library.

Bibliographic information published by the Deutsche Nationalbibliothek
The Deutsche Nationalbibliothek lists this publication in the Deutsche Nationalbibliografie; detailed bibliographic data are available on the Internet at <http://dnb.d-nb.de>.

© 2010 WILEY-VCH Verlag GmbH & Co. KGaA, Weinheim

All rights reserved (including those of translation into other languages). No part of this book may be reproduced in any form – by photoprinting, microfilm, or any other means – nor transmitted or translated into a machine language without written permission from the publishers. Registered names, trademarks, etc. used in this book, even when not specifically marked as such, are not to be considered unprotected by law.

Typesetting Toppan Best-set Premedia Limited
Cover Design Grafik Design Adam, Weinheim
ISBN: 978-3-527-31962-6

Contents

Preface to the Second Edition *XIII*
Preface to the First Edition *XV*

Introduction *1*

Part I Ozone in Overview

1 Toxicology *5*
1.1 Background *5*
1.2 Ozone in Gas *7*
1.2.1 Inhalation *7*
1.2.2 Skin Contact *7*
1.2.3 Eye Contact *7*
1.3 Ozone in Liquid *8*
1.4 By-products *9*
     References *11*

2 Reaction Mechanism *13*
2.1 Ozonation *13*
2.1.1 Indirect Reaction *13*
2.1.1.1 Initiation Step *14*
2.1.1.2 Radical Chain Reaction *15*
2.1.1.3 Termination Step *15*
2.1.1.4 Overall Reaction *16*
2.1.2 Direct Reaction *17*
2.2 Advanced Oxidation Processes (AOP) *20*
2.2.1 Ozone/Hydrogen Peroxide O_3/H_2O_2 *20*
2.2.2 Ozone/UV-Radiation O_3/UV *21*
2.2.3 Hydrogen Peroxide/UV-Radiation UV/H_2O_2 *22*
     References *24*

3 Ozone Applications 27
3.1 Historical Development 27
3.2 Overview of Ozone Applications 31
3.2.1 Ozone in the Gas Phase 32
3.2.2 Ozone in the Liquid Phase 34
3.3 Ozone in Drinking-Water Treatment 36
3.3.1 Disinfection 38
3.3.2 Oxidation of Inorganic Compounds 40
3.3.3 Oxidation of Organic Compounds 42
3.3.3.1 Natural Organic Matter (NOM) 42
3.3.3.2 Organic Micropollutants 43
3.3.4 Particle-Removal Processes 44
3.4 Ozonation in Waste-Water Treatment 46
3.4.1 Disinfection 47
3.4.2 Oxidation of Inorganic Compounds 48
3.4.3 Oxidation of Organic Compounds 49
3.4.3.1 Landfill Leachates – Partial Mineralization 52
3.4.3.2 Textile Waste Waters – Color Removal and Partial Mineralization 53
3.4.3.3 Other Applications 54
3.4.4 Particle-Removal Processes 55
3.5 Economical Aspects of Ozonation 56
References 59

Part II Ozone Applied

4 Experimental Design 69
4.1 Experimental Design Process 70
4.2 Experimental Design Steps 73
4.2.1 Define Goals 73
4.2.2 Define System 74
4.2.3 Select Analytical Methods and Methods of Data Evaluation 81
4.2.3.1 Ozone 83
4.2.3.2 Target Compound M 83
4.2.4 Determine Experimental Procedure 85
4.2.5 Evaluate Data 88
4.2.6 Assess Results 89
4.3 Reactor Design 94
4.3.1 Reactor Types 94
4.3.1.1 Operating Mode 94
4.3.1.2 Mixing 95
4.3.2 Comparison of Reactor Types 97
4.3.3 Design of Chemical Oxidation Reactors 100
4.3.3.1 Reaction System 102
4.3.3.2 Ancillary Systems 103
4.3.3.3 Process Integration 103

4.3.3.4 Controllability 104
4.3.3.5 Site Integration 104
4.4   Checklists for Experimental Design 104
4.4.1 Checklists for Each Experimental Design Step 105
4.5   Ozone Data Sheet 108
       References 111

5 Experimental Equipment and Analytical Methods 113
5.1   Materials in Contact with Ozone 113
5.1.1 Materials in Pilot- or Full-Scale Applications 116
5.1.1.1 Reactors 116
5.1.1.2 Piping 116
5.1.2 Materials in Lab-Scale Experiments 117
5.1.2.1 Reactors 117
5.1.2.2 Piping 118
5.2   Ozone Generation 118
5.2.1 Electrical Discharge Ozone Generators (EDOGs) 120
5.2.1.1 Chemistry 121
5.2.1.2 Engineering and Operation 123
5.2.1.3 Type of Feed Gas and its Preparation 125
5.2.1.4 Ozone Concentration, Production Capacity and Specific Energy Consumption 126
5.2.1.5 Use of EDOGs in Laboratory Experiments 126
5.2.2 Electrolytic Ozone Generators (ELOGs) 127
5.2.2.1 Use of ELOGs in Laboratory Experiments 130
5.3   Reactors Used for Ozonation 130
5.3.1 Overview of Hydrodynamic Behavior and Mass Transfer 131
5.3.2 Directly Gassed Reactors 135
5.3.2.1 Bubble Columns and Similar Reactors 135
5.3.2.2 Stirred-Tank Reactors 136
5.3.3 Indirectly and Nongassed Reactors 138
5.3.3.1 Tube Reactors 138
5.3.3.2 Membrane Reactors 139
5.3.4 Types of Gas Contactors 141
5.3.5 Mode of Operation 144
5.3.6 Experimental Procedure 145
5.3.6.1 Batch Experiments 145
5.3.6.2 Continuous-Flow Experiments 146
5.3.6.3 Process Combinations 146
5.4   Ozone Measurement 147
5.4.1 Methods 147
5.4.1.1 Iodometric Method (Gas and Liquid) 147
5.4.1.2 UV Absorption (Gas and Liquid) 147
5.4.1.3 Visible-Light Absorption (Gas and Liquid) 149
5.4.1.4 Indigo Method (Liquid) 149

5.4.1.5 N,N-diethyl-1,4 Phenylenediammonium – DPD (Liquid) 150
5.4.1.6 Chemiluminescence – CL (Liquid) 150
5.4.1.7 Membrane Ozone Electrode (Liquid) 151
5.4.2 Practical Aspects of Ozone Measurement 152
5.5 Safety Aspects 152
5.5.1 Vent Ozone Gas Destruction 152
5.5.2 Ambient Air Ozone Monitoring 154
5.6 Common Questions, Problems and Pitfalls 154
References 159

6 Mass Transfer 163

6.1 Theory of Mass Transfer 163
6.1.1 Mass Transfer in One Phase 164
6.1.2 Mass Transfer between Two Phases 166
6.1.3 Equilibrium Concentration for Ozone 167
6.1.4 Two-Film Theory 172
6.2 Parameters That Influence Mass Transfer 173
6.2.1 Mass Transfer with Simultaneous Chemical Reactions 175
6.2.1.1 Interdependence of Mass Transfer and Chemical Reaction 176
6.2.1.2 Effect of Kinetic Regime on Determination of Mass-Transfer Coefficients 180
6.2.2 Predicting the Mass-Transfer Coefficient 182
6.2.2.1 Theta Factor – Correction Factor for Temperature 183
6.2.2.2 Alpha Factor – Correction Factor for Water Composition 183
6.2.3 Influence of Water Constituents on Mass Transfer 184
6.2.3.1 Change in Bubble Coalescence 184
6.2.3.2 Changes in Surface Tension 185
6.3 Determination of Mass-Transfer Coefficients 186
6.3.1 Choice of Direct or Indirect Determination of $k_La(O_3)$ 187
6.3.2 General Experimental Considerations and Evaluation Methods 188
6.3.2.1 Equilibrium Concentration c_L^* 189
6.3.3 Nonsteady-State Methods Without Mass-Transfer Enhancement 191
6.3.3.1 Batch Model 191
6.3.3.2 Experimental Procedure 192
6.3.3.3 Continuous-Flow Model 192
6.3.3.4 Experimental Procedure 193
6.3.4 Steady-State Methods Without Mass-Transfer Enhancement 193
6.3.4.1 Semibatch and Continuous-Flow Models 194
6.3.4.2 Experimental Procedure 194
6.3.4.3 Simultaneous Determination of k_La and r_L 195
6.3.5 Methods with Mass-Transfer Enhancement 196
6.3.5.1 Experimental Procedure 197
6.3.6 Problems Inherent to the Determination of Mass-Transfer Coefficients 197

6.3.6.1 Nonsteady-State Method *197*
6.3.6.2 Steady-State Method *199*
References *200*

7   Reaction Kinetics *205*
7.1   Reaction Order *205*
7.1.1 Experimental Procedure to Determine the Reaction Order n *209*
7.1.1.1 Half-Life Method *210*
7.1.1.2 Initial Reaction Rate Method *210*
7.1.1.3 Trial and Error *210*
7.2   Reaction Rate Constants *211*
7.2.1 Determination of Rate Constants *213*
7.3   Parameters That Influence the Reaction Rate *215*
7.3.1 Concentration of Oxidants *216*
7.3.1.1 Direct Reactions *216*
7.3.1.2 Indirect or Hydroxyl Radical Reactions *216*
7.3.2 Temperature Dependency *217*
7.3.3 Influence of pH *217*
7.3.4 Influence of Inorganic Carbon *218*
7.3.5 Influence of Inorganic Salts *219*
7.3.6 Influence of Organic Carbon on the Radical Chain-Reaction Mechanism *220*
References *221*

8   Modeling of Ozonation Processes *225*
8.1   Ozone Modeling *228*
8.1.1 General Description of the Ozone Modeling Problem *228*
8.1.2 Chemical Model of Ozonation *231*
8.1.3 Mathematical Model of Ozonation *232*
8.1.3.1 Mass Balances *232*
8.1.3.2 Rate Equations *234*
8.1.3.3 Solving the Model *235*
8.1.4 Summary *236*
8.2   Modeling of Drinking-Water Oxidation *237*
8.2.1 Chemical and Mathematical Models *237*
8.2.2 Methods to Determine the Hydroxyl-Radical Concentration *240*
8.2.2.1 Indirect Measurement *241*
8.2.2.2 Complete Radical-Chain-Reaction Mechanism *244*
8.2.2.3 Semiempirical Method Based on Observable Parameters *245*
8.2.2.4 Semiempirical Method Based on Observed Hydroxyl Radical Initiating Rate *246*
8.2.2.5 Empirical Selectivity for Scavengers *247*
8.2.2.6 Summary of Chemical and Mathematical Models for Drinking Water *249*

Contents

- 8.2.3 Models Including Physical Processes 249
- 8.3 Modeling of Waste-Water Oxidation 250
- 8.3.1 Chemical and Mathematical Models 251
- 8.3.2 Empirical Models 260
- 8.3.3 Summary 261
- 8.4 Final Comments on Modeling 261
- References 263

9 Application of Ozone in Combined Processes 267
- 9.1 Advanced Oxidation Processes 268
- 9.1.1 Chemical AOPs 269
- 9.1.1.1 Principles and Goals 269
- 9.1.1.2 Existing Processes 269
- 9.1.1.3 Experimental Design 272
- 9.1.2 Catalytic Ozonation 273
- 9.1.2.1 Principles and Goals 273
- 9.1.2.2 Existing Processes and Current Research 275
- 9.1.2.3 Experimental Design 280
- 9.2 Three-Phase Systems 281
- 9.2.1 Principles and Goals 282
- 9.2.1.1 Gas / Water / Solvent Systems 283
- 9.2.1.2 Gas / Water / Solid Systems 284
- 9.2.2 Mass Transfer in Three-Phase Systems 284
- 9.2.3 Existing Processes and Current Research 287
- 9.2.3.1 Gas / Water / Solvent Systems 287
- 9.2.3.2 Gas / Water / Solid Systems 290
- 9.2.3.3 Change in the Solids 290
- 9.2.3.4 Change in Compounds Adsorbed on the Solids 292
- 9.2.3.5 Soil Ozonation 292
- 9.2.3.6 Regeneration of Adsorbents 293
- 9.2.4 Experimental Design 294
- 9.2.4.1 Define System 294
- 9.2.4.2 Select Analytical Methods 295
- 9.2.4.3 Determine Experimental Procedure 296
- 9.2.4.4 Evaluate Data and Assess Results 296
- 9.3 Chemical-Biological Processes (CBP) 297
- 9.3.1 Principles and Goals 297
- 9.3.2 Existing Processes and Current Research 301
- 9.3.2.1 Drinking-Water Applications 302
- 9.3.2.2 Waste-Water Applications 304
- 9.3.3 Experimental Design 314
- 9.3.3.1 Define System 315
- 9.3.3.2 Select Analytical Methods 318
- 9.3.3.3 Determine Experimental Procedure 320
- 9.3.3.4 Evaluate Data 323

9.3.3.5 Assess Results *324*
9.4 Applications in the Semiconductor Industry *327*
9.4.1 Production Sequence *327*
9.4.2 Principles and Goals *329*
9.4.3 Existing Processes for Cleaning and Oxidation *330*
9.4.4 Process and / or Experimental Design *333*
9.4.4.1 Define System *333*
9.4.4.2 Select Analytical Methods *333*
9.4.4.3 Determine Procedure *333*
9.4.4.4 Evaluate Data and Assess Results *334*
References *334*

Glossary of Terms *345*
Index *353*

Preface to the Second Edition

The first edition of this Practical Guide to Understanding Ozone and its Application was published in the year 2000 and was well received. It was even translated into Chinese in 2004. Since then much work on ozone has been carried out and a comprehensive new edition was needed. Changes in regulations for disinfection by-products in various parts of the world have driven the use of ozone in various water applications. Concern over micropollutants in wastewater effluents has spurred interest in its use in municipal waste water treatment as a polishing treatment before discharge. An important goal in updating the book was to summarize these recent developments. However, it remains focused on the basics necessary for someone getting started in ozone, mainly for the student, researcher and for people who are involved in process development and design.

The structure of the book – Part One: Ozone in Overview and Part Two: Ozone Applied – was maintained. Part One was expanded to cover disinfection by-products and provide an overview of the wider application areas of ozone as well as a short history of its use before reviewing the specific applications in water. In Part Two the practical aspects have been expanded and the differences in wastewater and drinking-water applications are dealt with more explicitly. Chapter 4 reviews experimental and reactor design concepts for those that need a quick refresher of the material learned in basic classes, while Chapters 6 and 8 build a firm basis to model the whole process. Chapter 6 has been expanded to include current mass-transfer theories and in Chapter 8 the description of modeling approaches deals explicitly with the differences in waste-water and drinking-water applications.

The update adds some "missing elements" that we have identified over the years of using the book. We hope the additions will increase the "practicality" of the guide for others as well.

Structure of the Book

The book consists of two parts: Part One: Ozone in Overview and Part Two: Ozone Applied. The first part is intended to provide a general background on ozonation, briefly reviewing the toxicology of ozone, its reaction mechanisms, and full-scale applications of ozonation. This provides motivation for experimental activity, applying ozone in the laboratory. The second part of the book tries to offer information on just how to go about it. The design of experiments and required equipment as well as analytical methods and data evaluation are first discussed. Then, the theoretical background needed to carry out these activities is explored. The goal here is to include the basics necessary for building a solid foundation, and to reference secondary sources, with which the reader can delve deeper into ozonation specifics. Part Two is rounded out with a discussion of applications that use ozone in combination with other treatment processes.

Acknowledgment

The authors would like to thank Kerstin, Rolf, Alan, Malte and Maarten for the patience and time they gave us to finish the second edition of this book.

Chris Gottschalk, Judy Libra, and Adrian Saupe
March 2009

Preface to the First Edition

The ozonation of compounds in water is a complex process. The mechanisms are very complicated, the parameters are many, but the possibilities of developing cost-effective treatment schemes for drinking water and waste water are large. To take advantage of this potential, it is important to know which parameters contribute to the process, which are important, and how they affect the process.

Because ozonation is so system dependent, most full-scale applications are first tried out bench-scale. That means designers and manufacturers of treatment systems, researchers, as well as potential industrial operators of ozonation must know not only the fundamentals of the mechanisms of ozonation, but also how to set up experiments so that the results can be interpreted, extrapolated, and applied.

Most books available today concentrate on either drinking-water or waste-water treatment, seldom dealing with both or explaining the essential differences. And only rare exceptions deal with the how-to of ozone experiments.

This guide fills the gap. It contains the cumulative knowledge gathered by the authors as researchers, teachers and ozone-system developers on experimental design, execution, interpretation and application. Drawing on experience gained from hours spent on laboratory research with drinking and waste waters, literature study, intensive discussion with leading experts, perplexed reflection and deep thought, the book offers practical help to avoid common pitfalls and unnecessary work.

This book is aimed at professionals in industry and research currently using ozonation who want to optimize their system, as well as students beginning work with ozonation. Much literature exists today about ozonation, but its practical use for beginners is limited by its specialization, and for the advanced by its magnitude and diversity.

The practical guide presents an overview of current theories and results from the specialized literature in short concise text, tables and figures accompanied with references to important secondary literature. It contains just enough information for beginners to start with, but goes rapidly to the detailed information that advanced readers need.

Introduction

Being poised on the edge of the third millennium in Berlin, the city of Siemens who constructed the first ozone generator almost 150 years ago, and writing a book on ozone applications in water can make one philosophical. Especially when one has been confronted with the puzzlement of most acquaintances about why anyone deliberately produces ozone and what it has to do with water. These two aspects of this book: ozone and water need some clarification.

Ozone can be present as a gas or dissolved in a liquid. The media reports almost exclusively on gaseous ozone. For example, that the beneficial ozone layer in our atmosphere is being depleted, allowing more damaging UV-radiation from the sun to reach us on earth. On the other hand, announcements in the media warn about too high ozone concentrations in our air on sunny days, causing damage to human health and the environment. That ozone can have beneficial and detrimental effects can be confusing if it is not made clear that the effect is dependent on the location of the ozone. Direct exposure is always detrimental.

The other aspect is water itself. A survey has shown that pure water contains only 98.1% H_2O, at least that is the common perception [1]. Although water pollution has not caused such an extreme change in the composition of our water resources, we have allowed many substances to enter the natural water cycle with detrimental effects on human health and the natural environment. The result is additional treatment processes are often necessary to prepare drinking water for every day use. And processes are necessary for waste-water treatment and groundwater remediation to prevent even larger contamination. Here, ozone comes into play. It is capable of oxidizing a large number of pollutants in an environmentally sound way, since normally no harmful end- or by-products are formed nor are secondary wastes produced. Unfortunately, we cannot make use of the ozone gas sometimes present in unacceptably high concentrations in our breathing air (e.g., $>240\,\mu g\,m^{-3}$), but we have to produce it in ozone generators from air or from pure oxygen using much energy, to reach concentrations higher by a factor of a million ($240\,g\,m^{-3}$).

The fact that almost 150 years after the production of the first ozone generator a book is still necessary on how to experiment with one, shows that ozonation is a complex subject. Some very good reference books and articles exist that explain the fundamentals of ozonation, the chemical reactions, the effect of some parameters. However, most work concentrates on either drinking-water treatment or waste-water treatment, seldom dealing with both, reflecting our personal experience that these are two separate "worlds". Since ozone applications in production processes are growing, even more people from various disciplines need access to information on ozone and how to make use of the results from the various applications. We have tried to bridge this information gap with this book, building on our diverse backgrounds in drinking-, waste- and process-water treatment.

Another area rarely dealt with in previous literature is the "how-to" of ozone experiments. This information is usually hard-earned by doctoral candidates and laboratory staff, and either not considered as appropriate information for a scientific treatise or considered as proprietary information that belongs to expertise. This lack of information has motivated us to write a book that contains not only fundamental information about the toxicology of ozone, its reaction mechanisms, and full-scale applications of ozonation (Part One: Ozone in Overview), but also information on how to set up experiments so that they produce results that can be interpreted and extrapolated (Part Two: Ozone Applied). The experimenter is provided with tools to improve his or her results and interpret results found in the literature. The required theoretical foundation is laid at the beginning of each chapter in Part Two, compact and tailored to ozone, followed by practical aspects. References are made to important literature sources to help direct the reader wishing for more in-depth information. A discussion of applications combining ozone with other processes illustrates how the oxidizing potential of ozone can be utilized.

Reference

1 Malt, B.C. (1994) Water is not H_2O. *Cognitive Psychology*, **27** (1), 41–70.

Part I Ozone in Overview

1
Toxicology

Toxicology examines the adverse effects of substances on living organisms. The effect on humans has been traditionally the subject of this study. The field of ecotoxicology has been developed to study the wider effects of substances on an ecosystem, not only on individual organisms, but also on the interactions between the elements in ecosystems. Both areas are important when evaluating the toxicology of ozonation applications. The species affected by a substance depends on the application – studies on drinking water concentrate on human toxicology and waste water on aquatic ecotoxicology.

This chapter will give a short overview of the toxicology of ozone. The types of toxicity and study subjects are briefly reviewed (Section 1.1), before the toxicological effects of exposure to ozone are presented. When talking about the effects of ozone, one has to differentiate between the routes of exposure and the type of compound being examined. The exposure can take place with:

- ozone in gas (Section 1.2);
- ozone in liquid (Section 1.3); and
- by-products formed by ozone reactions (Section 1.4).

1.1
Background

In the description of the effect of a substance on an organism, consideration of the length of exposure necessary for the effect is essential. Toxicity is usually differentiated into three types according to the exposure. Acute toxicity describes a fast harmful effect after only a short-term exposure (<4 d) or exposure in limited amounts, for example, a fast-reacting poison. Subchronic reactions from chemicals are mostly determined by biochemical changes as well as changes in growth, behavior and other factors over a time period of several months. For chronic toxicity, the harmful effect of a substance is measured over a much longer time period, from years to a lifetime. The harmful effect could be reversible or irreversible, cause benign or malignant tumors, mutagenic or teratogenic effects, bodily injury or death [1].

Ozonation of Water and Waste Water. 2nd Ed. Ch. Gottschalk, J.A. Libra, and A. Saupe
Copyright © 2010 WILEY-VCH Verlag GmbH & Co. KGaA, Weinheim
ISBN: 978-3-527-31962-6

Human toxicology employs a variety of testing methods to evaluate effects on human health. The tests can be ordered according to a hierarchy of significance. If available, epidemiological studies of humans exposed to a particular environmental situation are preferred because their results are usually directly applicable to human health risk. However, in most cases experiments with animals or cultured cells are generally necessary to gain information. Aquatic ecotoxicology evaluates the probability of an adverse impact of a substance on the aquatic environment at the present as well as in the future, considering the total flow into the system [2]. It encompasses laboratory ecotoxicity tests on appropriate test organisms to explore relationships between exposure and effect under controlled conditions as well as studies of the effects of substances or effluents under a variety of ecological conditions in complex field ecosystems [3].

Ecotoxicity tests or bioassays measure the responses induced by the substances under controlled conditions in the laboratory, generally using cultured organisms in the tests. The laboratory test organisms should be representative of the four groups: microorganisms, plants, invertebrates, and fish. Common test organisms for invertebrate toxicity are the water fleas, *Daphnia* and *Ceriodaphnia*, brine shrimp *Artemia salina*. Various microorganisms can be used: for example, the green microalgae *Selenastrum capricornutum* or marine microorganisms that exhibit bioluminescence such as *Vibrio fischeri* (formerly known as *Photobacterium phosphoreum*). The results are often reported as a lethal dose or concentration (LD or LC) with LC50 the concentration where 50% of the test organisms survived. The effective dose or concentration (ED or EC) is defined analogously where EC50 is used to describe adverse effects in 50% of the test organisms within the prescribed test period [4].

Standardized bioassays have been developed and optimized over the last decades to quantify effects on bacteria, daphnia and fish [5]. These tests are designed to assess the toxicity of specific compounds as well as whole effluents on aquatic organisms. They are quick to perform, easy to handle and comparatively inexpensive, with the goal of allowing the toxicity of a complex water matrix to be estimated. They have been incorporated into regulatory practice in various countries. Extensive reviews of bioassay use and international experience can be found in [6, 7].

In general, the test results are usually not directly applicable to risks in human health or in the aquatic environment and must be interpreted by toxicologists [8]. It is their responsibility to provide risk assessments based on the test results and to derive guidelines or standards for water quality below which no significant health risk is encountered [9].

Although much progress has been made in laying a scientific basis for ecotoxicology and interpreting bioassay results [6], there are still many problems associated with predicting effects in complex ecosystems [10]. The results from bioassays are in general matrix-specific and usually give no hint to the compounds responsible for any adverse effects. Moreover, ozone with its ability to oxidize a wide spectrum of compounds increases the complexity of the problem. Due to its extreme reactivity and high redox potential, ozone can directly oxidize compounds as well as produce highly reactive, short-lived free radicals that can further react.

While ozone itself rapidly decomposes in water, leaving oxygen as the only residual, decomposition by-products may be left behind. By-products from both types of reactions can be formed. This poses the problem that not only must it be determined if there are measurable toxicological effects, but also which compounds are responsible for the measured effects. Especially in drinking water and foods, the by-products from ozone reactions with organics and inorganics are of concern for their potential chronic toxicity.

1.2 Ozone in Gas

Ozone is a highly toxic, oxidizing gas. The routes of entry are inhalation, skin and eyes.

1.2.1 Inhalation

Acute Effects: Ozone concentrations in excess of a few tenths of a ppm (1 ppm = 2 mg m^{-3}, 20 °C, 101.3 kPa) cause occasional discomfort to exposed individuals in the form of headache, coughing, dryness of throat and mucous membranes, and irritation of the nose following exposures of short duration. The odor threshold is about 0.02 ppm, however, a desensibilization occurs over time. Exposure to higher concentrations can also produce delayed lung edema in addition to lassitude, frontal headache, sensation of substernal pressure, constriction or oppression, acid in mouth, and anorexia. More severe exposures have produced dyspnea, coughing, choking sensation, tachycardia, vertigo, lowering of blood pressure, severe cramping chest pain, and generalized body pain. It is estimated that 50 ppm for 30 min would be fatal.

Chronic Exposures: chronic exposure symptoms are similar to acute exposures with pulmonary lung function decrements depending on concentrations and duration of exposure. Asthma, allergies, and other respiratory disorders have been observed. Breathing disorders, tumorgenic, direct and indirect genetic damage have been found in animal and/or human tissue studies.

Carcinogenicity: Justifiably suspected of having carcinogenic potential (group B).

1.2.2 Skin Contact

Contact with ozone may irritate the skin, burns and frostbite can also occur.

1.2.3 Eye Contact

Exposed persons may sense eye irritation at or above 0.1 ppm ozone.

The severity of injury depends on both the concentration of ozone and the duration of exposure, which is in some regulations included in threshold values concerning workplace exposure.

Workplace exposure limits differ depending on the regulatory agency. The following list gives some examples of regulations in the United States [11]:

- OSHA (Occupational Safety and Health Administration): The legal airborne permissible exposure limit (PEL) is 0.1 ppm averaged over an 8-hour workshift.
- ACGIH (American Conference of Governmental Industrial Hygienists): TIME-WEIGHTED AVERAGE (TLV-TWA): Heavy work 0.05 ppm; Moderate work 0.08 ppm; Light work 0.1 ppm; for two hours or less exposure time, heavy/moderate/light work loads 0.2 ppm.
- NIOSH (National Institute for Occupational Safety and Health): The recommended airborne exposure limit is 0.1 ppm, which should not be exceeded at any time. Immediately Dangerous to Life or Health Concentration IDLH: 5 ppm.
- EPA (Environmental Protection Agency) National Ambient Air Quality Standard for ozone is a maximum 8-hour average outdoor concentration of 0.08 ppm.

In the MAK-list in Germany (maximal allowable workplace concentration) ozone has been categorized as IIIb, which means a substance being justifiably suspected to be carcinogenic. The older MAK value of $200\,\mu g\,m^{-3}$ (= 0.1 ppm) was therefore suspended until it is known if ozone shows carcinogenic effects [12].

Note: For safety reasons ozone should always be used with an ambient air ozone monitor (measuring ranges 0–1 ppm) with a safety shutdown procedure.

1.3
Ozone in Liquid

No health hazard data are available and no limits for workplace exist for ozone in liquid. Ozonated water in high concentrations can lead to eye and skin irritation. Langlais (1991) summarize some LC_{50}-values (concentration that is lethal to half of the test animals) found in fish tests [8]:

- Bluegills (*Lepomis macrochius*) for 24 h: $0.06\,mg\,l^{-1}$
- Rainbow trout (*Salmo gairdneri*) for 96 h: $0.0093\,mg\,l^{-1}$
- White perch (*Morone americana*) for 24 h: $0.38\,mg\,l^{-1}$

It is important to note that the differentiation between ozone and its by-products in such tests is often not possible.

Most of the possible toxic effects from ozone in gas can also occur when using liquid ozone, due to the potential risk of it gassing-out. Consequently, liquid ozone has a strong odor and should always be used in closed piping and vessels.

1.4
By-products

Ozone is highly reactive and can oxidize compounds directly or indirectly via hydroxyl radicals, so that a multitude of by-products can be produced in ozone applications. In this section we mainly look at by-products produced by reactions in water. Such by-products can be of concern not only in drinking water, but in any application associated with human exposure, such as disinfection in swimming pools, food processing and waste-water reuse too. These by-products are often referred to as disinfection by-products, DBP.

The dilemma with chemical disinfectants is that in order to achieve the goal of deactivating microorganisms, they must be highly reactive compounds. This carries with it the drawback that they react with most organics and many inorganics in the water producing by-products that may be harmful. The original concern over DBP began with the discovery in the 1970s that chlorine used for drinking-water disinfection could react with natural organics in the water to produce chloroform. It was soon found that chlorination can produce other organochlorine by-products and, in the presence of other halogen ions, for example, bromide and iodide, a variety of halogenated organics that are grouped together according to their structures: trihalomethanes (THMs) and haloacetic acids (HAAs), haloketones (HKs), haloacetonitriles (HANs), and chloral hydrate (CH) as well as bromate and chlorate [13].

Concern over DBP has led to increased use of alternative disinfection methods such as ozone, chlorine dioxide and chloramines. They too can produce DBP. Continuous analytical developments make it possible to detect more polar compounds at very low concentrations. For example, in their nationwide study on DBP in drinking waters, the US EPA was able to quantitatively analyze for over 50 DBPs [14].

Unfortunately, DBP formation is very complex and highly dependent on water quality as well as on the treatment processes and operating conditions used. It is influenced by the water constituents present (e.g., TOC, bromide, ammonia, carbonate alkalinity), the treatment train (type and order of treatment stages, e.g., removal of NOM before ozonation) and operating conditions (e.g., pH, temperature, disinfectant dose, contact time), so that seasonal variations are possible at one location. This makes it difficult to compare disinfection methods used at various plants. This is true for drinking water as well as other water types. If disinfectants are used in combination, the interplay between the effects of each disinfectant must be considered. For instance, the type of DBP found should be differentiated according to whether ozone is used in combination with other disinfectants and treatment processes. In some countries, legal mandates of a chlorine residual in the drinking water distribution network necessitate the use of chlorine-containing disinfectants as a final stage, even though ozone may be used as the major disinfection process. Disinfection combinations are also often used in the treatment of swimming-pool waters. In such combinations, the ozonation stage itself may not produce harmful DBP, but the final chlorination of the oxidized products may.

Normally, ozonation results in the formation of organic by-products since complete mineralization seldom occurs. The types of by-products formed depend on the organic precursors in the water, which can be highly variable. The organic composition of natural organic matter (NOM) ozonation found in surface water is different from the composition of organic matter found in waste-water effluents. In general, though, organic compounds such as organic acids, aldehydes and ketones are formed [15]. This organic DBP can cause increased bacterial regrowth in drinking-water distribution systems.

Furthermore, the dissolved organic carbon concentration in swimming pools and waste water is typically greater than in drinking or surface water, resulting in faster ozone decomposition. As a result, a higher ozone dose is required to meet the water-treatment goals, potentially leading to increased DBP formation.

On the inorganic side, the formation of bromate, iodate and chlorate may be of concern [16]. Bromate – a regulated DBP – has received the most attention due to its potential carcinogenic effect. The mechanism of formation from bromide is described in Chapter 2. If both NOM and bromide are present, brominated organohalogen compounds can be formed. However, in his comprehensive review of ozonation DBPs, von Gunten [17] reports that these reactions are of minor importance. In addition, research has shown that under drinking-water conditions chloride is not oxidized during ozonation. Chlorate is only formed if ozonation has been preceded by the addition of chlorine and/or chlorine dioxide.

In order to evaluate the toxicity of ozonation by-products, their effects on target organisms (human, animals, fish, etc.) need to be determined. Normally, whole-effluent testing is carried out since identifying all the substances that compose the TOC of a ground-, drinking- or waste-water sample can rarely be achieved. Controlled testing with synthetic mixtures of such matrices may not contain important trace DBPs. Furthermore, the toxicity of specific compounds in a complex mixture may also depend on the background matrix and cause synergistic or antagonistic interactions with other substances. Good overviews of toxicological methods and results have been published for DBP in drinking water [8, 9], swimming pools [18] and waste water [19–21]). In general, the reviews show that test results are variable with indications that ozone treatment can either increase or decrease toxicity and mutagenicity. Therefore, since the results are site-specific and seasonal, before adopting a particular disinfection method, the mutagenic and toxic effects at various doses and seasons with the real water should be studied. As Langlais et al. [8] pointed out this variability is due to the fact that many reactions with ozone are dose and pH dependent.

When DBPs are of concern, there appear to be three possible ways of reduction:

- Remove precursors that react with the disinfectants to form the unwanted DBP. Since the level of harmful by-products can be substantially reduced by the removal of organic substances prior to ozonation, NOM or other organics can be removed by GAC absorption and membrane filtration or coagulation.

- Optimize the water treatment to control the DBPs formation. The ozonation stage can be operated to reduce the formation of bromate by controlling of the pH and/or dissolved ozone concentration (see Chapter 3 for further details).

- Remove DBPs that are formed, for example, with GAC filtration and membrane processes. However, since most DBP are difficult to remove, avoidance is the best policy.

References

1 Wentz, C.A. (1998) *Safety, Health, and Environmental Protection*, WCB/McGraw-Hill, Boston.

2 Klein, W. (1999) Aquatische Ökotoxikologie: Stoffeigenschaften und ökologisches Risiko, Preprints Band 1, 4. GVC-Abwasser-Kongress 1999, Bremen.

3 Chapman, J.C. (1995) The role of ecotoxicity testing in assessing water quality. *Australian Journal of Ecology*, 20 (1), 20–27.

4 Novotny, V. and Olem, H. (1994) *Water Quality: Prevention, Identification, and Management of Diffuse Pollution*, Van Nostrand Reinhold, New York.

5 US Environmental Protection Agency (1999) *Toxicity Reduction Evaluation Guidance for Municipal Wastewater Treatment Plants*, EPA-833B-99-002, Office of Water, Washington, D.C.

6 Tonkes, M., den Besten, P.J. and Leverett, D. (2005) Bioassays and tiered approaches for monitoring surface water quality and effluents, in *Ecotoxicological Testing of Marine and Freshwater Ecosystems: Emerging Techniques, Trends, and Strategies* (eds P.J. den Besten and M. Munawar), CRC Press, Boca Raton.

7 Power, E.A. and Boumphrey, R.S. (2004) International trends in bioassay use for effluent management. *Ecotoxicology*, 13 (5), 377–398.

8 Langlais, B., Reckhow, D.A. and Brink, D.R. (1991) *Ozone in Water Treatment: Application and Engineering*, American Water Works Association Research Foundation, Denver, Lewis Publishers Inc., Michigan.

9 van Leeuwen, F.X.R. (2000) Safe drinking water: the toxicologist's approach. *Food and Chemical Toxicology*, 38 (Suppl. 1), S51–S58.

10 Schäfers, C. and Klein, W. (1998) *Ökosystemare Ansätze in der Ökotoxikologie, Zittau 18.–19.5.1998* (eds J. Oehlmann and B. Markert), Ecomed Verlag, Landsberg.

11 OSHA (2004) Chemical sampling information-ozone, http://www.osha.gov/dts/chemicalsampling/data/CH_259300.html (accessed December 2008).

12 DFG Deutsche Forschungsgemeinschaft: MAK- und BAT-Werte-Liste (2008) *Senatskommission zur Prüfung gesundheitsschädlicher Arbeitsstoffe, Mitteilung 44*, Wiley-VCH Verlag GmbH, Weinheim.

13 WHO (2000) *Environmental Health Criteria 216: Disinfectants and Disinfectant By-Products*, World Health Organisation, Geneva (Updated 2004), http://www.who.int/ipcs/publications/ehc/216_disinfectants_part_2.pdf (accessed January 2009).

14 US EPA (2002) Nationwide DBP occurrence study. EPA/600/R-02/068, US EPA Ecosystems Research Division, http://www.epa.gov/athens/publications/DBP.html (accessed January 2009).

15 Hammes, F., Sahli, E., Köster, O., Kaiser, H.-P., Egli, T. and von Gunten, U. (2006) Mechanistic and kinetic evaluation of organic disinfection by-products and assimilable organic carbon (AOC) formation during the ozonation of drinking water. *Water Research*, 40, 2275–2286.

16 von Gunten, U. (2003) Ozonation of drinking water: Part I. Oxidation kinetics and product formation. *Water Research*, 37, 1443–1467.

17 von Gunten, U. (2003) Ozonation of drinking water: Part II. Disinfection and by-product formation in presence of bromide, iodide or chlorine. *Water Research*, **37** (7), 1469–1487.

18 Zwiener, C., Richardson, S.D., De Marini, D.M., Grummt, T., Glauner, T. and Frimmel, F.H. (2007) Drowning in disinfection by-products? Swimming-pool water quality reconsidered. *Environmental Science & Technology*, **41** (2), 363–372.

19 Leong, L.Y.C., Kuo, J. and Tang, C.-C. (2008) Disinfection of wastewater effluent–comparison of alternative technologies. WERF Report 04-HHE-4, Water Environment Research Foundation, Alexandria, Virginia.

20 Monarca, S., Feretti, D., Collivignarelli, C., Guzzella, L., Zerbini, I., Bertanza, G. and Pedrazzani, R. (2000) The influence of different disinfectants on mutagenicity and toxicity of urban waste water. *Water Research*, **34** (17), 4261–4269.

21 Paraskeva, P. and Graham, N.J. (2002) Ozonation of municipal wastewater effluents. *Water Environment Research*, **74** (6), 569–581.

2
Reaction Mechanism

Two of the strongest chemical oxidants are ozone and hydroxyl radicals. Ozone can react directly with a compound or it can produce hydroxyl radicals that then react with a compound. These two reaction mechanisms are considered in Section 2.1. Hydroxyl radicals can also be produced in other ways. Advanced oxidation processes are alternative techniques for catalyzing the production of these radicals (Section 2.2).

2.1
Ozonation

Ozone is an unstable gas that has to be produced at the point of use. In order to bring ozone into contact with a target substance in the water phase, it must first be transferred into water via a gas–liquid contactor. The subsequent reaction is not straightforward, since many chemical reactions can occur simultaneously.

Ozone can react with substances in two different ways, indirect and direct. These two reaction pathways lead to different oxidation products and are controlled by different types of kinetics. Figure 2.1 gives an overview of the indirect and direct pathways, and their interaction.

2.1.1
Indirect Reaction

The indirect reaction pathway involves radicals, which are molecules that have an unpaired electron. The unpaired electron is represented in this book by the "°" next to the chemical structure. Most radicals are highly unstable and immediately undergo a reaction with another molecule in order to obtain the missing electron.

The ozone radical chain mechanism can be divided into three different steps, the initiation, chain propagation, and termination. The first step is the decay of ozone, accelerated by initiators, for example, OH-, to form secondary oxidants such as hydroxyl radicals (OH°). They react nonselectively and immediately ($k = 10^8$–10^{10} M^{-1} s^{-1}) with target molecules [1, 2]. For example, the hydroxyl radical

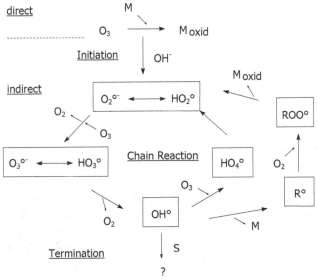

Figure 2.1 Model for the indirect and direct ozonation, S: Scavenger, R: Reaction product, M: Micropollutant (modified from [3]).

regains its missing electron by removing a hydrogen electron from the target molecule to form a water molecule. In losing an electron, the target molecule itself becomes a radical, which will react further, propagating the chain reaction. However, if instead a radical reacts with a second radical, thus, each pairing its unpaired electron, the chain reaction is terminated. The radicals have neutralized each other.

The radical pathway is very complex and is influenced by many substances. The major reactions and reaction products of the radical pathway based on the two most important models are discussed below [3, 4].

Only the main reactions that are necessary to describe the mechanism are explained. Further reactions can be found in Staehelin and Hoigné [5], Buxton et al. [6], Bühler et al. [7] and Chelkowska et al. [8].

2.1.1.1 Initiation Step

The reaction between hydroxide ions and ozone leads to the formation of one superoxide anion $O_2^{\circ -}$ and one hydroperoxyl radical HO_2°.

$$O_3 + OH^- \rightarrow O_2^{\circ -} + HO_2^{\circ} \qquad k_1 = 70 \text{ M}^{-1} \text{ s}^{-1} \tag{2.1}$$

The hydroperoxyl radical is in acid–base equilibrium with the superoxide anion.

$$HO_2^{\circ} \leftrightarrow O_2^{\circ -} + H^+ \qquad pK_a = 4.8 \tag{2.2}$$

2.1.1.2 Radical Chain Reaction

The superoxide anion $O_2^{\circ-}$ then reacts with ozone to form an ozonide anion $\left(O_3^{\circ-}\right)$. This decomposes immediately via hydrogen trioxide HO_3° to an OH° radical.

$$O_3 + O_2^{\circ-} \to O_3^{\circ-} + O_2 \quad k_2 = 1.6 \times 10^9 \ M^{-1} \ s^{-1} \tag{2.3}$$

$$HO_3^{\circ} \leftrightarrow O_3^{\circ-} + H^+ \quad pK_a = 6.2 \tag{2.4}$$

$$HO_3^{\circ} \to OH^{\circ} + O_2 \quad k_3 = 1.1 \times 10^5 \ s^{-1} \tag{2.5}$$

The OH° can react with ozone in the following way [9]:

$$OH^{\circ} + O_3 \to HO_4^{\circ} \quad k_4 = 2.0 \times 10^9 \ M^{-1} \ s^{-1} \tag{2.6}$$

$$HO_4^{\circ} \to O_2 + HO_2^{\circ} \quad k_5 = 2.8 \times 10^4 \ s^{-1} \tag{2.7}$$

With the decay of HO_4° into oxygen and a hydroperoxyl radical, the chain reaction can start anew (see 2.2 and 2.3). Overall for the chain reaction two moles of ozone are consumed (see 2.1 and 2.3). Substances that convert OH° into superoxide radicals $O_2^{\circ-}/HO_2^{\circ}$ promote the chain reaction; they act as chain carriers, the so-called promoters.

Organic molecules, R, can also act as promoters. Some of them contain functional groups that react with OH° and form organic radicals R°.

$$H_2R + OH^{\circ} \to HR^{\circ} + H_2O \tag{2.8}$$

If oxygen is present, organic peroxy radicals ROO° can be formed. These can further react, and so enter again into the chain reaction.

$$HR^{\circ} + O_2 \to HRO_2^{\circ} \tag{2.9}$$

$$HRO_2^{\circ} \to R + HO_2^{\circ} \tag{2.10}$$

$$HRO_2^{\circ} \to RO + OH^{\circ} \tag{2.11}$$

The experimental proof of the existence of HO_4°, necessary for the verification of this reaction pathway proposed by Hoigné, is missing [10]. These radicals are not found in the radical chain cycles of the model from Tomiyasu et al. [4]. However, the result of both models is the same.

2.1.1.3 Termination Step

Some organic and inorganic substances react with OH° to form secondary radicals that do not produce superoxide radicals $HO_2^{\circ}/O_2^{\circ-}$. These inhibitors (or scavengers) generally terminate the chain reaction and inhibit ozone decay.

$$OH^{\circ} + CO_3^{2-} \to OH^- + CO_3^{\circ-} \quad k_6 = 4.2 \times 10^8 \ M^{-1} \ s^{-1} \tag{2.12}$$

$$OH^{\circ} + HCO_3^- \to OH^- + HCO_3^{\circ} \quad k_7 = 1.5 \times 10^7 \ M^{-1} \ s^{-1} \tag{2.13}$$

Another possibility to terminate the chain reaction is the reaction of two radicals:

$$OH° + HO_2° \rightarrow O_2 + H_2O \qquad k_8 = 3.7 \times 10^{10} \text{ M}^{-1} \text{ s}^{-1} \qquad (2.14)$$

2.1.1.4 Overall Reaction

The combination of the Equations 2.1–2.7 for the overall reaction shows that three ozone molecules produce two OH°.

$$3\,O_3 + OH^- + H^+ \rightarrow 2\,OH° + 4\,O_2 \qquad (2.15)$$

Thus, the decay of ozone initiated by the hydroxide ion leads to a chain reaction, producing fast-reacting and nonselective OH–radicals. The OH° reacts with the target molecule at the position with the highest electron density due to its electrophilic properties. Detailed information on OH° reactions can be found in von Sonntag [11], which gives a good overview of the degradation mechanism of aromatics by OH° in water. Due to their reactivity OH-radicals have a very short half-life, for example, less than 10 ms at an initial concentration of 10^{-4} M. Buxton et al. [6] showed that the reaction rate constants for hydroxyl radicals and aromatic compounds are close to the diffusion limit. This means they react as soon as they come into contact with each other.

Many substances exist that initiate, promote or terminate the chain reaction. Table 2.1 gives some examples.

Staehelin and Hoigné [13] found that even phosphate, which is known to react only slowly with OH°, can act as an efficient scavenger when used in concentrations typically found in buffer solutions (50 mM).

Obviously, the action of humic acid is contradictory. It can react as either scavenger or promoter, depending on its concentration [12].

The classical OH° scavenger *tert*–butyl alcohol is often used to suppress the chain reaction. This substance reduces the ozone consumption rate by a factor of seven when the initial concentration is about 50 µM [13].

Table 2.1 Typical initiators, promoters and scavengers for decomposition of ozone in water [3, 12].

Initiator	Promoter	Scavenger
OH^-	Humic acid	HCO_3^-/CO_3^{2-}
H_2O_2/HO_2^-	*aryl*-R	PO_4^{4-}
Fe^{2+}	Primary and secondary alcohols	Humic acid *alkyl*-R *tert*-butyl alcohol (TBA)

Bicarbonate and carbonate play an important role as scavengers of OH° radicals in natural systems. The reaction rate constants are relatively low but the concentration range in natural systems is comparatively high, so that this reaction cannot be ignored. A comparison of the reaction rate constants ($k_6 = 4.2 \times 10^8 \, M^{-1} s^{-1}$ for CO_3^{2-} and $k_7 = 1.5 \times 10^7 \, M^{-1} s^{-1}$ for HCO_3^-) shows that carbonate is a stronger scavenger than bicarbonate. This means that the reaction rate with 100% of the inorganic carbon being present as bicarbonate is comparable to that with 3.6% as carbonate. Hoigné and Bader [14] assumed that the reaction products from bicarbonate and carbonate ions with OH° do not interact further with ozone.

By adding carbonate to ozonated water, the half-life of ozone can be increased. Even the addition of a few μmoles decreases the decay rate of ozone by about a factor of ten or more [14]. Increasing the concentration of bicarbonate/carbonate up to a concentration of 1.5 mM increases the stability of ozone. Thereafter, no further stabilization occurs [15].

2.1.2
Direct Reaction

The direct oxidation (M + O_3) of organic components by ozone is a selective reaction with slow reaction rate constants, typically being in the range of ($k_D = 1.0-10^6 \, M^{-1} s^{-1}$). The ozone molecule reacts with the unsaturated bond due to its dipolar structure and leads to a splitting of the bond, which is based on the so-called Criegee mechanism (see Figure 2.2). The Criegee mechanism itself was developed for nonaqueous solutions.

In general, ozone reacts the faster with organic water contaminants, the higher their electron density, that is, their degree of nucleophilicity. Ozone will react faster with certain types of aromatic and aliphatic compounds, for example, those carrying electron-supplying substituents such as hydroxyl or amine groups. If there is no such substituent the rate of ozonation is much slower.

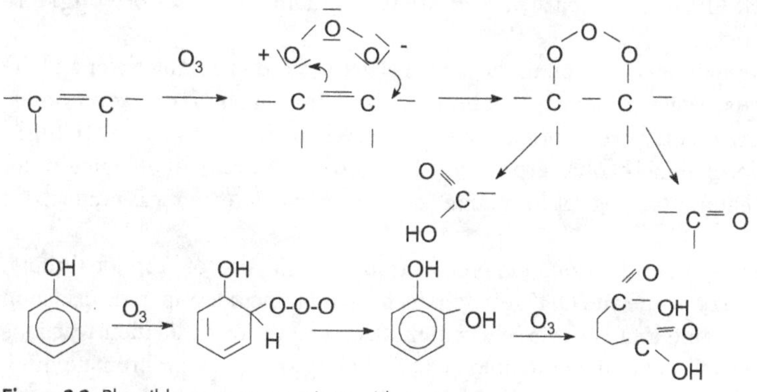

Figure 2.2 Plausible aqueous reactions with ozone.

Table 2.2 Oxidation of organic compounds by ozonation [1, 2, 16].

Compound	Type	k_D (M^{-1}s^{-1})
	Aliphatic: saturated, alkanes	10^{-2}
	Aliphatic: e$^-$ supplying substitutes, alcohols	10^{-2}–1
	Aliphatic: unsaturated, alkenes	1–10^4
	Aromatics: nonsubstituted	1–10^2
Benzene	Aromatic ring: nonsubstituted	2
Chlorobenzene	Aromatic ring: e$^-$ detracting substitutes	0.8
Phenol	Aromatic ring: e$^-$ supplying substitutes undissociated	1.3×10^3
Phenol	Aromatic ring: e$^-$ supplying substitutes dissociated	1.4×10^9

The following order of reactivity toward ozone can be used as a rule of thumb for the various target compound groups:

Saturated Aliphatic < Aromatic Ring < Unsaturated Aliphatic
e$^-$-detracting substitutes < non substituted < e$^-$-supplying substitutes
Undissociated < Dissociated

Table 2.2 gives some general and specific examples of the reactivity of organic compounds toward ozone.

Inorganic compounds can react much faster with ozone than organic compounds. The variation in direct rate constants for inorganic compounds, however, spans a much wider range – more than 12 orders of magnitude compared to 6 – with exceptions – for organic compounds (k_D = 1.0–10^6 M^{-1}s^{-1}). The reaction rate increases with the increasing degree of nucleophilicity of the compound, similar to the trend with organic compounds [17]. Likewise, the reaction is faster with ionized or dissociated inorganic compounds.

Table 2.3 provides an overview of selected inorganic compounds, their reaction products including the rate of direct and radical oxidation.

A critical reaction here is the formation of bromate, a potential carcinogen, from bromide in the water source. The WHO [21] standard, the European Union [22] and the USEPA [23] established a maximum contaminant level of 10 µg l^{-1} in drinking water.

The bromide concentration in the source water used for drinking water and the treatment used determines whether bromate formation will be a concern. Bromide levels below 20 µg l^{-1} are unproblematic [17]. Levels between 50 µg l^{-1} to 100 µg l^{-1} bromate formation become a problem. Levels of bromide in the source water can have natural (geological formation) or anthropogenic (chemical production) causes.

Bromate is formed in ozonation process from the oxidation of bromide through combination of ozone and OH° reaction. Its formation includes up to six oxidation states of bromine and is extremely complicated. The various oxidation reactions are discussed in detail in von Gunten [20] including strategies for bromate minimization (e.g., ammonia dosage, lowering pH) (Figure 2.3).

Table 2.3 Oxidation of inorganic compounds by ozonation [17–20].

Compound	Products	k_D (M^{-1}s^{-1})	k_R (M^{-1}s^{-1})
Fe^{2+}	$Fe(OH)_3$	8.2×10^5	3.5×10^8
Mn^{2+}	MnO_2	1.5×10^5	2.6×10^7
	MnO_4^-		
NO_2^-	NO_3^-	3.7×10^5	6×10^9
NH_4^+/NH_3	NO_3^-	0/20	$-/9.7 \times 10^9$
CN^-	CO_2, NO_3^-	$10^3–10^5$	8×10^9
H_2S/S^{2-}	SO_4^{2-}	$3 \times 10^4/3 \times 10^9$	$1.5 \times 10^{10}/9 \times 10^9$
Br^-	$HOBr/OBr^-$	160	
$HOBr/OBr^-$	$HBrO_2/BrO_2$	<0.01/530	
$HBrO_2/BrO_2$	BrO_3^-	$-/<10^5$	
Cl^-	$HOCl$	~0.003	
I^-	HOI/OI^-, IO_3^-	1.2×10^9	
$HOCl/OCl^-$	ClO_3^-	>0.002/120	
$HClO_2$	ClO_3^-	>10^4	
ClO_2	ClO_3^-	1100	

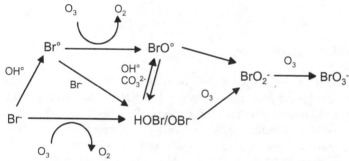

Figure 2.3 Reaction scheme for bromate formation (after von Gunten 2003 [17]). Bold lines show the main pathway.

Mizuno et al. [24] developed a model to predict the formation of bromate ion as well as hypobromous acid/hypobromide ion through the radical. This model could be used for drinking-water treatment processes to simulate the formation of these species.

Analogous to bromate formation from bromide, it is possible for other halogens to be oxidized to their respective oxyhalide forms, for example, chlorate, iodate. Iodate, however, is not considered a carcinogenic risk because it is quickly reduced to iodide in the body [20]. In addition, iodide concentrations in natural waters are usually fairly low (<10 µg l^{-1}). For chlorate the WHO gives no guideline value for drinking water, however, it is suggested that it should be kept as low as possible. Toxicological reports have shown that chlorate can have toxic effects [25]. In Switzerland, the drinking water standard for chlorate is 200 µg l^{-1} [26], while the state of California has set an action level of 800 µg l^{-1} for chlorate in drinking water [27].

Further information about the reaction rates as well as the rate constants with ozone and OH° can be found in many publications, for example, Glaze, 1987 [28]; Yao and Haag, 1991 [29]; Haag and Yao, 1992 [30]; Hoigné and Bader, 1983 [1, 2] as well as Hoigné and Bader, 1985 [19].

Generally speaking, direct ozonation is important if the radical reactions are inhibited. Such is the case if the water either does not contain compounds that initiate the chain reaction (initiators) or if it contains many that terminate the chain reaction very quickly (scavengers). With increasing concentrations of scavengers the mechanism of oxidation tends to the direct pathway. Therefore, both inorganic carbons as well as organic compounds play an important role.

Normally, under acidic conditions (pH < 4) the direct pathway dominates, above pH = 10 it changes to the indirect.

In ground and surface waters (pH ≅ 7) both pathways–direct and indirect–can be of importance [3]. In special wastewaters even at pH = 2 the indirect oxidation can be of importance, depending much on the contaminants/promoters present [31]. Both pathways should always be considered when developing a treatment scheme.

2.2
Advanced Oxidation Processes (AOP)

Advanced oxidation processes (AOPs) have been defined by Glaze et al. [28] as processes that "involve the generation of hydroxyl radicals in sufficient quantity to effect water purification". AOPs are used to specifically generate OH°. The OH° have a higher oxidation potential (2.8 eV) than molecular ozone (2.07 eV) and can attack organic and inorganic molecules rapidly and nonselectively. The most common processes are O_3/H_2O_2, O_3/UV and H_2O_2/UV. Research continues on AOP development as more cost-effective methods are sought. For example, using catalytic ozonation with dissolved or solid catalysts can reduce operating costs for pH adjustment since it is equally effective under both highly acidic and highly alkaline conditions or substituting solar energy for UV lamps can reduce energy costs. Each of these processes involves chemistry similar to that discussed above in the previous section. An overview of the reactions involved in the most common processes is presented below.

2.2.1
Ozone/Hydrogen Peroxide O_3/H_2O_2

Hydrogen peroxide reacts with ozone when present as an anion, HO_2^-. The reaction rate is dependent on the initial concentration of ozone/hydrogen peroxide.

$$H_2O_2 \leftrightarrow HO_2^- + H^+ \quad pK_a = 11.8 \quad (2.16)$$

$$HO_2^- + O_3 \rightarrow HO_2^\circ + O_3^{\circ-} \quad k_{11} = 2.2 \times 10^6 \ M^{-1} \ s^{-1} \quad (2.17)$$

The reaction of ozone with the undissociated hydrogen peroxide, which would lead to a loss of ozone and hydrogen peroxide, is negligible [32]:

$$H_2O_2 + O_3 \rightarrow H_2O + 2O_2 \quad k_{10} < 10^{-2} \ M^{-1} \ s^{-1} \quad (2.18)$$

The hydroperoxyl radical HO_2° and the ozonide anion $O_3^{\circ-}$ produced in Equation 2.17 then enter the chain reaction of the indirect pathway (2.2–2.7) to produce OH° [5, 7].

Comparison of the initial reaction of ozone with HO_2^- ($k_9 = 2.2 \times 10^6 \ M^{-1} s^{-1}$) and to that of ozone with OH^- ($k_1 = 70 \ M^{-1} s^{-1}$) shows that in the O_3/H_2O_2 system the initiation step by OH^- is negligible. Whenever the concentration of hydrogen peroxide is above $10^{-7} M$ and the pH-value less than 12, HO_2^- has a greater effect than OH^- has on the decomposition rate of ozone in water.

The combination of Equations 2.2–2.7, 2.17 and 2.18 shows that two ozone molecules produce two OH°:

$$2O_3 + H_2O_2 \rightarrow 2OH^\circ + 3O_2 \quad (2.19)$$

2.2.2
Ozone/UV-Radiation O_3/UV

The advanced oxidation process with ozone and UV-radiation is initiated by the photolysis of ozone. The photodecomposition of ozone leads to hydrogen peroxide [33].

$$O_3 + H_2O \xrightarrow{h\nu} H_2O_2 + O_2 \quad (2.20)$$

Guittoneau et al. [34] confirmed that one mole of H_2O_2 is formed from one mole of ozone at 254 nm and pH < 1.8, this ratio decreases with the increase of pH.

This system contains three components to produce OH° and/or to oxidize the pollutant for subsequent reactions:

⇒ UV-radiation
⇒ ozone
⇒ hydrogen peroxide

Therefore, all removal mechanisms should be considered when evaluating this AOP. Direct oxidation by hydrogen peroxide can usually be neglected under normal conditions (pH between 5 to 10 and ambient temperature). However, direct photolysis of the pollutant can occur if it absorbs the wavelength used. Depending on the conditions, direct and indirect ozone reactions with the pollutant are possible. In addition, combinations of the three components – ozone/hydrogen peroxide or UV-radiation/hydrogen peroxide – also produce OH° that contribute to the overall results.

Ultraviolet lamps must have a maximum radiation output at 254 nm for an efficient ozone photolysis.

2.2.3
Hydrogen Peroxide/UV-Radiation UV/H_2O_2

The direct photolysis of hydrogen peroxide leads to $OH°$:

$$H_2O_2 \xrightarrow{h\nu} 2\,OH° \qquad \varepsilon_{254\,nm} = 18.6\ M^{-1}\ cm^{-1} \qquad (2.21)$$

In addition, the ionized form of hydrogen peroxide HO_2^-, which is in acid–base equilibrium (see 2.16), also absorbs the wavelength 254 nm, and decomposes to produce one $OH°$ and an oxygen anion radical $O^{°-}$, which can produce with H_2O another $OH°$:

$$HO_2^- \xrightarrow{h\nu} OH° + O^{°-} \qquad \varepsilon_{254\,nm} = 240\ M^{-1}\ cm^{-1} \qquad (2.22)$$

$$O^{°-} + H_2O \rightarrow OH° + OH^- \qquad (2.23)$$

Propagation:

$$H_2O_2 + OH° \rightarrow H_2O + HO_2° \qquad (2.24)$$

$$HO_2° + H_2O_2 \rightarrow H_2O + O_2° + OH° \qquad (2.25)$$

Termination of this radical chain reaction includes the subsequent radical–radical recombination:

$$OH° + OH° \rightarrow H_2O_2 \qquad (2.26)$$

$$HO_2° + OH° \rightarrow H_2O + O_2 \qquad (2.27)$$

Figure 2.4 gives an overview of the reactions involved in the AOPs. The reaction rate constants for all these reactions (2.1–2.23) have been taken from several references (e.g., [31, 35, 36]) and are presented to illustrate the various orders of magnitude. It is possible to find different values for one reaction. If a model is going to be used to describe the measured data, the appropriate reaction constant has to be chosen to achieve the best fit.

From the chemical point of view, the effect of O_3/UV is comparable to that of O_3/H_2O_2 if direct photolysis is negligible [28, 37]. Table 2.4 summarizes the chemistry involved in the generation of hydroxyl radicals from the described four processes.

In principle, the most direct method for the generation of hydroxyl radicals is through the combination of hydrogen peroxide with UV. By the photolysis, 100% of the hydrogen peroxide is transformed into $OH°$. But the extinction coefficient ε of ozone at the wavelength of 254 nm is much higher ($\varepsilon_{254\,nm} = 3300\ M^{-1}\ cm^{-1}$)

2.2 Advanced Oxidation Processes (AOP)

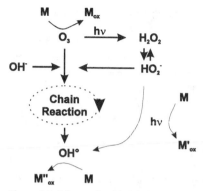

Figure 2.4 Advanced oxidation processes.

Table 2.4 Theoretical amount of oxidants and UV required for the formation of hydroxyl radicals in ozone-peroxide–UV systems [28].

System	Moles of oxidants consumed per mole of OH° formed		
	O_3	UV[a]	H_2O_2
Ozone–Hydroxide Ion[b]	1.5	–	–
Ozone–UV	1.5	0.5	(0.5)[c]
Ozone–Hydrogen Peroxide[b]	1.0		0.5
Hydrogen Peroxide–UV	–	0.5	0.5

a Moles of photons (Einsteins) required for each mole of OH° formed.
b Assumes that superoxide O_2^- is formed that yields one OH° per O_2^-, may not be the case in certain waters.
c Hydrogen peroxide formed *in situ*.

than that of hydrogen peroxide ($\varepsilon_{254\,nm} = 18.6\,M^{-1}cm^{-1}$) [38], so the photolysis of ozone yields more OH° than that from hydrogen peroxide for the same energy input (see Table 2.5). Therefore, to achieve comparable treatment results, higher dosages of hydrogen peroxide or longer treatment time are necessary with UV/H_2O_2.

This comparison shows the theoretical advantages of the various reactions. In reality, a high production of OH° can lead to a low reaction rate because the radicals recombine and are not useful for the oxidation process. Also not considered are the effects of different inorganic and/or organic compounds in the water that can either accelerate or decelerate the reaction.

Table 2.5 Theoretical formation of hydroxyl radicals from the photolysis of ozone and hydrogen peroxide [27].

	ε_{254nm} in $M^{-1} cm^{-1}$	Stoichiometry	OH° formed per incident photon[a]
H_2O_2	20	$H_2O_2 \rightarrow 2\ OH°$	0.09
O_3	3–300	$O_3 \rightarrow 2\ OH°$	2.00

a Assumes 10 cm path length; $c(O_3) = c(H_2O_2) = 10^{-4}$ M.

Further information concerning the parameters that influence the concentration of hydroxyl radicals is given in Section 7.4, as well as a short overview about AOPs in Section 9.1 Various models to calculate the actual OH-radical concentration can be found in the literature, some are described in Chapter 8.

References

1 Hoigné, J. and Bader, H. (1983) Rate constants of reactions of ozone with organic and inorganic compounds in water–I. Non dissociating organic compounds. *Water Research*, **17**, 173–183.
2 Hoigné, J. and Bader, H. (1983) Rate constants of reactions of ozone with organic and inorganic compounds in water–II. Dissociating organic compounds. *Water Research*, **17**, 185–194.
3 Staehelin, J. and Hoigné, J. (1983) Reaktionsmechanismus und kinetik des ozonzerfalls in wasser in gegenwart organischer stoffe. *Vom Wasser*, **61**, 337–348.
4 Tomiyasu, H., Fukutomi, H. and Gordon, G. (1985) Kinetics and mechanisms of ozone decomposition in basic aqueous solutions. *Inorganic Chemistry*, **24**, 2962–2985.
5 Staehelin, J. and Hoigné, J. (1982) Decomposition of ozone in water: rate of initiation by hydroxide ions and hydrogen peroxide. *Environmental Science and Technology*, **16**, 676–681.
6 Buxton, G.V., Greenstock, C.L., Helman, W.P. and Ross, A.B. (1988) Critical view of rate constants for oxidation of hydrated electrons, hydrogen atoms and hydroxyl radicals (OH°/O°⁻) in aqueous solutions. *Journal of Physical and Chemical Reference Data*, **17**, 513–884.
7 Bühler, R.E., Staehelin, J. and Hoigné, J. (1984) Ozone decomposition in water studies by pulse radiolysis. 1. HO_2/O_2^- and HO_3/O_3^- as intermediates. *Journal of Physical Chemistry*, **88**, 2560–2564.
8 Chelkowska, K., Grasso, D., Fabian, I. and Gordon, G. (1992) Numerical simulation of aqueous ozone decomposition. *Ozone: Science and Engineering*, **14**, 33–49.
9 Hoigné, J. (1982) Mechanisms, rates and selectivities of oxidations of organic compounds initiated by ozonation of water, in *Handbook of Ozone Technology and Applications*, Vol. 1 (eds R.G. Rice and A. Netzer), Ann Arbor Science Publishers, Ann Arbor, MI, pp. 341–379.
10 Mizuno, T., Tsuno, H. and Yamada, H. (2007) Development of ozone self-decomposition model for engineering design. *Ozone: Science and Engineering*, **29**, 55–63
11 Von Sonntag, C. (1996) Degradation of aromatics by advanced oxidation processes in water remediation: some basic considerations. *Journal Water Supply Research and Technology-Aqua*, **45**, 84–91.
12 Xiong, F. and Graham, N.J.D. (1992) Research note: removal of atrazine through ozonation in the presence of humic substances. *Ozone: Science and Engineering*, **14**, 283–301.

13 Staehelin, J. and Hoigné, J. (1985) Decomposition of ozone in water in the presence of organic solutes acting as promoters and inhibitors of radical chain reactions. *Environmental Science and Technology*, **19**, 1206–1213.
14 Hoigné, J. and Bader, H. (1977) Beeinflussung der oxidationswirkung von ozon und OH-radikalen durch carbonat. *Vom Wasser*, **48**, 283–304.
15 Forni, L., Bahnemann, D. and Hart, E.J. (1982) Mechanism of the hydroxide ion initiated decomposition of ozone in aqueous solution. *Journal of Physical Chemistry*, **86**, 255–259.
16 Francis, P.D. (1987) Oxidation by UV and ozone of organic contaminants dissolved in deionized and raw main water. *Ozone: Science and Engineering*, **9**, 369–390.
17 Von Gunten, U. (2003) Ozonation of drinking water: Part I. oxidation kinetics and product formation. *Water Research*, **37**, 1443–1467.
18 Langlais, B., Reckhow D.A. and Brink D.R. (eds) (1991) *Ozone in Water Treatment – Application and Engineering*. Cooperative Research Report. American Water Works Association Research Foundation: Company Générale des Eaux and Lewis Publisher, Chelsea, MI, ISBN: 0-87371-477-1.
19 Hoigné, J. and Bader, H. (1985) Rate constants of reactions of ozone with organic and inorganic compounds in water – III. Inorganic compounds and radicals. *Water Research*, **19**, 993–1004.
20 Von Gunten, U. (2003) Ozonation of drinking water: Part II. Disinfection and by-product formation in presence of bromide, iodide or chlorine. *Water Research*, **37**, 1469–1487.
21 WHO (2004) *Guidelines of Drinking Water Quality: Recommondations*, Vol. 1, 3rd edn, World Health Organization, Geneva, http://www.who.int/water_sanitation_health/dwq/GDWQ2004web.pdf.
22 EU (1998) *Official Journal of the European Community*, **L330**, Directive 98/83/EG.
23 USEPA (1989) National primary drinking water regulations: final rule. *Federal Register*, **54**, 27485–27541

24 Mizuno, T., Tsuno, H. and Yamada, H. (2007) A simple model to predict formation of bromate ion and hypobromous acid/hypobromite ion through hydroxyl radical pathway during ozonation. *Ozone: Science and Engineering*, **29**, 3–11.
25 Snyder, S.A., Vanderford, B.J. and Rexing, D.J. (2005) Trace analysis of bromate, chlorate, iodate, and perchlorate in natural and bottled waters. *Environmental Science and Technology*, **39**, 4586–4593.
26 Swiss Federal Department of Interior (2000) *Ordinance on Contamination in Food*, Swiss Federal Department of Interior, Bern, Switzerland. Verordnung des EDI über Fremd – und Inhaltsstoffe in Lebensmitteln (FIV) vom 26. Juni 1995 (Stand 25. Mai 2009) 817.021.23.
27 Shane, A., Snyder, R., Pleus, C., Vanderford B.J., Holady J.C. (2006) Perchlorate and chlorate in dietary supplements and flavor enhancing ingredients, *Analytica Chimica Acta*, **567**(1), 26–32.
28 Glaze, W.H. and Kang, J.-W. (1987) The chemistry of water treatment processes involving ozone, hydrogen peroxide and ultraviolet radiation. *Ozone: Science and Engineering*, **9**, 335–352.
29 Yao, C.C.D. and Haag, W.R. (1991) Rate constants for direct reactions of ozone with several drinking contaminants. *Water Research*, **25**, 761–773.
30 Haag, W.R. and Yao, C.C.D. (1992) Rate constants for reaction of hydroxyl radicals with several drinking water contaminants. *Environmental Science and Technology*, **26**, 1005–1013.
31 Beltrán, F.J., Encinar, J.M. and García-Araya, M. (1993) Oxidation by ozone and chlorine dioxide of two distillery wastewater contaminants: gallic acid and epicatechin. *Water Research*, **27**, 1023–1028.
32 Taube, H. and Bray, W.C. (1940) Chain reactions in aqueous solutions containing ozone, hydrogen peroxide and acid. *Journal of the American Chemical Society*, **62**, 3357–3373.
33 Peyton, G.R. and Glaze, W.H. (1987) Mechanism of photolytic ozonation, *ASC Symposium–Series 327*, American

Chemical Society, Washington DC, pp. 76–88.

34 Guittonneau, P., De Laat, J. and Dore, M. (1990) Kinetic-study of the photodecomposition of aqueous ozone by UV irradiation at 253.7 nm. *Environmental Technology*, **11**, 477–490.

35 Paillard, H., Brunet, R. and Doré, M. (1988) Optimal conditions for applying an ozone/hydrogen peroxide oxidizing system. *Water Research*, **22**, 91–103.

36 De Laat, J., Tacc, E. and Doré, M. (1994) Etude de l'oxydation de chloroethanes en milieu aqueux dilue par H_2O_2/UV. *Water Research*, **28**, 2507–2519.

37 Prados, M., Paillard, H. and Roche, P. (1995) hydroxyl radical oxidation processes for the removal of triazine from natural water. *Ozone: Science and Engineering*, **17**, 183–194.

38 Guittonneau, P., Glaze, W.H., Duguet, J.P. and Wable, O. (1991) Characterization of natural waters for potential to oxidize organic pollutants with ozone. Proceedings 10th Ozone World Congress and Exhibition, Monaco, International Ozone Association, Zürich, Switzerland.

3
Ozone Applications

Ozone is applied in a wide range of processes due to its strong oxidizing properties. Its ability to oxidize compounds can be used for their destruction or for the synthesis of new chemicals. Ozone application has increased enormously both in number and diversity since its discovery by Schönbein in 1839. This historical development and the diversity of ozone applications will be explored in the beginning of this chapter (Sections 3.1–3.2).

While the oxidative properties of ozone can be used on reaction partners in all three phases: gas/liquid/solid, the main focus of this book as well as this chapter is on the use of ozone in water applications. Ozone has been integrated into production processes that utilize its oxidizing potential, for example, bleaching in the pulp and paper industry, metal oxidation in the semiconductor industry. However, its main application is in the treatment and purification of many types of water: ground and surface waters for drinking-water use, domestic and industrial waste waters for reuse or discharge to natural water bodies, as well as waters in swimming pools and cooling-tower systems. These are discussed in the last two sections of the chapter (Sections 3.3–3.4). This is rounded off by a brief discussion of the economical aspects of ozone use in Section 3.5.

3.1
Historical Development

Ozone is produced naturally by the discharge of lightening as well as artificially by the discharge of electricity in the presence of oxygen. It owes its name and discovery to its distinctive smell. The Dutch scientist van Marum described the "odor of electricity" in 1785 in his experiments with electrical discharges in oxygen from the then-largest static electricity machine [1, 2]. However, it was not until 1839 that Schönbein, noting the smell at the anode of an electrolytic cell, postulated the production of a new substance as its source. He named it after the Greek word for smell *"ozein"*. Although the smell of lightening has been noted since ancient times, it was usually associated with sulfur until Schönbein identified the two smells as being one and the same in 1840. In fact, Schönbein had postulated early that ozone is present in the atmosphere based on the smell. Already in 1857 enough measurements on ambient ozone concentrations in larger cities had been made, that the daily and seasonal variations in the ozone concentration over urban areas were recognizable from analysis of the available data [3]. The main source of ozone production in the troposphere remains atmospheric photochemical UV-induced reactions with atmospheric pollutants rather than man-made generators.

Schönbein's discovery spurred intensive research on ozone's properties in the ensuing years. Its strong oxidizing properties, similar to chlorine, were reported already in 1845 by Schönbein, who tested its ability to bleach colors and its danger to living beings through inhalation on a mouse [3]. One of the first methods to measure ozone was based on its ability to oxidize potassium iodide to give elementary iodine, leading to the starch-iodide reaction as a test for ozone. It was recognized early on that ozone was an allotrope of oxygen and in 1865 Soret concluded from experimental observations that O_3 was the molecular formula for ozone [2].

The development of the silent discharge apparatus by Siemens in 1857 and its ability to produce larger quantities of ozone, spurred widespread investigations of its multipurpose capabilities. These investigations quickly turned up a wide range of applications – from killing microorganisms (1873), artificial ageing of liquors (1890) [4], food-preservation agent (1909) [5], improving ventilation of buildings (1912) [6], to cleaning and sterilizing drinking water (1886) [7] and waste water.

At the turn of the nineteenth century ozone units started to be installed for disinfecting surface water for use as drinking water; it was first used in the Netherlands in 1893 in Oudshoorn. The technology initially spread quickly, reaching approximately 50 installations by 1915. The first ones were small ($12\,000\,m^3\,d^{-1}$ in 1898, Paris, France; $6000\,m^3\,d^{-1}$ in 1901, Wiesbaden, Germany) and grew in size ($90\,000\,m^3\,d^{-1}$ in 1909 Paris, France; $47\,000\,m^3\,d^{-1}$ in 1910, Petersburg, Russia) [4].

While disinfection was the initial purpose, it was soon realized that multiple positive effects could be achieved with ozone such as taste and odor improvement, and color removal. It was mainly the last stage of treatment until the 1960s (post-ozonation), when in order to utilize ozone's ability to oxidize manganese and iron, its coagulating effects and even more recently micropollutant oxidation, it was moved up in the treatment train [8]. Before the coagulation and sedimentation stage, the term pre-ozonation is used. When it is placed before the filtration stage it is called intermediate ozone. The filtration stage is then often a combination of filtration to remove suspended particles and biological activity to remove the biodegradable compounds produced by ozonation. This ability of ozone to breakdown biorefractory compounds into biodegradable products is exploited in chemical biological processes that are mainly used in waste-water applications. Often, a biological stage to remove biodegradable compounds is added before the ozonation stage to reduce the amount of ozone needed, in addition to the biological stage after. These effects will be discussed further in Sections 3.3, 3.4, and 9.3. The placement of the ozone stage in the treatment train and the terms used for the various configurations are illustrated in Figure 3.1.

Ozonation use increased in France after the 1920s, reaching over 200 DWTP in the 1960s and has remained steady since then. In contrast, ozonation of drinking water supplies declined in importance in the USA and Germany in the 1920s as

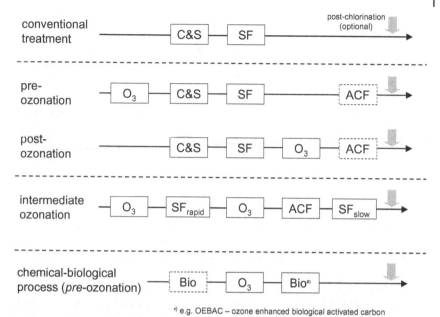

Figure 3.1 Various configurations of multistage drinking-water treatment trains. Stages included are: ozonation (O_3), coagulation and sedimentation (C&S), sand filtration (SF) activated-carbon filtration (ACF), and biodegradation (Bio).

high-quality groundwater replaced surface waters as the source for drinking water and applications for surplus chlorine production were sought [4]. Since chlorine is a waste product from the production of sodium hydroxide, which was in great demand for industrial processes in the early part of the twentieth century, its price was low, as were the investment costs for the dosing equipment. A major advantage of chlorine in its many forms was that it could be produced offsite and stored onsite. Moreover, it could be dosed to maintain residual disinfection effects in the distribution network. The development of high-volume pumping capabilities and well techniques also played an important role. High-quality groundwater requiring little or no treatment replaced surface water as the drinking-water source. Furthermore, improved treatment processes using a strategy of multiple barriers made the use of disinfectants unnecessary [9].

The high investment costs and low service life of ozone generators at that time brought about the decline of ozonation in drinking-water treatment, especially in the United States. So that in 1980 there were less than 10 known water-treatment plants using ozone in the USA. The rapid increase in ozone use in the USA to over 300 now [10] was caused partly by improvement in ozone technology and equipment performance, but mainly by changes in drinking-water regulations due to concerns over disinfectant by-products. In 1998 both the US EPA and the EU promulgated regulations that, by limiting halogenated disinfection by-products in drinking water, encouraged the use of alternative disinfection methods [11, 12]. For example, the US and EU standards for total trihalomethanes are 80 and

100 µg l^{-1} respectively. In California alone there are as many as 44 DWTPs using ozone with a total ozone-generation capacity of 32 000 kg d^{-1} (70 000 ppd [pounds per day]) and it is expected that by 2013 the number of large installations will increase by 21, resulting in a doubling of the ozone-generation capacity and an increase in the total amount of water treated with ozone to as much as 20.5 Mio m^3 d^{-1} (5.4 billion gallons per day) [13].

Similar concerns over disinfection by-products spurred ozone's growth in the food-processing industry as an antimicrobial and disinfection agent. Ozone applications in food processing, water disinfection, packaging cleansing and equipment sterilization increased after ozone was given the status as generally regarded as safe (GRAS) for bottled water (US FDA [14]) and approved in general as a antimicrobial agent in food processing in the USA [15] and in Japan [16]. Examples of treatment conditions for some applications can be found in [17]. A further spread in ozone applications may be prompted by the EU Integrated Pollution Prevention and Control (IPPC) Directive 96/61/EC and activities to evaluate ozone's potential to be considered a best available technology (BAT) for cleaning and disinfection in food-processing plants [18].

Industrial ozone generators span a wide range of production capacity, for example, from 0.1 kg O$_3$ h^{-1} up to 200 kg O$_3$ h^{-1} from a single unit [19]. In general, the specific energy consumption for ozone production is dependent on the type of ozone generator and on the type of feed gas. Approximately 70% of the total energy costs in ozone processes are for ozone generation, while the rest is used for process-gas preparation, the gas contactor, off-gas treatment and auxiliary processes. Since as much as 90% of the energy used in ozone production is lost as heat, the large ozone generators used in municipal drinking-water preparation require water cooling to remove the heat. Smaller sized units are usually air-cooled. General types of generators for the various applications are listed in Table 3.1.

Table 3.1 Overview of applications and type of ozone generator used.

Area	Application	Typical generator	Feed gas	Generator cooling
Municipal	Drinking water	0.5–50 kg h^{-1}	Air/LOX/PSA	Drinking water
	Waste water	2–200 kg h^{-1}	LOX/(V)PSA	Chilled water
Industrial	Waste water	1–50 kg h^{-1}	Air/LOX/PSA	Chilled water
	Process water	0.2–20 kg h^{-1}	Air/LOX	Chilled water
	Food processing	0.1–1 kg h^{-1}	Air/LOX	Chilled water
	Pulp bleaching	50–500 kg h^{-1}	PSA/VPSA	Chilled water
Residential	Drinking water	<0.5 kg h^{-1}	Air	Drinking water
	Pools/Spas	0.1–2 kg h^{-1}	Air	Chilled water
Lab-scale	Experiments	1–10 g h^{-1}	O$_2$/Air	Air cooled

The ozone concentration that can be achieved depends on the feed gas. As a general rule, twice as much ozone can be produced with oxygen than with air using the same amount of energy [8]. With air, ozone concentrations between 1–3% (wt.) using roughly up to $20\,kWh\,kg\,O_3^{-1}$ can be reached, while with oxygen higher concentrations 6–13% (wt.) using ~$10\,kWh\,kg\,O_3^{-1}$ are usual. However, the energy to handle the pure oxygen delivered as liquid oxygen (LOX), approx. $0.5\,kWh\,N\,m^{-3}\,O_2$, or produce it onsite via pressure swing absorption (PSA) or vacuum pressure swing absorption (VPSA) is not considered in this comparison.

Improved ozone generators with longer service life have pushed the expansion of ozone applications, in addition to the regulatory changes. However, while ozone's oxidative power and multiple effects make it a good choice for many industrial processes, its remains underutilized in many areas due to its relatively high capital and operating costs. The multiple benefits of ozonation are often hard to quantify and calculate into cost comparisons. Where health concerns and regulatory aspects play a role though, ozone use is growing. For example, compared with traditional chlorine treatment ozonation still is four times more expensive, however, it yields higher treatment efficiency and thus is gaining interest in the USA again [20].

In addition, improved understanding and control of the reaction system can lead to reduced operating costs. Great strides have been made in process automation and control so that optimized reactor operation can reduce costs. Furthermore, improved tools for reactor assessment and optimization such as computational fluid dynamics (CFD) are also available. CFD has often been used in recent years to assess and optimize the hydrodynamics of full-scale ozonation reactors for drinking-water treatment. For instance, modeling helped determine how mass transfer could be improved in an existing plant, resulting in the doubling of the ct-value as well as the inactivation efficiency [21]. Its use has greatly improved the possibilities and ease of modeling nonideal systems [22] so that if the disinfection efficiency has to be improved, the effect of constructive measures to improve hydrodynamics can be evaluated, instead of only considering increased ozone concentrations [23].

3.2
Overview of Ozone Applications

The applications of ozone can be broken down into destruction and synthesis. The destructive uses of ozone are more widespread, although ozone's ability to react selectively at an unsaturated carbon center is also used in organic chemistry to transform molecules to produce desired end-products or intermediates in the synthesis of natural and non-natural products. For example in medical chemistry, molecules can be modified by ozone to produce aldehydes or ketones that can be further transformed to biologically active alcohols, carboxylic acids or amines, depending on the conditions used in further steps [24]. The first step, oxidation with ozone, is usually carried out with ozone dissolved in a solvent at low temperatures. Further applications are in the area of polymer production. Ozone can be used to modify natural or synthetic polymers, functionalizing them with acid or hydroxyl groups in a variety of applications [25, 26] – for example, modification of industrial polypropylene microfiltration membranes for improving performance through a higher hydrophilicity [27] or for synthesizing designed functional polymers in biotechnology [28] – via reactions leading to polymer peroxides, which are useful initiators of radical polymerization. In addition, ozone's selective reactions have made it an extremely useful tool for structure determination of natural products.

Most destructive applications rely on the modification of the compound (e.g., a change in the oxidation state of a metal or surface structure, destruction of odoriferous sidechains) or inactivation of a pathogen, only a few rely on complete oxidation (e.g., mineralization of an organic micropollutant to CO_2 and H_2O). Both direct and indirect (via hydroxyl radicals) reactions can play a role.

The destructive properties of ozone can be used on reaction partners in all three phases: gas/liquid/solid. Ozone itself is a gas at standard conditions. It can be used as such or dissolved in a liquid. In general, ozone applications can be grouped into four main areas:

- disinfection (or pathogen control);
- oxidation of inorganic compounds;
- oxidation of organic compounds, including taste, odor, color removal; and
- particle removal.

These areas are explored briefly in Sections 3.2.1 and 3.2.2 for both gas- and liquid-phase applications, before expanding on them for drinking- and waste-water applications in Sections 3.3 and 3.4.

3.2.1
Ozone in the Gas Phase

While this book concentrates on the use of ozone in waters, the uses of ozone in the gas phase are manifold. A variety of uses are based on ozone's disinfectant properties, its ability as a biocide to kill living organisms, viruses and spores. For instance, in agriculture it can be used as a grain fumigant for insect and fungal control in storage with no detrimental effect on grain quality [29, 30] as well as an herbicide in weed control under plastic covering in the field [31] or a pesticide for treatment of soilborne pathogens through root-zone injection [32, 33].

Ozone gas has also found application in pathogen control in food preservation and packaging [5], in sterilization of medical instruments and equipment as well as in therapeutic uses in medicine – externally (perfusions for wounds, lesions, etc.), or internally (insufflation for ulcerative colitis, etc.) [34]. Water content in the gas plays a crucial role in its effectiveness in many applications. In general, inactivation rates of various pathogens increase with increasing relative humidity [35,

36]. However, a minimum relative humidity is often required. For example, no significant inactivation of Bacillus spores was attained at a relative humidity below 50% [36]. Since the reaction rate of ozone in the gas phase is usually much slower than in the liquid phase, residence times must be chosen appropriately [37].

Further uses of ozone in the gas phase depend on its ability to oxidize organic and inorganic compounds. For example, it can be used in *in-situ* soil and groundwater remediation where it is injected into the soil or dosed directly into the groundwater [38]. Other uses range from continuous applications such as treating off-gases from industrial processes or agricultural facilities (e.g., animal waste lagoons), and odor control in animal confinements [39] or in human buildings, to one-time treatment in commercial applications such as fire restoration or deodorization of housing and automobiles [40].

The air may contain hazardous compounds (e.g., phenols, cyanides, dioxans, etc.) to be destroyed or odorous compounds (e.g., ammonia, hydrogen sulfide, mercaptans, fatty acids, aldehydes and amines) to be neutralized. Predicting the efficacy of ozone in particular systems is difficult to impossible. This is especially the case with odors, since the identification of their chemical composition and the conditions that contribute to their dispersion are complex [41]. Therefore, testing is usually necessary to determine the required doses and cost effectiveness.

In the case of off-gas treatment in reactors, the dose or residence time can be increased in the reaction system until treatment is effective. In addition, ozonation can be combined with a solid catalyst [42] or a liquid scrubber [43] in order to increase the efficiency of the gas treatment. In direct applications dosing ozone to the ambient air, the maximum dose allowable is restricted by safety considerations for the exposed materials, animals or humans. In fire restoration or deodorizing applications where odors are eliminated on surfaces in unoccupied enclosures, the dose is limited by the sensitivity of the material being treated.

In controlling indoor air pollution in human buildings or odor in animal confinements, however, the concentration of ozone allowed in the ambient air is severely limited by public health standards. The US EPA concluded from the available scientific evidence that the concentration of ozone would have to greatly exceed health standards to be effective in removing most indoor air contaminants. The reaction products are an additional concern, since they themselves can be irritating, odorous or corrosive [44]. Therefore, current applications for improving indoor air quality rely on applying ozone in higher concentrations either at times when the building is unoccupied, with the goal that ozone has reacted to below public health standards before people enter the building or in the air-recirculation system, dosing so that the ozone reacts within the duct system [40].

While higher concentrations of ozone may be tolerable in animal confinements, similar concerns over effectiveness at allowable doses and oxidation by-products also apply here. Some odor contributors such as ammonium and volatile fatty acids are not effectively oxidized by ozone [45, 46]. Furthermore, a recent study found that oxidation of dust particles can increase the level of fine/ultrafine particles and aerosols [45].

3.2.2
Ozone in the Liquid Phase

There are a myriad of applications using dissolved ozone. While aqueous ozone can be produced *in-situ* electrolytically, in most cases ozone is produced as a gas and must be transferred into the liquid. Ozone can be dissolved in organic solvents to exploit its higher solubility for the synthesis of compounds, as already mentioned above on synthetic organic chemistry [24, 26], or for the destruction of organic compounds in oil emulsions in waste-water treatment [47]. In medical uses, oils can be ozonated to increase their antimicrobial activity for ointments [48, 49] and blood can be ozonated externally and reintroduced into the patient [50].

The most widely used liquid for the absorption of ozone, however, is water. Generally, the target substance to be oxidized is present in the water into which the ozone is absorbed. But some applications exist in which water is used as a carrier to transport dissolved ozone to the target substance. Here again the applications can be broken down into the four groups according to the primary goal of ozonation: disinfection, oxidation of inorganics, organics, and particle removal. Ozone is usually chosen for one or two major purposes, although it can produce multiple effects. An overview of various ozone applications is found in Table 3.2 with the primary and secondary goals associated with them.

The disinfectant properties of aqueous ozone can be used to treat the water into which it is transferred (e.g., drinking water, process water, swimming pools [51], cooling-tower systems [52, 53], ballast water in ships [54]) or ozonated water can be used as a delivery system to bring dissolved ozone in contact with pathogens in a variety of applications. For example, in dental and medical applications ozonated water can be rinsed or applied topically (caries-causing bacteria [55, 56]), swallowed (gastritis) or irrigated (chronic intestinal or bladder inflammation [57]). In agriculture, application of aqueous ozone to the soil has reduced soilborne pathogens [58] and in the food and beverage industry, ozonated water is used for clean-in-place (CIP) applications [18] to inactivate a variety of common molds, viruses and bacteria [59].

In some applications a mixture of both methods occurs. For instance, in food processing where water is used as a transport medium, the water as well as food surfaces in contact with the water are disinfected. Similarly in bottled water, the water is contacted with ozone to safely ensure its disinfection and a residual ozone level is maintained at the time of bottling to provide an additional safety factor to disinfect the bottles.

The ability of aqueous ozone to oxidize inorganics and organics has found many applications in industrial processing. Its ability to oxidize metal surfaces is exploited in the semiconductor industry and in coating processes. It has been integrated into production processes that utilize its bleaching ability, for example, in the pulp and paper industry and textile finishing, as well as in laundries. Ozone in laundries allows washing to be conducted using cold water, thereby saving considerable heat energy and water consumption [60]. In pulp-bleaching

Table 3.2 Overview of primary (++) and secondary (+) goals in some water-ozonation applications.

Removal goals	Drinking water	Food industry	Swimming pools	Cooling water	Industrial effluents	WWTP effluents
Disinfection/biocide	++	++	++	++		++
Improved particle separation	+	+				+
Iron & manganese elimination	+	++		+		
Removal of organics						
Natural organic matter (NOM)	+	+				
Micropollutants (pesticides, pharmaceuticals, etc.)	++				++	+
AOX removal	++				++	
Taste enhancement	+	++				
Deodorization	+	++	+			
COD or DOC removal	+		+		+	+
Improved biodegradability	+				++	+
Reuse goals						
Water recycling (in-house)					++	
Water reuse (e.g., irrigation or groundwater infiltration)					+	++

processes, ozonation in combination with oxygen and hydrogen peroxide can fully replace chlorine-containing chemicals. Thus, the production of high concentrations of AOX is avoided and the plant may be run according to the "closed mill" concept. The typical ozone consumption is in the range of 2–4 kg ozone per ton (Mg) of pulp [61]. One of the world-largest industrial ozonation systems is in operation in a pulp-bleaching process in Finland with a production capacity 420 kg O_3 h^{-1} (22 000 ppd) [62]. An additional advantage is that the remaining oxygen in the off-gas is used for several other oxygen-consuming processes in the bleaching sequence [62]. Another of the world's largest industrial ozonation systems is located in Cincinnati (USA), where 90 ozone generators produce as much as $570\,kg\,O_3\,h^{-1}$ (30000 ppd). Here, ozone converts animal fats to esters, acids, etc. [63]. Further applications in the chemical and paper industries are summarized in [64].

While industrial use of the versatile and powerful oxidant are more often found in production processes where value is added, rather than in waste treatment, due to the costs and operation requirements (skilled personnel and safety considerations), there still are some applications. Ozone can be found especially in areas requiring space-saving compact installations that avoid waste streams (waste water or ballast water treatment on ships); in process-water treatment and recycling schemes, in detoxification/removal of hazardous or persistent compounds (hazardous waste or industrial waste-water treatment), or where regulations and/or toxicological considerations have restricted use of chlorine-based oxidants as disinfectants in municipal waste-water treatment plants (MWWTP) [65–67]. In addition, interest in ozonation for the removal of persistent compounds from secondary effluent from MWWTP is increasing, especially with the intention of water reuse after groundwater infiltration or aquifer storage [20, 68].

In the following two sections, applications of ozone for drinking-water and waste-water treatment are examined with special attention to full-scale systems. While the source of the water plays an important role in the treatment required, in both sections, the discussion is arranged according to the four main treatment goals:

- disinfection (or pathogen control);
- oxidation of inorganic compounds;
- oxidation of organic compounds, including taste, odor, color removal; and
- particle removal.

3.3
Ozone in Drinking-Water Treatment
Martin Jekel

Producing high-quality drinking water is a constant challenge since the quality requirements continue to rise as more and more chemical pollutants and microorganisms, such as the cysts and oocysts of parasites (*Giardia, Cryptosporidium*) are identified in source waters and concern over disinfection by-products increases. These concerns have induced renewed interest in ozonation and ozone-based advanced oxidation processes. Their effectiveness is based upon the multiple effects produced by the oxidative and disinfective activity of ozone and ozone-derived oxidizing species such as OH-radicals. These effects can be utilized for disinfection, oxidation of inorganic or organic compounds, including taste, odor and color removal as well as for particle removal.

Typically, ozone is chosen for one or two major purposes, but several side-effects may exist, inducing positive and/or negative effects that need thorough consideration. For instance, care has to be taken in the application of ozonation, since research has shown that hazardous by-products can be formed, for example, bromate in the ozonation of waters containing bromide. Further discussion of this aspect is found in Chapters 1 and 2 and in von Gunten [69].

Drinking-water supplies can be based on natural ground waters (the source with highest priority), on artificially recharged ground waters or bank-filtered surface waters, on lakes and dam reservoirs and on river waters. Most applications of ozone are found in water-treatment systems dealing with polluted surface waters and contaminated underground waters, whereas pure ground waters are either not treated or require only removal of ferrous and manganese ions and/or stabilization. In addition, ozone can be used to treat waters in swimming pools, cooling-tower systems or for other commercial purposes. The design of the treatment process and choice of operating conditions have to be made to utilize or reduce the multiple effects of ozonation.

Nearly all ozonation effects and their respective extent and kinetic pattern depend on the amount of ozone consumed in the ozone contactor and subsequent reactors. This requires the search for and definition of optimal operational parameters for an ozonation stage, such as a concentration–time-value (ct-value) for a given degree of disinfection or the ratio of ozone mass consumed per mass of organic compounds initially present.

Another consideration connected with these multiple effects is the optimum placement of the ozonation stages within a whole treatment scheme. Ozonation can be used as an early stage in the treatment train (pre-ozonation), where its ability to oxidize manganese and iron as well as micropollutants and DBP precursors, and/or its coagulating effects is used as a pretreatment to improve removal in the following stages. It can be used as the primary disinfection stage where it is placed in the middle (intermediate ozonation) or near the end of the train (post-ozonation) (see Figure 3.1). However, ozone cannot be used as a secondary disinfectant to maintain water quality throughout the distribution system up to the tap because it decays too rapidly. The efficiency of every ozonation unit and the ozone demand depend on the water quality produced by the preceding process units (e.g., particle removal or biodegradation). Ozonation will also have pronounced effects downstream in the treatment sequence, for example, improved biodegradation of dissolved organics.

While the source of the water plays an important role in the treatment required, the following overview of ozonation in full-scale drinking-water treatment is not arranged according to the water source, but rather according to the four main effects of ozone on the water constituents and treatment goals. In addition, attention is given to the appropriate combination of the ozonation process with preceding and subsequent treatment steps. More detailed information on the mechanisms and goals of ozone and associated oxidation processes in drinking-water treatment is provided by Camel and Bermond (1998) [70], while extensive coverage of the design and operation of drinking-water ozonation plants is found in Langlais et al. (1991) [8]. Practical experience and extensive knowledge on how to run a drinking-water treatment plant (DWTP) employing ozone is reported in the work of Rakness (2005) [71].

3.3.1
Disinfection

The introduction of ozone in water treatment started about a century ago and was directed at the disinfection of microbiologically polluted water in order to stop waterborne disease. Later, chlorine and also chlorine dioxide were introduced and have been used successfully to control pathogenic pollution, excluding the parasitic organisms. In view of this problem and the well-known formation of halogenated disinfection by-products (especially *tri*-halomethanes, THMs) by chlorine, there is renewed interest in the use of ozone for disinfection, but in the intermediate stages of treatment, not as the last step. The short lifetime of dissolved ozone and the production of biodegradable organics (BDOC, sometimes also called assimilable organic carbon, AOC) from natural organic matter (NOM) do not favor its use as a final process, but rather its positioning before a rapid filtration/activated-carbon filtration/slow sand filtration or underground passage (see Figure 3.1). Ozone is an essential part of the multiple-barrier principle against pathogenic organisms (and organic pollution). It is strongly recommended to remove most particulate material before ozonation to prevent encapsulated microorganisms from escaping an ozone attack and to reduce the ozone demand, which helps in establishing a "free ozone residual", that is, a residual concentration of dissolved ozone, for a certain time.

In the design of chemical disinfection, the concept of ct (free disinfectant concentration c multiplied by the available contact time t) is frequently applied, based on the law of Chick (1908) [72] and Watson (1908) [73]. ct-values have been reported for various microorganisms to achieve a given degree of inactivation, like a two or three log reduction in the concentration of microorganisms (99 or 99.9% removal), but careful application of these data is recommended with respect to different water sources. For example, a study of the main parameters of process design for waste-water disinfection by ozonation has shown that the transferred ozone dose is the critical parameter and that the ct concept should not be used in waste-water ozonation [67].

Contaminated water can contain a multitude of bacteria, viruses and parasites excreted in animal and human faeces. It is generally accepted today that molecular ozone is a very effective and promising disinfectant, often better than free chlorine, chlorine dioxide, chloramines or hydroxyl radicals. The relative resistance of microorganisms follows roughly the order: bacteria, viruses and parasite cysts. Very often, a ct value of $1.6-2\,\mathrm{mg\,l^{-1}\,min}$ (e.g., $0.4\,\mathrm{mg\,l^{-1}}$ ozone for 4 or 5 min) is considered to be sufficient for effective disinfection during postozonation [74], that is, after particulate matter is removed down to low turbidities (less than ca. 0.2 NTU). Research has shown that parasites require a much higher ct-value for a two or three log removal than viruses, for example, for a two-log removal at 5 °C *Cryptosporidium parvum* oocysts can require at least $33\,\mathrm{mg\,l^{-1}\,min}$ [75], while *Giardia muris* cysts require a ct of $1.8-2.0\,\mathrm{mg\,l^{-1}\,min}$ compared with 0.02 and $0.2\,\mathrm{mg\,l^{-1}\,min}$ for *E. coli* and *Poliovirus 1*, respectively [76]. In addition, the water temperature plays a large role in determining the required ct-values.

The function between ct and temperature depends on the organism. For example, for every 10 °C decrease in temperature the ct for a two-log removal of *C. parvum* oocysts increases threefold [75], while it generally increases twofold for most viruses and Giardia [77].

An important secondary effect to be minimized in ozone disinfection is the formation of bromate (BrO_3^-) in waters that contain bromide (Br^-), which is usually the case as concentrations in the range of 10 to 1000 µg l^{-1} are found in natural waters. Since bromate is a potential carcinogen, its concentration in drinking water is limited to 10 µg l^{-1} in many countries (e.g., EU, USA) and this value has also been set as provisional guideline concentration by the World Health Organization (WHO) [78]. As a rule of thumb bromate is formed when the residual concentration of dissolved ozone in the DWTP is about 0.1 mg l^{-1} or higher, and in general its concentration increases linearly with higher ozone exposures (ct-values) [78]. Therefore, if a high degree of disinfection is required involving a high ct-value, low bromate concentrations may be difficult to achieve, depending on the bromide concentration in the source water. This is especially valid if a 2-log (99%) inactivation of *Cryptosporidium parvum* oocysts has to be achieved for which a temperature-dependent range of ct of 3.1–48.0 mg l^{-1} min is recommended by the USEPA 2006 [79]. Since the chemical processes behind bromate formation are very complex, but nevertheless well understood [69], drinking-water treatment with ozone to reduce DBP formation is often a difficult optimization problem.

Various strategies can be used to control bromate production such as lowering the ozone concentration or the pH-value to less than six, or dosing ammonia or hydrogen peroxide. A good overview of the matter with special focus on bromate modeling and control strategies is given by Jarvis (2007) [78]. Another area to be considered to reduce ozone concentrations in full-scale applications is the optimization of the contactors using computational fluid dynamics (CFD). CFD has greatly improved the possibilities and ease of modeling nonideal systems [22] so that if the disinfection efficiency has to be improved, the effect of constructive measures to improve hydrodynamics can be evaluated, instead of only considering increased ozone concentrations [23].

In swimming-pool water clean-up ozone is normally used with the objective to minimize the formation of halogenated disinfection by-products (DBPs), for example, *tri*-halomethanes (THMs). In a typical pool-water ozonation system ozone is applied in a recirculation stream, prior to subsequent chlorine disinfection. Further treatment steps in pool water treatment are flocculation and sand filtration. According to the German standard DIN 19643 ozonation of pool water is performed with a ct-value of 2.4–15 mg l^{-1} min (0.8–1.5 mg l^{-1} for 3–10 min) [80]. A comparative study on the application of ozone or the AOPs, O_3/H_2O_2 and O_3/UV, has shown that O_3/H_2O_2 treatment was superior to O_3/UV and ozonation alone and resulted in a net THM reduction at a reasonable increase in the operational costs [81].

Typical ct-values in drinking- and swimming-pool-water ozonation are summarized in Table 3.3.

Table 3.3 Typical ct-values and ozone dosages (c) for disinfection in drinking- and swimming-pool-water ozonation.

Application	Typical values			Reference/remarks
	ct	c	t	
	g O_3 m^{-3} min	g O_3 m^{-3}	min	
Drinking-water treatment				
Preozonation	5–10	–	–	[82]
Preozonation in intermediate ozonation treatment train	–	1.1	50	[82] DWTP Lengg, Zurich, CH, 70 000 m^3 d^{-1}
Postozonation	1.6–2	0.4	4–5	low turbidity <0.2 NTU
Swimming-pool water				
Typical system	2.4–15	0.8–1.5	3–10	[80] German Standard (DIN)
Typical system	–	1.0 (28 °C) 1.5 (35 °C)	–	[62] example of temperature dependence of c

3.3.2
Oxidation of Inorganic Compounds

Whereas the use of ozonation to oxidize inorganic surfaces in the semiconductor industry is growing (see Section 9.4), ozonation for the oxidative removal or transformation of inorganic constituents of drinking and waste waters is a rather rare application, because other methods exist for most of the target compounds. However, inorganic compounds may be oxidized as a secondary effect of ozonation for other purposes (particle removal, organics oxidation). Table 3.4 provides an overview of the target and product compounds and a qualitative indication of the rate of oxidation in drinking waters.

As was already mentioned above a critical ozone reaction with inorganics is the formation of bromate, a potential carcinogen, from bromide in the water source. The WHO standard is set at 10 µg l^{-1} and is applied in all DWTPs in Europe and the United States of America. If bromate formation is a problem, possible measures to limit bromate formation are: adjusting the ozone dosage, lowering the pH to less than 6 or dosing a small amount of ammonia or hydrogen peroxide ([78, 84]). To remove bromate is difficult, but could occur in activated-carbon filters ([83, 85–87]).

As shown in Table 3.4, ozone can destroy other disinfectants. This should be avoided by dosing them not ahead of ozonation stages, but rather at the end of the total treatment before the distribution of water to the supply net (see Figure

Table 3.4 Oxidation of inorganic compounds by ozonation ([8, 82]).

Compound	Products	Rate of oxidation	Remarks
Fe^{2+}	$Fe(OH)_3$	Fast	Filtration of solids required; application in the beverage industry
Mn^{2+}	$MnO(OH)_2$	Fast	Filtration of solids required; application in the beverage industry
	MnO_4^-	Fast	At higher residual ozone conc., reduction and filtration required
NO_2^-	NO_3^-	Fast	Nitrite is a toxic compound
NH_4^+/NH_3	NO_3^-	Slow at pH < 9 Moderate at pH > 9	Not relevant
CN^-	CO_2, NO_3^-	Fast	Application in waste water
H_2S/S^{2-}	SO_4^{2-}	Fast	Not relevant
As-III	As-V	Fast	Preoxidation for subsequent As removal
Cl^-	HOCl	Near zero	Not relevant
Br^-	$HOBr/OBr^-$ BrO_3^-	Moderate	Bromination of organic compounds possible; bromate as toxic by-product
I^-	HOI/OI^-, IO_3^-	Fast	Not relevant
$HOCl/OCl^-$	ClO_3^-	Slow	Loss of free chlorine
Chloramines, Bromamines		Moderate	Loss of combined chlorine
ClO_2 ClO_2^-	ClO_3^- ClO_3^-	Fast Fast	Loss of free chlorine dioxide
H_2O_2	$OH°$	Moderate	Basis of O_3/H_2O_2 process, (AOP)

3.1). A special case is the reaction of ozone with H_2O_2 (correctly with the species HO_2^-). Here, the "destruction" of ozone is intended and the combination of ozone and hydrogen peroxide (sometimes called the *"peroxone"* process) is used as an advanced oxidation process (AOP) for intensified formation of hydroxyl radicals and their oxidative attack on persistent organic target compounds (persistent against ozone in the direct reaction mechanism) (see Chapter 2).

3.3.3
Oxidation of Organic Compounds

3.3.3.1 Natural Organic Matter (NOM)
All water sources may contain natural organic matter, but concentrations (usually measured as dissolved organic carbon, DOC) differ from 0.2 to more than $10\,\mathrm{mg\,l^{-1}}$. NOM is a direct quality problem due to its color and odor, but more important are indirect problems, such as the formation of organic disinfection-by-products (DBPs, e.g., *tri*-halomethanes (THMs) due to chlorination), support of bacterial regrowth in the distribution system, disturbances of treatment efficiency in particle separation, elevated requirements for coagulants and oxidants or reductions in the removal of trace organics during adsorption and oxidation, etc.

Removal of NOM or its alteration to products less reactive to chlorine is a priority task in modern water treatment. Various processes can be used such as chemical oxidation by ozone, biodegradation, adsorption, enhanced coagulation or even membrane technologies. A DOC-level of approximately $1\,\mathrm{mg\,l^{-1}}$ appears to be the lower limit of ozone applications, but a few cases exist where waters with lower concentrations of NOM (ground water) have been treated.

The position for NOM-oxidation in water treatment schemes often is an intermediate one, for example, between settling/flotation and rapid filtration or between rapid filtration and activated-carbon filters or other post-treatment units (see Figure 3.1). A decisive operational parameter for organic carbon removal is the specific ozone consumption $D(O_3)*$, which is the ratio of $g\,O_3$ consumed per $g\,DOC$ initially present.

The tasks of NOM-ozonation are [70]:

- removal of color and UV-absorbance;
- increase in biodegradable organic carbon ahead of biological stages;
- reduction of potential disinfection-by-product formation, including *tri*-halomethanes;
- direct reduction of DOC/TOC-levels by mineralization.

The first three treatment goals are much more relevant and applicable to full-scale plants than the last one. The reason is that the ozone demand for direct chemical mineralization is usually very high, typically requiring a specific ozone consumption $D(O_3)*$ of more than $3\,\mathrm{g\,O_3\,g^{-1}\,DOC}$ to achieve a removal efficiency of 20% or more.

The removal of color and UV-absorbance is one of the easier tasks due to quick reactions and comparatively low specific ozone consumption requirements in the range below $1\,\mathrm{g\,O_3\,g^{-1}\,DOC}$. Thus, this effect is observed in preozonation steps for improved particle separation. Color can be removed by 90% or more, while UV-absorbance at 254 nm is commonly reduced to 20–50% of the initial value. The reaction mechanism here is primarily the direct ozone attack on C double bonds in aromatic and chromophoric molecules leading to the formation of "bleached" products, like aliphatic acids, ketones and aldehydes. This oxidative reaction with UV/VIS-active substances induces molecular changes, but not mineralization. These changes are also the basis for the production of biodegradable metabolites and the formation of smaller molecules with a higher hydrophilicity, that tend to form less DBPs with the chlorine disinfectant.

For optimal production of biodegradable DOC (AOC) specific O_3-consumptions of about 1–2 g g^{-1} are advised. Higher ratios lead to an enhanced oxidation of intermediates to carbon dioxide (direct mineralization). The AOC/DOC ratio after oxidation may be 0.1 to 0.6, and is frequently found to be 0.3–0.5. The AOC-formation prohibits the direct supply of ozonated NOM waters to the distribution system, due to severe bacterial regrowth after ozone decay. It is essential to add a treatment step with high bacterial activity (rapid filters, activated-carbon filters, underground passage, slow sand filters) to remove AOC and achieve a microbiologically "stable" water.

The reduction in DBP-formation also depends on the specific ozone consumption. Typical reductions are in the range of 10 to 60% (compared to nonozonated water), at specific ozone dosages between 0.5 to 2 g O_3 g^{-1} DOC initially present. If bromide is present, brominated organic DBPs and bromate formation may occur.

Complete mineralization of NOM does not appear to be economical, compared with partial oxidation and biodegradation. Typical intermediate organic metabolites (like oxalic acid) are difficult to oxidize by molecular ozone; nonselective OH-radicals are better suited. Therefore, AOP processes, which are designed to produce OH-radicals through the addition of H_2O_2 or via UV-irradiation, may achieve better mineralization (see Section 9.1).

If ozone is used not only for NOM treatment, but also for disinfection, then the necessary ozone dosage has to be chosen based on the highest requirement for either DOC removal or the ct-value for disinfection. The latter objective may be the dominating one in the case of raw waters with microbiological contamination.

3.3.3.2 Organic Micropollutants

Organic micropollutants are found in surface and ground waters, always in conjunction with more or less NOM, but at low concentrations in the range of 0.01 µg l^{-1} to 100 µg l^{-1} (in water sources of sufficient quality for a water supply). Their degradation by ozone to oxidized metabolites or even to mineral products is a complex process, due to the influences of various water-quality parameters (pH, inorganic and organic carbon, etc.) on the two known major reaction pathways: direct electrophilic ozone reaction and the oxidation via the nonselective, fast-reacting OH-radicals.

In practical ozone applications, micropollutant (or trace organic) oxidation has not been a primary task, but was considered to be a positive side-effect. However, due to the development of modern analytical tools and the detection of a large number of micropollutants in source waters (some of them being potentially health-hazardous) interest in trace organic oxidation has been growing continuously. During the 1980s the detection of micropollutants such as pesticides was the major reason to introduce and study AOP techniques, because most were only poorly accessible to a direct ozone attack. Since the mid 1990s concern about endocrine-disrupting chemicals (EDC) has increased. In the meantime a great number of chemicals such as pharmaceuticals, cosmetic or personal care products (PPCP) has been detected in very low concentrations in raw water sources in Europe and the USA [68, 88].

It must be kept in mind that in nearly all cases, the target compounds will not be mineralized, but rather transformed to by-products, which are typically more polar in nature and smaller in molecular weight. Quite often some of the products formed do not react further with ozone, the so-called dead-end products. A complete removal of organic products does not occur and it is essential to have a subsequent treatment unit, such as biological filtration systems (if the by-products are degradable) or an adsorption on activated carbon. In the latter case, the oxidized trace pollutant may be less adsorbable, due to the increase in hydrophilicity. If oxidized products are left in the water, their toxicological evaluation is recommended and should be compared to the risk of the original pollutant.

Numerous publications are available on ozonation and ozone-based AOPs for the oxidation of specific organic compounds, either in pure waters, model waters or in full-scale systems (see, e.g., [70, 89, 90]). Kinetic constants k_D and k_R for the direct (molecular) and indirect (radical) oxidation are reported and provide good insight into the range of oxidation rates, however, other water-quality parameters exhibit a strong influence, especially on the indirect oxidation by OH-radicals. This may result in considerably differing observed rate constants, which depend on the individual "water matrix". Von Gunten (2003) has given a comprehensive review of this matter [69, 84].

The list of organic groups in Table 3.5 provides a qualitative presentation of expected degrees of removal in full-scale drinking-water treatment plants.

3.3.4
Particle-Removal Processes

All surface waters contain particles of different origin, sizes and materials, which must be removed efficiently before water distribution. There is a renewed interest in improved particle separation due to the hygienic problems with infectious cysts and oocysts of parasites (*Giardia, Cryptosporidium*), which are particles in the size range of 3–12 mm. Depending on the raw-water quality, particle separation may typically be accomplished by:

- rapid or slow filtration;
- coagulation/flocculation/deep-bed filtration or coagulation/flocculation/floc separation;
- settling or flotation and rapid filtration.

Table 3.5 Degree of removal of trace organics during ozonation in full-scale drinking-water treatment plants.

Substances	Typical range of reaction rate constants k_D or k_R in l mol^{-1} s^{-1}		Degree of removal by O_3, range in %	Remarks
	O_3	OH·		
Taste and odor	n.a.	10^6–10^9	20–90	Water source specific
(S-organics)	10–10^3	10^9–10^{10}	n.a.	
2-Methylisoborneol, Geosmin	<10	10^9–10^{10}	40–95	Improvement by AOPs
Alkanes	10^{-2}	10^6–10^9	<10	AOPs support
Alkenes	10^3–10^5	10^9–10^{11}	10–100	oxidation; chlorinated
Chlorinated alkenes		10^7–10^9		alkenes are more difficult to oxidize
Aromatics	1–10^2	10^8–10^{10}	30–100	Highly halogenated
Phenols	10^2–10^4	10^9–10^{10}	n.a.	phenols are more difficult to oxidize by ozone
Aldehydes	10^{-1}–1	10^9	low	Typical products of
Alcohols	10^{-2}–1	10^9–10^{10}		ozonation, easily
Carboxylic acids	10^{-5}–10	10^7–10^9		biodegradable
N-containing aliphatics and aromatics	10^{-2}–10^2	10^8–10^{10}	0–50	AOPs may increase oxidation rate
Pesticides	10^{-2}–10^2	10^9–10^{10}	0–80	Very substance specific
Atrazine	10	10^9–10^{10}		triazines (e.g., atrazine) require AOPs
Polyaromatic hydrocarbons	10^3–10^5	10^9–10^{10}	High, up to 100	AOPs do not increase oxidation rate

n.a. = not available.

Also, membrane processes such as nano-, micro- and ultrafiltration have been studied and introduced for near-to-complete particle removal.

It has been observed for more than 40 years that "preozonation" ahead of particle-removal units can improve the efficiency significantly, can induce a lower coagulant demand or allow higher flow rates, for example, in deep-bed filtration. A typical scheme for a surface-water treatment including preozonation for particle removal is shown in Figure 3.1. Ozone gas is added either before or together with the coagulant (ferric or aluminum salts or cationic polymers) at rather low dosages of 0.5–2 mg l^{-1}. The terms "microflocculation" or "ozone-induced particle destabilization" are used in practice [90].

The mechanisms involved appear to be rather complex and several mechanistic models have been described (for a review see Jekel [90]). Results from the references therein as well as from additional pilot and full-scale applications indicate that an optimal ozone dosage exists, typically in the lower range of 0.5–2 mg l^{-1} or, related to the DOC, 0.1–1 mg mg^{-1}. The optimal point must be determined by tests in the combined treatment.

The relative improvements (reference is particle removal without ozone) are quite variable, but were reported to be about 20–90% lower turbidities and/or lower particle counts in the filtered water. The presence of dissolved organic matter is frequently essential and DOC should be at least 1 mg l^{-1}. The preozonation effects depend strongly on the presence of alkaline-earth cations, especially calcium.

Positive effects of preozonation are found in algae removal, usually a difficult task. Preozonation may be combined with flotation, an effective technique for separating coagulated algae. The algae cells are not destroyed at the low dosages required. In addition, ozonation is an effective process for destruction of both intracellular and extracellular algal toxins.

The preozonation effects are detected in treatment for particle removal with and without coagulants. Reduced species, like Fe^{2+}, Mn^{2+} or NO_2^- are oxidized quickly and may precipitate ($Fe(OH)_3$, $MnO(OH)_2$), also supporting coagulation. Depending on the amount of oxidizable material present, the preozonation dosage may be insufficient to establish an ozone residual in the water, but in cleaner waters the ct-value necessary to disinfect the water effectively may be met, meaning that the dissolved ozone concentration remains high enough during a certain reaction time.

The ozonation reactors for preozonation have to deal with the particle content of raw waters and are sometimes combined with the coagulant mixing tank. Suitable transfer devices for the ozone-containing gas are injector systems, radial diffusers or turbines with blades.

3.4
Ozonation in Waste-Water Treatment

The application of ozonation for the treatment of waste waters can be found in almost all branches of industry, treating a variety of waste waters and contaminants, as well as in municipal waste-water treatment plants for purposes ranging from disinfection and micropollutant oxidation of the effluent for water reuse to the reduction in the mass of biosolids. Ozone-generation capacity in the various applications can vary to a large extent. It may range from 0.1 kg h^{-1} to 500 kg h^1 (see Table 3.1) depending on the amount of waste water, its content and the treatment goal, so that there can be no general definition of what "full-scale" is in waste-water applications. Furthermore, the required ozone dose for waste-water ozonation depends on the type of industry and the kind of waste water. Important aspects that influence these requirements are:

- the removal goal: oxidative transformation of specific compounds due to their toxicity or color, decrease in lumped parameters ($SUVA_{254}$, DOC or COD), disinfection, or particle removal;
- the discharge requirements: in-house pretreatment for water recycling or indirect discharge to WWTPs, end-of-pipe treatment for direct discharge to rivers or bays;
- the overall treatment scheme: only chemical processes or combinations: chemical/biological and/or physical.

Since variable costs for energy and oxygen can play a decisive role in the choice of a waste-water treatment system, waste-water ozonation is often embedded into a multistage chemical-biological process (CBP) to reduce ozone consumption. Most often a CBP system employs biodegradation at least before and also often after the chemical oxidation step (see Figure 3.1). In general, savings in the amount of ozone used and other costs have often been achieved by the application of such combined ozonation/biodegradation systems.

The most frequently used contactors in large-scale waste-water ozonation systems are bubble-column reactors equipped with diffusers or Venturi injectors, mostly operated in a reactor-in-series countercurrent continuous mode. Many full-scale ozone reactors are operated at elevated pressure (2–6 bar_{abs}) in order to achieve a high ozone-transfer efficiency, which also increases the overall process efficiency.

The discussion of ozonation systems for waste waters in the following sections is grouped according to the main removal goal of the application, analogous to that used in drinking-water ozonation systems.

3.4.1
Disinfection

Disinfection of waste waters may be carried out when treated waste-water effluent is directly reused for irrigation or process-water applications or may be required before discharge of effluent into receiving waters to meet water-quality standards in some countries. This is true for municipal waste waters in the United States, where discharge permits normally include standards for fecal coliform. The standards are part of the National Pollutant Discharge Elimination System (NPDES) permit required for each WWTP. In Europe, bacteriological standards are not part of the EU Urban Wastewater Treatment Directive. Disinfection of municipal effluents is being implemented in some areas, however, due to the need to conform to the standards in the EU Bathing Water Directive for surface waters used for bathing.

Chlorine is most often used in the USA, for example, in approximately 75% of the WWTP installations in the USA [66]. However, use of chlorine can cause formation of chlorinated disinfection by-products (DBPs) such as tri-halomethanes (THM). Therefore, in analogy with the situation in drinking-water treatment, chlorine-free processes such as UV treatment or ozonation, either alone or in combination with H_2O_2, are treatment alternatives. While the use of ozone in more than 40 municipal waste-water treatment plants was reported for the year 1994 in the United States [71], UV applications are more widespread. A survey of 4450 WWTP by the Water Environment Research Foundation (2008) showed that ozonation of waste-water effluents from municipal WWTPs is very seldom applied today, for example, in as few as 0.2% of the utilities (7 out of 4450) while UV is applied in 918 cases (20.6%) and no disinfection in 173 plants (3.9%) [66]. The study lists several reasons for the discontinued use of ozone: changes in legislation, operational and maintenance problems in first-generation ozone facilities, inability to attain performance objectives without major modifications, and high ozone-dosage requirements at some facilities. However, due to advances in ozone-generation technology over the last decades, ozone has become more economically competitive. Moreover, the secondary effects from ozone disinfection such as the oxidation of compounds responsible for odor or color or improved removal of suspended solids as well as micropollutant removal also cause increased interest in its use.

This highlights a potential problem for direct water reuse. Although microorganism reactivation after ozonation is not likely to occur, biodegradable by-products (BDOC) like aldehydes, carboxylic acids, and ketones are formed during ozonation from the remaining DOC in the WWTP effluent, bacterial growth in distribution systems and applications using the recycled water is possible [67]. The removal of BDOC due to biodegradation by immobilized bacteria can be achieved in a subsequent treatment stage employing rapid sand filtration or in a granular activated-carbon unit (sometimes referred to as the ozone-enhanced biological activated-carbon process OEBAC). Further discussion of such combinations can be found in Section 9.3 (see also Figure 3.1). Another concern with respect to indirect potable-water reuse applications, that is, after groundwater infiltration, is the formation of bromate from bromide, which is caused by applying ozone doses greater than the immediate ozone demand resulting in low but measurable dissolved ozone concentrations [67]. Similar observations have been reported in [68], that is, bromate was only formed if dissolved ozone was present.

3.4.2
Oxidation of Inorganic Compounds

Ozonation of inorganic compounds in waste waters with the aim to destroy very toxic substances is mostly restricted to cyanide removal. For example, it is used to remove cyanide in the waste waters of gold and/or silver extraction from mineral ores [62]. Conversely, ozonation has also been found to be a potentially cost-efficient process for the recovery of cyanide from thiocyanate in such applications in pilot-scale testing [91]. Cyanide is also frequently used in galvanic processes in the metal processing and electronics industry, where it not only appears as free cyanide (CN^-) but more often occurs in complexed forms associated with iron or copper. While ozone reacts very fast with free cyanide, so that the process is likely to be mass transfer controlled at cyanide concentrations above $5\,mg\,l^{-1}$ [92], complexed cyanides are more stable to the attack of molecular ozone [93]. Consequently, in order to oxidize both forms of cyanide, the application of nonselective hydroxyl radicals is more promising. The H_2O_2/UV process is an efficient and easy-to-handle treatment technology for this application [94].

Nitrite (NO_2^-) as well as sulfide (H_2S/S^{2-}) removal from waste waters is sometimes performed by ozonation. Both substances react fast with ozone [83]. Biological treatment, though may be a cost-efficient treatment alternative for these substances, for example, biological denitrification or biological sulfide removal. Another important inorganic oxidation reaction is the above-mentioned undesired oxidation of bromide. This topic was dealt with in detail in Section 3.3.2.

3.4.3
Oxidation of Organic Compounds

The majority of problematic substances in industrial waste waters and also in WWTP effluents are organic compounds. Often, a complex mixture, composed of many individual substances present in a wide range of concentrations (from ng to $g\,l^{-1}$) has to be treated. The predominant tasks associated with ozone treatment of waste waters containing organic compounds are:

- the transformation of toxic compounds that often occur as micropollutants in a complex matrix, for example, in very low concentrations in the μg or ng per liter range;

- the partial oxidation of the biologically refractory part of the DOC, mostly applied with the aim to generate biodegradable DOC (BDOC) in order to improve subsequent biodegradation;

- the removal of color.

Ozonation is rarely used as a standalone treatment for waste water. Similar to the treatment of drinking waters, a near-to-complete mineralization of the DOC cannot be achieved economically and the combination of ozonation with other processes is recommended. The ozonation stage may be a pretreatment before discharge to an offsite biological WWTP or integrated into an onsite biological combination. The success of the whole treatment train should be measured by the overall DOC removal.

Full-scale ozonation systems have been used to treat waste waters, such as landfill leachates, waste waters from the textile, pharmaceutical and chemical industries as well as effluents from municipal WWTP (Table 3.6). These are often used in conjunction with biological treatment stages. For example, waste water comprised of a mixture of domestic sewage (20–60%) and industrial (80–40%) waste waters predominantly from pharmaceutical production is first treated in a conventional activated sludge process in the WWTP in Kalundborg (Denmark) before ozonation. Using an ozone dose of $0.18\,kg\,O_3\,m^{-3}$ approximately $1200\,m^3\,h^{-1}$

Table 3.6 Overview of technological features, operating parameters and treatment costs of full-scale plants for waste-water ozonation.

Ref.	Type of WW	Type of treatment system	No. and type of ozone reactors (operating pressure)	Ozone production capacity	Nominal // real liquid flow rate	Ozone yield coefficient $Y(O_3/M)$ $(M = COD)$	Specific costs (without annuity)	Remarks
				kg O_3 h^{-1}	m^3 d^{-1}	kg O_3 kg^{-1} ΔM	EUR m^{-3}	
[64, 95]	Industrial & municipal (80 & 20%)	Bio-O_3	2 × 3 BC in series	2 × 90	1000	~3.5	0.1–0.3	MWWTP Kalundborg (DK); industrial WW from pharmaceutical production
[96]	Landfill leachate	Bio-O_3-Bio (sequential)	3 BC (1 bar$_{abs}$)	12	n.a. // 40–108	1.6–2.0	3-5 (energy only) 31 (total)	Controlled Ca-oxalate precipitation and recycling to 1st Bio; only 6% variable costs, low potential for optimization
[97]	Landfill leachate	Bio-NF-O_3-Bio (recycle)	1 BC with fixed bed catalyst (approx. 4 bar$_{abs}$)	6	50 ≡ 10 // n.a.	1.5–1.8 0.8	n.a.	ECOCLEAR®; ozone used for NF-concentrate; O_3-transfer at <2.0 kWh kg^{-1} O_3
[98]	Landfill Leachate	Bio-O_3-Bio (integrated)	1 BC	1–8	n.a. // 1.0–20	0.9–1.2	2.5–7.5	BioQuint® process, 12 plants in operation since 1995

Ref	Application	Process	Reactor configuration	Size	Flow	Ozone dose	Energy	Notes
[99, 100]	Landfill leachate	Bio-O$_3$-Bio (sequential)	2 or 3 BC + 1 IZR (1 bar$_{abs}$)	12	70–140 // ≤250	BC: 2.3–3.2 IZR: 1.8–2.5 (depending on HCO$_3^-$ content)	BC and IZR: 15–50 (depending on COD load)	Conventional BC-Venturi system, reactor made from PVC; IZR made from duplex steel (1.4462) for high chloride content; no foaming, O$_3$-transfer at 2.0 kWh kg^{-1}
[64]	Textile	Bio-O$_3$	4 BC (1 bar$_{abs}$)	4 × 40	120000	n.a.	0.2	Mainly decolorization, oxidation of surfactants to <1.5 mg l^{-1}, water reuse in textile factories
[64]	Textile	Bio-O$_3$	n.a.	2 × 14	2500	n.a.	~0.2	Decolorization and tenside removal
[101]	Industrial	Bio-O$_3$-Bio	2 BC in series (1 bar$_{abs}$)	10–15	n.a. // 144–600	~1.5	9	Nitrite, nitroaromatics and polyether-alcohols in influent foaming,
[98]	Pulp & paper	Bio-O$_3$	n.a.	40–100	n.a. // 300–1000	n.a.	0.05–0.25	final polishing: removal of odor, color, AOX and COD

n.a. = not available.

of the biologically treated effluent is oxidized, effectively reducing the COD from approx. 110 to 70 mg l^{-1} before discharge. In this full-scale application the operating costs amount to roughly 0.30 € per m^3 water treated [102]. According to Ried *et al.* [64] more than 40 combined ozonation and biodegradation applications exist in Europe, but many of them are not published. Only the data of 16 utilities, mostly installed in Germany, are covered in this paper. This is also the reason why Table 3.6 also contains some older data.

The main pollutants associated with these waters are refractory xenobiotic organics, which are sometimes called micropollutants, and include [103, 104]:

- humic compounds (brown or yellow colored) and adsorbable organic halogens (AOX) in the landfill leachates;
- colored (poly-)aromatic compounds often incorporating considerable amounts of metal ions (Cu, Ni, Zn, Cr) in textile waste waters;
- endocrine-disrupting chemicals (EDCs) or pharmaceutical or personal care products (PPCPs) from the pharmaceutical and/or chemical industry;
- toxic or biocidal substances (e.g., pesticides);
- detergents from the cosmetic and other industries;
- COD and colored compounds in solutions of the pulp and paper production.

Advanced micropollutant removal with ozonation and/or AOPs in conjunction with biologically enhanced activated-carbon filtration is not only technically but also economically feasible. This could drastically reduce the micropollutant load discharged to the environment after (centralized) biological treatment in MWWTP. The increase in specific costs was estimated at 0.05 to 0.20 € m^{-3}, however, widespread full-scale testing has still to be performed [105]. Pilot-scale testing of the ozonation of waste-water effluent as a pretreatment step before groundwater infiltration resulted in good treatment efficiency with regard to disinfection, removal of color and DOC. It was found to be practically feasible in terms of specific costs of 0.01 to 0.05 € m^{-3} in the case of a hypothetical full-scale application of 200 000 m^3 d^{-1} in Berlin, Germany [68].

The most frequent operational problems in waste-water ozonation systems are foaming and the formation and precipitation of calcium oxalate, calcium carbonates and ferrous hydroxide (Fe(OH)$_3$) that may easily clog the reactor, piping and valves and also damage the pumps.

Goals, technology and results of full-scale applications on some types of waste waters are discussed in more detail in the following sections. An overview of technological features, operating parameters and treatment costs of full-scale plants for waste-water ozonation is given in Table 3.6.

3.4.3.1 Landfill Leachates – Partial Mineralization

Ozonation systems are used to oxidize the highly biorefractory organic compounds in landfill leachate. Over thirty systems were located in Germany in 1999 [62]. The comparatively widespread application in Germany is mainly due to the legal requirements on the treatment of leachate before discharge [106], which have spurred the development of efficient treatment systems. Commonly, the effluent of such plants is directly discharged to the receiving water, which requires the effluent meet limits on COD below 200 mg l^{-1}, an AOX level below 500 µg l^{-1} and a toxicity factor to fish lower than $G_F = 2$.

Due to the high concentration and complex nature of the organic content of landfill leachates, the ozonation stage is mostly operated in combination with biological processes, that is, often between two biological systems (Bio-O$_3$-Bio) [64]. In the first biological stage almost all easily biodegradable organic compounds are removed. The biological stage can be designed to remove the nitrogen-species as well; nitrification/denitrification can remove ammonium, nitrite and nitrate to low levels. Remaining is often a considerable amount of biorefractory organic compounds. Here, ozone is used in order to partly oxidize these substances, with the goal of increasing their biodegradability in the subsequent biological treatment stage. An important advantage of this process combination is that no secondary wastes are produced, as would be the case if the O$_3$-Bio process steps were substituted by an activated-carbon treatment.

Triggered by comparatively high treatment costs, an integrated (cyclic) chemical/biological process (CBP) named BioQuint®, was developed and implemented in full-scale treatment of landfill leachates during the 1990s [64, 98]. In this integrated CBP, the effluent from the biological system is ozonated in an ozone reactor to achieve partial oxidation and then recycled several times to the bioreactor. This combination reduces the specific ozone consumption due to a higher amount of compounds being biodegraded, instead of mineralized by extended ozonation, for example, from 2.5 kg O$_3$ kg^{-1} COD removed to less than 1 kg ozone per kg eliminated COD ([64]; see also Section 9.3 on CBPs). In five full-scale landfill leachate treatment plants (LLTPs) with flow rates between 6.25 and 25 m^3 h^{-1} the values of the ozone dose range from 0.5 to 1.9 kg O$_3$ m^{-3} [64].

Also, other advanced reaction systems with improved oxidation efficiency like heterogeneous catalytic ozonation [97] or improved mass-transfer efficiency such as an impinging zone reactor (IZR) [99], came into operation then (Table 3.6). In heterogeneous catalytic ozonation, radicals such as O$^{-°}$, O$_2^{-°}$ and O$_3^{-°}$ – but not OH-radicals – are the main oxidizing species. Here, the oxidation of previously adsorbed pollutants develops at the surface of the special grade activated carbon that acts as a promoter for the generation of highly reactive radicals but not as catalyst [107]. The main oxidation products are low weight organic acids like formic or acetic acid ([97, 107, 108]; see also Section 9.2 on three-phase systems). The IZR showed superior performance over a Venturi ejector-type of reactor by using 30% less ozone (1.8 vs. 2.5 kg O$_3$ kg^{-1} COD being removed), 50% less power (2 vs. 4 kWh kg^{-1} O$_3$ and seven times smaller reactor volume (20 m^3 vs. 150 m^3) [99].

All systems proved capable of reaching the required effluent limits at comparatively low specific ozone absorptions of 0.5 to 1.8 g O$_3$ g^{-1} COD in the influent of the ozonation stage (for further details see Table 3.6).

3.4.3.2 Textile Waste Waters – Color Removal and Partial Mineralization

The finding that ozone is very effective in color removal has increased interest in ozone application in the textile processing industry as legislative regulations on waste-water discharges have grown tighter. For example, national standards for handling dyebaths and regulating color in textile waste waters for direct discharges were established in Germany in the 1990s [109] and best available technologys (BAT) for the textile industry in the European Union through the IPPC Directive in 2003 [110]. Preceding these national standards were local regulations for dealing with color and foaming problems from direct discharges to natural waters. Most solutions were end-of-the-pipe systems treating the full-stream of waste water mixed from all textile processes for direct discharge [111]. Later, more sophisticated processes like the in-house treatment of segregated streams, small in volume and high in color, were found to be effective and sometimes more economical treatment alternatives for pretreatment before indirect discharge or in-house biological treatment ([112, 113]; see also the examples in Table 3.6).

While the main goal of treatment may be to remove nonbiodegradable (residual) color from the waste streams, removal of surfactants or partial oxidation of the DOC in order to achieve improved biodegradability may be secondary goals. Where ozone is used to treat full-stream waste waters, it is normally applied after a biodegradation stage (postozonation). In three full-scale applications, treating 160, 2500 and 5000 $m^3 h^{-1}$ of mixed waste water, the removal of color (and surfactants) is achieved with comparatively low ozone doses in the range of 0.032 to 0.112 kg $O_3 m^{-3}$ and, if specific costs of 0.15 to 0.2 € g^{-1} O_3 are assumed, at low specific operation costs between 0.05 to 0.25 € m^{-3} ([64]; see also Table 3.6).

In contrast, the treatment of segregated streams of dyebaths waste water with high concentrations of DOC and color requires much higher ozone doses per m^3, resulting in costs from 70–170 € m^{-3}. This is offset, though, by the vastly reduced amount of water that has to be treated, often thousands of times smaller. The large differences and inverse relationship in specific costs and flow rates lead to lower total costs for segregated versus full-stream treatment [112, 113]. Furthermore, the lower costs for segregated waste-stream treatment are accompanied by higher color removal and overall DOC removal. They are designed to remove more of the contaminants not just dilute them. This demonstrates that costs for segregated and full-stream treatment processes cannot be compared indiscriminately. Treatment schemes based on segregated waste streams coupled with process-water recycling can save quite a bit of money.

3.4.3.3 Other Applications

An interesting example of full-scale ozonation of a waste water from the chemical industry is the plant at BASF Schwarzheide GmbH, Germany, which is equipped with an ozone-generation capacity of 25 kg $O_3 h^{-1}$ (1333 ppd) capacity [62]. Again, the whole treatment train consists of a combination of chemical and biological processes. A Bio-O_3-Bio system is applied for the treatment of waste water coming from the manufacture of polyurethane foams. Here, the main goal of ozonation is the removal of toxic and refractory nitroaromatic compounds, so that the degree of COD removal amounted to only 4% of the COD load to the whole treatment train [101]. In this application considerable operational difficulties were caused by ozone consumption for nitrite oxidation due to insufficient biological denitrification as well as intense foaming in the ozone reactor.

Additional full-scale applications exist where ozonation is used with the objective to remove COD from industrial waste waters. For example, plants are operating with flow rates ranging from 30 to $100\,m^3\,h^{-1}$ and ozone doses of 0.67 and $0.43\,kg\,O_3\,m^{-3}$ in the chemical industry as Bio-O_3 or Bio-O_3-Bio processes, and as Bio-O_3-Bio treatment trains with flow rates of 580 and $1100\,m^3\,h^{-1}$ at ozone doses of 0.07 and $0.17\,kg\,O_3\,m^{-3}$ in the paper industry [64].

3.4.4
Particle-Removal Processes

As mentioned in Section 3.3.4, preozonation of surface waters can improve particle removal due to microflocculation caused by ozone-induced particle destabilization. In waste-water applications these effects may also occur, but primarily as a side-effect. Applications of ozonation solely with the goal to achieve a high degree in particle removal are not reported for (industrial) waste waters. In municipal WWTP, however, ozonation can be used to change the surface properties of activated sludge flocs to control bulking in secondary sedimentation basins, as well as to improve settleability or dewatering of waste sludge (also termed biosolids).

Furthermore, the application of ozone can be used to disrupt the cells of biosolids from biotreatment processes to increase their biodegradability in a subsequent biological stage, reducing or even eliminating excess sludge production. This reduction or disintegration of sludge with ozonation involves the sequential processes of floc disintegration by ozone, solubilization and subsequent mineralization of the released organics by bacteria. In a study on sludge ozonation at $0.05\,g\,O_3\,g^{-1}$ TSS (total suspended solids) it was found that 8% of the COD was mineralized, 22% was solubilized and 70% remained as (small) particulate organic matter [114]. The following biological treatment step can either be the main aeration basin or an anaerobic digestion tank, depending on which solid stream is being ozonated. A schematic overview of the options to integrate sludge ozonation into WWTPs is shown in Figure 3.2.

The suitability of sludge ozonation to reduce the production of excess sludge by 40 to 100% in biological waste-water treatment processes has been demonstrated in full-scale applications. The recommended ozone dose ranges from 0.03 to $0.05\,g\,O_3\,g^{-1}$ TSS, which is appropriate to achieve a balance between sludge reduction efficiency and cost [115]. However, this balance is not easy to find, since the process is very site-specific. Results from sludge ozonation studies are difficult to generalize because of the complexity of sludge, the numerous parameters that influence the process and the different treatment schemes used. Furthermore, the local legal regulations as well as technological and economic boundary conditions for sludge disposal determine its feasibility (see also Section 9.3.2). In general, sludge ozonation may be economical in WWTP that have high sludge disposal costs and operational problems such as sludge foaming and bulking [115].

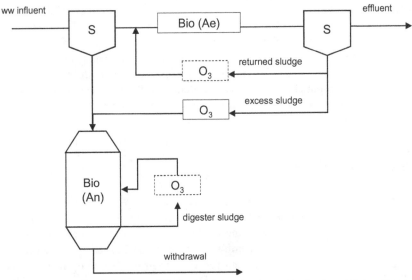

Figure 3.2 Various configurations of sludge ozonation in waste-water treatment trains. Stages included are: ozonation (O_3), aerobic biology (Bio (Ae)), anaerobic biology (Bio (An)) and settlers (S) (adapted from [115]).

Considering the amount of approximately 10 million tons of organic dry matter from waste-water treatment plants in each the EU, the USA and China, it seems reasonable to increase efforts to design and optimize an economic sludge-reduction process [115].

3.5
Economical Aspects of Ozonation

Within the twenty years between 1987 and 2007, the cost efficiency of ozone production has improved considerably due to the following technological advances [10, 13]:

- higher ozone yield per unit of electrode area due to medium frequency, high-efficiency technology;
- ozone generators operating at 10% to 12% wt in oxygen or higher;
- increase in the unitary ozone production capacity by a factor of more than two, now being between 8 and 14 kWh kg^{-1} O_3;
- improved operational reliability due to process control and automation as well as optimized mixing conditions and ozone transfer methods in the reactor chambers.

3.5 Economical Aspects of Ozonation

All in all, the operational costs for the production of 1 kg ozone are typically in the range of 1.5 to 2.0 € [64]. For drinking water this translates to operating costs ranging from less than one to approximately 5 €-cent per m^3, dependent on the treatment objective. Similar costs can be assumed for the ozonation of WWTP effluent if disinfection and/or the removal of easily oxidizable micropollutants is the goal, for example, many endocrine-disrupting chemicals (EDCs) and pharmaceuticals. The costs for the treatment of industrial effluents depend heavily on their composition and range from approximately 20 €-cent per m^3 for the treatment of waste waters from the textile and paper industry to as much as 4.0 € for the removal of DOC from a landfill leachate. An overview of the operational costs in the treatment of drinking and waste water is given in Table 3.7.

Considering the sum of the investment and operation costs, ozonation still is not an inexpensive technology. Although safe operation is not a problem, ozonation systems require considerable safety precautions, which also contribute to the relatively high investment costs. Especially, in smaller applications the investment and capital costs cannot be neglected, since they can considerably lengthen the payback times [111, 112].

In a pilot study in Berlin (Germany) a WWTP effluent was treated with the goal of reuse as drinking water after a subsequent underground passage. A specific ozone consumption of $1.9\,g\,O_3\,g^{-1}$ DOC was required. In this case the specific treatment costs were calculated to be approximately $0.05\,€\,m^{-3}$ for the ozonation of $200\,000\,m^3\,d^{-1}$ and were considered as not prohibitive for a possible full-scale treatment [68]. Joss et al. (2008) estimated similar total costs for such an application, between 0.05 and $0.20\,€\,m^{-3}$ [105]. While these costs per m^3 may be small, when large volumes are treated, total costs can run into millions of Euros [116]. The benefits of treatment must be weighed against the environmental and financial costs, especially of the high energy consumption. In the case of Berlin, such a process would be to protect its own drinking-water supply. This goal was also at the root of the recent decision by the Orange County Wastewater Authority to install large ozonation systems with treatment capacities between 300–750 MGD for the purpose of water reuse [20].

In 2003 the world market volume for ozone equipment was about 250 million US$ [25], thus being a rather small market. Nevertheless, there has been continued increase in installed ozonation capacity throughout the last two decades [10, 13]. While only about 60 larger water works were using ozone in the USA in 1992 [117], in the year 2007 there were already more than 300 major water works (above 19 MLD/5 MGD) using ozone [10]. Currently, ozone is used at 44 California drinking-water treatment plants (DWTPs) representing approximately one-fourth of the current 35 950 MLD (9500 MGD) in water-treatment capacity in the United States. The 21 DWTPs in California scheduled to add an ozone system or upgrade an existing ozone system by 2013 will increase the water-treatment capacity that includes ozone to 20 450 MLD (5400 MGD) and the ozone-generation capacity to about $2650\,kg\,h^{-1}$ (140 000 ppd) [13].

This chapter has given a short overview of the various areas in which ozone is being applied and which goals can be achieved with it. Some of the technical and

Table 3.7 Overview of the operational costs in the treatment of drinking and waste water depending on the treatment goal.

Application	Primary treatment objective of ozonation	Ozone dose / (typical values) kg O_3 m^{-3}	Operational costs € m^{-3a}	Reference/ Remark
Drinking-water treatment				
	Disinfection	~0.010	~0.015–0.02	At 10 mg^{-1} DOC and $D(O_3)$* reported in Section 3.3.4
	Color (e.g., SUVA$_{254}$)	<0.010	<0.015–0.02	Ditto
	DOC → BDOC[b]	~0.010–0.020	~0.015–0.04	Ditto
	DOC (~20%)	>0.030	>0.045–0.06	Ditto
	Particle removal	0.0005–0.002	0.00075–0.004	See Section 3.3.4
Waste-water treatment				
MWWTP effluent				
	Disinfection + endocrines and pharmaceuticals	0.005–0.020	0.0075–0.04	[64]
	Disinfection + endocrines and pharmaceuticals	0.012	0.018–0.024	[68]/p. 100
(Filtered and nitrified)	Disinfection	0.005–0.010	0.0075–0.02	[76]
	Disinfection	0.015–0.020	0.02–0.04	[76]
	COD	0.030–0.100	0.045–0.20	[64]
Landfill leachate	COD	0.500–2.000	0.75–4.00	[64]/5 examples
Textile (+sewage)	Color + tenside	0.030–0.110	0.045–0.22	[64]/3 examples
Chemical industry	COD	0.430, 0.666	0.65–1.33	[64]/2 examples
Paper industry	COD	0.068, 0.172	0.10–0.34	[64]/2 examples

a At 1.5–2.0 € kg^{-1} O_3 [64].
b Formation of biodegradable DOC (BDOC).

economical criteria that are used to evaluate whether to choose an ozonation system in drinking- and waste-water treatment are:

- achieve the treatment goal, for example, reduce the contamination below the legal limits;
- use a minimum of ozone (oxygen);
- keep the investment and energy costs low;
- be safe to operate;
- produce no toxic by-products.

The following chapters in this book provide an understanding on how to design and operate an ozonation system so that these criteria can be met.

References

1 Biography of Martinus Van Marum, http://profiles.incredible-people.com/martinus-van-marum/ (accessed November 2008).

2 Rubin, M.B. (2001) The history of ozone. The Schönbein Period, 1839–1868. *Bulletin for the History of Chemistry*, **26** (1), 40–56.

3 Lemmerich, J. (1990) *Die Entdeckung des Ozons und die ersten 100 Jahre der Ozonforschung*, ISBN 9783928068017, Sigma, Berlin.

4 Kurzmann, G.E. (1993) Ozonanwendung in der Wasseraufbereitung, in *Ozonanwendung in der Wasseraufbereitung*, 2. Ed. (eds G.E. Kurzmann et al.), Expert-Verlag, Ehingen bei Böblingen.

5 Rice, R.G., Farquhar, J.W. and Bollyky, L.J. (1982) Review of the applications of ozone for increasing storage times of perishable foods. *Ozone: Science and Engineering*, **4**, 147–163.

6 Hamor, W.A. (1912) The industrial uses of ozone. *Industrial and Engineering Chemistry*, **4** (6), 459–460.

7 Rideal, E.K. (1920) *Ozone*, D. Van Nostrand Co., New York, http://ia331402.us.archive.org/2/items/ozoneeric00riderich/ozoneeric00riderich.pdf (accessed November 2008).

8 Langlais, B., Reckhow, D.A. and Brink, D.R. (eds) (1991) *Ozone in Water Treatment – Application and Engineering*, Cooperative Research Report, American Water Works Association Research Foundation: Company Générale des Eaux and Lewis Publisher, Chelsea, MI, ISBN 0-87371-477-1.

9 German Federal Environment Agency (2006) Qualität von Wasser für den menschlichen Gebrauch (Trinkwasser) in Deutschland. [Quality of Water for human consumption (drinking water) in Germany], http://www.umweltdaten.de/publikationen/fpdf-l/3012.pdf (accessed November 2008).

10 Rakness, K.L. (2007) Twenty (20) years of advances in ozone operations and performance at drinking-water treatment plants. *Ozone News*, **35** (6), 17–25.

11 US EPA (1998) National primary drinking water regulations: disinfectants and disinfection by-products. *Federal Register* **63** (241), 69389–69476, http://www.epa.gov/ogwdw000/mdbp/dbpfr.html (accessed November 2008).

12 EU (1998) Council Directive 98/83/EC of 3 November 1998 on the quality of water intended for human consumption, http://eur-lex.europa.eu/LexUriServ/LexUriServ.do?uri=CELEX:31998L0083:EN:NOT (accessed November 2008).

13 Thompson, C., Drago, J. and Yamamoto, G. (2008) 20 years of ozone system installations in California. *Ozone News*, **36** (2), 14–22.

14 FDA (1982) 21 CFR 184.1563- Ozone. Code of Federal Regulations – Title 21: Food and Drugs (Edition 2006),

http://edocket.access.gpo.gov/cfr_2006/aprqtr/pdf/21cfr184.1583.pdf (accessed December 2008).
15 FDA (2001) 21 CFR 173.368 – Ozone. Code of Federal Regulations – Title 21: Food and Drugs (Edition 2002), http://edocket.access.gpo.gov/cfr_2002/aprqtr/pdf/21cfr173.368.pdf (accessed December 2008).
16 Naito, S. and Takahara, H. (2006) Ozone contribution in food industry in Japan. *Ozone: Science and Engineering*, **28** (6), 425–429.
17 Rice, R.G. and Graham, D.M. (2001) US FDA regulatory approval of ozone as an antimicrobial agent – what is allowed and what needs to be understood. *Ozone News*, **29** (5), 22–31.
18 Pascual, A., Llorca, I. and Canut, A. (2007) Use of ozone in food industries for reducing the environmental impact of cleaning and disinfection activities. *Trends in Food Science and Technology*, **18** (Suppl. 1), S29–35.
19 Ozonia (2007) Engineered Systems, http://www.ozonia.com/ozone/engineered.html (accessed June 2007).
20 Metropolitan Water District of Southern California (2008) Ozone at a glance, www.mwdh2o.com (accessed December 2008).
21 Cockx, A., Do-Quang, Z., Liné, A. and Roustan, M. (1999) Use of computational fluid dynamics for simulating hydrodynamics and mass transfer in industrial ozonation towers. *Chemical Engineering Science*, **54**, 5085–5090.
22 Ta, C.T. and Hague, J. (2004) A two-phase computational fluid dynamics model for ozone tank design and troubleshooting in water treatment. *Ozone: Science and Engineering*, **26**, 403–411.
23 Smeets, P.W.M.H., van der Helm, A.W.C., Dullemont, Y.J., Rietveld, L.C., van Dijk, J.C. and Medema, G.J. (2006) Inactivation of Escherichia coli by ozone under bench-scale plug-flow and full-scale hydraulic conditions. *Water Research*, **40**, 3239–3248.
24 McGuire, J., Bond, G. and Haslam, P.J. (2003) An Overview on the use of Ozonolysis in Synthetic Chemistry (1978–2003), http://www.ozonolysis.com/ozonolysis/web.nsf/web (accessed October 2008).
25 Suske, W. (2004) Spezial: Analytik/Umwelt, Ozontechnologien im grossen Stil – gewusst wie. *Chemische Rundschau*, Nr. **8**, 24.
26 Robin, J.J. (2004) Overview of the use of ozone in the synthesis of new polymers and the modification of polymers. *Advances in Polymer Science*, **167**, 35–79.
27 Wang, Y., Kim, J.-H., Choo, K.-H., Lee, Y.-S. and Lee, C.-H. (2000) Hydrophilic modification of polypropylene microfiltration membranes by ozone-induced graft polymerization. *Journal of Membrane Science*, **169** (2), 269–276.
28 Rimmer, S. and Collins, S. (2006) The use of radical polymerization in the synthesis of telechelic oligomers. *Reactive and Functional Polymers*, **66** (1), 177–186, http://biopolymer.group.shef.ac.uk/research/documents/OzoneinPolymerSynthesis.pdf.
29 Mendez, F., Maier, D.E., Mason, L.J. and Woloshuk, C.P. (2003) Penetration of ozone into columns of stored grains and effects on chemical composition and processing performance. *Journal of Stored Products Research*, **39**, 33–44.
30 Raila, A., Lugauskas, A., Steponavicius, D., Railiene, M., Steponaviciene, A. and Zvicevicius, E. (2006) Application of ozone for reduction of mycological infection in wheat grain. *Annals of Agricultural and Environmental Medicine*, **13** (2), 287–294, http://www.aaem.pl/pdf/13287.pdf (accessed December 2008).
31 Organic Materials Review Institute (OMRI) (2002) Ozone – Crops. National Organic Standards Board Technical Advisory Panel Review Compiled by for the USDA National Organic Program, http://www.omri.org/ozone_crops.pdf (accessed October 2008).
32 Pryor, A. (1999) Results of 2 years of field trials using ozone gas as a soil treatment, in *1999 Annual International Conference on Methyl Bromide Alternatives and Emissions Reductions* (ed. G.L. Obenauf), US EPA and USDA,

http://mbao.org/1999airc/32pryora.pdf (accessed December 2008).

33 Larson, L.E. (1999) Integrated agricultural technologies demonstrations. Public Interest Energy Research (PIER), Rpt. No. P600-00-012, California Energy Commission, Sacramento, CA, p. 100, http://www.energy.ca.gov/reports/2002-01-10_600-00-012.PDF.

34 Bocci, V., Aldinucci, C., Borrelli, E., Corradeschi, F., Diadori, A., Fanetti, G. and Valacchi, F. (2001) Ozone in medicine. *Ozone: Science and Engineering*, 23, 207–217.

35 Aydogan, A. and Gurol, M.D. (2006) Application of gaseous ozone for inactivation of Bacillus subtilis spores. *Journal of the Air and Waste Management Association*, 56 (2), 179–185.

36 Ishizaki, K., Shinriki, N. and Matsuyama, H. (1986) Inactivation of Bacillus spores by gaseous ozone. *Journal of Applied Bacteriology*, 60, 67–72.

37 Razumovskii, S.D. (1995) Comparison of reaction rates for ozone-alkene and ozone-alkane systems in the gas phase and in solution. *Russian Chemical Bulletin*, 44 (12), 2287–2288.

38 Siegrist, R.L., Urynowicz, M.A. and West, O.R. (2000) In situ chemical oxidation for remediation of contaminated soil and ground water. *Ground Water Currents*, 37, EPA 542-N-00-006.

39 Ullman, J. (2005) Remedial activities to reduce atmospheric pollutants from animal feeding operations. *Agricultural Engineering International*, VII, Invited Overview Paper No. 9, Website: cigr-ejournal.tamu.edu.

40 Rice, R.G. (2002) Century 21–pregnant with ozone. *Ozone: Science and Engineering*, 24 (1), 1–15.

41 Yuwono, A. and Lammers, P. (2004) Odor Pollution in the environment and the detection instrumentation. *Agricultural Engineering International*, VII, Invited Overview Paper, Vol. 6, Website: cigr-ejournal.tamu.edu.

42 Tamm, U. (2002) *Beseitigung von organischen Schadstoffen in Abgasen durch Oxidation mit Ozon*, Dissertation Uni Halle, Germany, http://sundoc.bibliothek.uni-halle.de/diss-online/02/02H232/prom.pdf (accessed December 2008).

43 Meuli, K. and Miller, O. (1993) Ozon zur Abluftreinigung, in *Ozonanwendung in der Wasseraufbereitung*, 2. Ed. (eds G.E. Kurzmann et al.), Expert-Verlag, Ehingen bei Böblingen.

44 U.S. EPA (2008) Ozone generators that are sold as air cleaners, http://www.epa.gov/iaq/pubs/ozonegen.html (accessed January 2009).

45 Wang, L., Oviedo-Rondón, E., Small, J., Li, Q. and Liu, Z. (2008) Ozone application for mitigating ammonia emission from poultry manure: field and laboratory evaluations. Proceedings of Mitigating Air Emissions from Animal Feeding Operations Conference, Iowa State University, http://www.ag.iastate.edu/wastemgmt/Mitigation_Conference_proceedings/CD_proceedings/Animal_Housing-Treatment/Wang-Ozone_application.pdf (accessed January 2009).

46 Kim-Yang, H., von Bernuth, R.D., Hill, J.D. and Davies, S.H. (2005) Effect of ozonation on odor and concentration of odorous organic compounds in air in a swine housing facility. *Transactions of the ASAE*, 48 (6), 2297–2302.

47 Kornmüller, A. and Wiesmann, U. (2003) Ozonation of polycyclic aromatic hydrocarbons in oil/water-emulsions: mass transfer and reaction kinetics. *Water Research*, 37 (5), 1023–1032.

48 Sechi, L.A., Lezcano, I., Nunez, N., Espim, M., Duprè, I., Pinna, A., Molicotti, P., Fadda, G. and Zanetti, S. (2001) Antibacterial activity of ozonized sunflower oil (Oleozon). *Journal of Applied Microbiology*, 90 (2), 279–284.

49 Díaz, M., Gavín, J.A., Gómez, M., Curtielles, V. and Hernández, F. (2006) Study of ozonated sunflower oil using 1H NMR and microbiological analysis. *Ozone: Science and Engineering*, 28 (1), 59–63.

50 Bocci, V. (2005) *Ozone: A New Medical Drug*, Springer, The Netherlands.

51 Zwiener, C., Richardson, S.D., DeMarini, D.M., Grummt, T., Glauner, T. and Frimmel, V. (2007) Drowning in

disinfection by-products? Assessing swimming-pool water. *Environmental Science and Technology*, **41**, 363–372.

52 Strittmatter, R.J., Yang, B. and Johnson, D.A. (1993) A comprehensive investigation on the application of ozone in cooling water systems – correlation of bench–top, pilot scale and field application data. *Ozone: Science and Engineering*, **15** (1), 47–80.

53 Puckorius, P.R. (1993) Ozone use in cooling-tower systems – current guidelines – where it works. *Ozone: Science and Engineering*, **15** (1), 81–93.

54 Sassi, J., Viitasalo, S., Rytkönen, J. and Leppäkoski, E. (2005) *Experiments with Ultraviolet Light, Ultrasound and Ozone Technologies for Onboard Ballast Water Treatment*, Espoo 2005. VTT Tiedotteita. Research Notes 2313, VTT Technical Research Centre of Finland, Espoo, p. 80. ISBN 951.38.6748.X, http://www.vtt.fi/inf/pdf/tiedotteet/2005/T2313.pdf (accessed December 2008).

55 Baysan, A. and Lynch, E. (2005) The use of ozone in dentistry and medicine. *Primary Dental Care*, **12** (2), 47–52.

56 Brazzelli, M., McKenzie, L., Fielding, S., Fraser, C., Clarkson, J., Kilonzo, M. and Waugh, N. (2006) Systematic review of the effectiveness and cost-effectiveness of HealOzone for the treatment of occlusal pit/fissure caries and root caries. *Health Technology Assessment*, **10** (16), iii–iv, ix–80.

57 Bocci, V.A. (2006) Scientific and medical aspects of ozone therapy. State of the art. *Archives of Medical Research*, **37** (4), 425–435.

58 Fujiwara, K., Kadoya, M., Hayashi, Y. and Kurata, K. (2006) Effects of ozonated water application on the population density of fusarium oxysporum f. sp. lycopersici in soil columns. *Ozone: Science and Engineering*, **28** (2), 125–127.

59 Ozonecip Project (2007) Study of the ozone technology: public report, http://www.ozonecip.net/pdf/Ozonetechnology.pdf (accessed January 2009).

60 Cardis, D., Tapp, C., DeBrum, M. and Rice, R.G. (2007) Ozone in the laundry industry – practical experiences in the United Kingdom. *Ozone: Science and Engineering*, **29** (2), 85–99.

61 Wedeco, I.T.T. (2008) Ozone in bleaching processes, http://itt.wedeco.de/Bleaching_Processes.535.html (accessed April 2008).

62 Böhme, A. (1999) Ozone technology of German industrial enterprises. *Ozone: Science and Engineering*, **21**, 163–176.

63 Ciufa, V. (2007) personal communication.

64 Ried, A., Mielcke, J., Wieland, A., Schaefer, S. and Sievers, M. (2006) An overview of the integration of ozone systems in biological treatment steps. 4th International Conference on Oxidation Technologies for Water and Waster and Wastewater Treatment, May 15–17, 2006 in Goslar, Germany. Cutec Serial Publication No 69.

65 Takahara, H., Nakayama, S. and Tsuno, H. (2006) Application of ozone to municipal sewage treatment. Wasser Berlin – International Conference Ozone and UV, April 3rd 2006.

66 Leong, L.Y.C., Kuo, J. and Tang, C.-C. (2008) Disinfection of wastewater effluent – comparison of alternative technologies, WERF Report 04-HHE-4. Water Environment Research Foundation, Alexandria, VA.

67 Wert, E.C., Rosario-Ortiz, F., Drury, D.D. and Snyder, S.A. (2007) Formation of oxidation by-products from ozonation of wastewater. *Water Research*, **41**, 1481–1490.

68 Schuhmacher, J. (2006) Ozonung zur weitergehenden Aufbereitung kommunaler Kläranlagenabläufe. Doctoral Thesis. Technical University of Berlin, Germany, http://opus.kobv.de/tuberlin/volltexte/2006/1218 (accessed December 2008).

69 von Gunten, U. (2003) Review: ozonation of drinking water: Part II. Disinfection and by-product formation in presence of bromide, iodide or chlorine. *Water Research*, **37**, 1469–1487.

70 Camel, V. and Bermond, A. (1998) The use of ozone and associated oxidation processes in drinking-water treatment. *Water Research*, **32**, 3208–3222.

71 Rakness, K.L. (2005) *Ozone in Drinking Water Treatment: Process Design, Operation, and Optimization*, American

References

Water Works Association, ISBN 1-58321-379-1.

72 Chick, H. (1908) An investigation of the laws in disinfection. *Journal of Hygienic*, 8, 92–158.

73 Watson, H.E. (1908) A note on the variation of the rate of disinfection with the change in the concentration of disinfectant. *Journal of Hygienic*, 8, 536–542.

74 Xu, P., Janex, M.-L., Savoye, P., Cockx, A. and Lazarova, V. (2002) Wastewater disinfection by ozone: main parameters for process design. *Water Research*, 36, 1043–1055.

75 Rennecker, J.L., Mariñas, B.J., Owens, J.H. and Rice, E.W. (2000) Inactivation of *Cryptosporidium parvum* oocysts with ozone. *Water Research*, 33, 2481–2488.

76 Koltunski, E. and Plumridge, J. (2000) *Ozone as a Disinfecting Agent in the Reuse of Wastewater*, Ozonia Ltd, Duebendorf, http://www.degremont-technologies.com/IMG/pdf/tech_ozonia_disinfecting-agent.pdf (accessed March 2009).

77 EPA (1999) *Alternative Disinfectants and Oxidants Guidance Manual*, Chapter 3 Ozone. 815-R-99-014, EPA, http://www.epa.gov/safewater/mdbp/pdf/alter/chapt_3.pdf.

78 Jarvis, P., Parsons, S.A. and Smith, R. (2007) Modeling bromate formation during ozonation. *Ozone: Science and Engineering*, 29 (6), 429–442.

79 USEPA (2006) Long term 2 enhanced surface-water treatment rule (LT2), Volume 71 (3), January 2006.

80 German Standard DIN (1997) 19643. *Part 1-5 Treatment of Water of Swimming Pools and Baths*. Normenausschuss Wasserwesen (NAW) im DIN, Deutschen Institut für Normung e.V., Beuth Verlag, Berlin.

81 Glauner, T. and Frimmel, F.H. (2006) Advanced oxidation – a powerful tool for pool water treatment. Wasser Berlin – International Conference Ozone and UV, April 3rd 2006.

82 Hammes, F., Salhi, E., Köster, O., Kaiser, H.-P., Egli, T. and von Gunten, U. (2006) Mechanistic and kinetic evaluation of organic disinfection by-product and assimilable organic carbon (AOC) formation during the ozonation of drinking water. *Water Research*, 40, 2275–2286.

83 Hoigné, J., Bader, H., Haag, W.R. and Staehelin, J. (1985) Rate constants of reactions of ozone with organic and inorganic compounds in water–III. Inorganic compounds and radicals. *Water Research*, 19, 173–183.

84 von Gunten, U. (2003) Review: ozonation of drinking water: Part I. Oxidation kinetics and product formation. *Water Research*, 37, 1443–1467.

85 Haag, W.R. and Hoigné, J. (1983) Ozonation of bromide-containing waters: kinetics of formation of hypobromous acid and bromate. *Environmental Science and Technology*, 17, 261–267.

86 Von Gunten, U. and Hoigné, J. (1994) Bromate formation during ozonation of bromide-containing waters: interaction of ozone and hydroxyl radical reaction. *Environmental Science and Technology* 28, 1234–1242.

87 Koudjonou, B.K., Croué, J.P. and Legube, B. (1994) Bromate Formation during ozonation of bromide in the presence of organic matter. Proceedings of the first International Research Symposium on Water Treatment By-Products, Poitier, France, 29–30. September 1:8.1–8.14.

88 Snyder, S.A., Wert, E.C., Lei, H., Westerhoff, P. and Yoon, Y. (2007) *Removal of EDCs and Pharmaceuticals in Drinking and Reuse Treatment Processes*, American Water Works Assn.

89 Hoigné, J. and Bader, H. (1983) Rate constants of reactions of ozone with organic and inorganic compounds in water.-II. Dissociating organic compounds. *Water Research*, 17, 185–194.

90 Jekel, R.M. (1998) Effects and mechanisms involved in preoxidation and particle separation processes. *Water Science and Technology*, 37, 1–7.

91 Botz, M.M., Dimitriadis, D., Polglase, T., Phillips, W. and Jenny, R. (2001) Processes for the regeneration of cyanide from thiocyanate. *Minerals and Metallurgical Processing*, 18 (3), 126–132.

92 Zeevalkink, J.A., Vlisser, D.C., Arnolds, P. and Boelhouwer, C. (1980) Mechanism and kinetics of cyanide ozonation. *Water Research*, 14, 1375–1385.

93 Gurol, M.D. et al. (1985) Oxidation of cyanides in industrial wastewaters. *Environmental Progress*, 4, 46–51.

94 Oppenländer, T. (2003) *Photochemical Purification of Water and Air, Advanced Oxidation Processes (AOPs): Principles, Reaction Mechanisms, Reactor Concept*, Wiley-VCH Verlag GmbH, Weinheim.

95 Schaefer, S., Ried, A., Wieland, A. and Mielcke, J. (2006) Cost-effective water-reuse with ozone as an oxidant. 4th International Conference on Oxidation Technologies for Water and Waster and Wastewater Treatment, May 15–17, 2006 in Goslar, Germany Cutec Serial Publication No 69.

96 Stegmanns, R., Maurer, C. and Kraus, S. (1995) Betriebserfahrungen bei der Sickerwasserbehandlung auf der Deponie Fernthal, in *2. Fachtagung naßoxidative Abwasserbehandlung*, CUTEC Schriftenreihe Nr. 23 (ed. A. Vogelpohl), Clausthaler Umwelttechnik-Institut GmbH (CUTEC), International Conference Oxidation Technology for Water and Wastewater Treatment.

97 Kaptijn, J.P. (1997) The Ecoclear® process. Results from full-scale installations. *Ozone: Science and Engineering*, 19, 297–305.

98 Ried, A. and Mielcke, J. (1999) The state of development and operational experience gained with processing leachate with a combination of ozone and biological treatment. Proceedings of the 14th Ozone World Congress.

99 Rüütel, P.I.L., Lee, S.-Y., Barratt, P. and White, V. (1998) *Efficient Use of Ozone with the CHEMOX™-SR Reactor*, Knowledge Paper No. 2, Air Products and Chemicals Inc., Walton on Thames, England.

100 Barratt, P.A., Baumgartl, A., Hannay, N., Vetter, M. and Xiong, F. (1996) CHEMOX™: advanced waste-water treatment with the impinging zone reactor, in *Clausthaler Umwelt-Akademie: Oxidation of Water and Wastewater* (ed. A. Vogelpohl), (Hrsg.), Goslar, pp. 20–22. Mai.

101 Krost, H. (1995) Ozon knackt CSB. *WLB Wasser, Luft und Boden*, 5, 36–38.

102 Ried, A., Mielcke, J. and Kampmann, M. (2006) The right treatment step: ozone and ozone/H_2O_2 for the degradation of non-biodegradable COD. Wasser Berlin 2006, International Conference on Ozone and AOP, April 3rd 2006, pp. 25–33.

103 Masten, S.J. and Davies, S.H.R. (1994) The use of ozonation to degrade organic contaminants in wastewaters. *Environmental Science and Technology*, 28, A181–A185.

104 Keisuke, I., Gamal El-Din, M. and Snyder, S.A. (2008) Ozonation and advanced oxidation treatment of emerging organic pollutants in water and wastewater. *Ozone: Science and Engineering*, 30, 21–26.

105 Joss, A., Siegrist, H. and Ternes, T.A. (2008) Are we about to upgrade waste-water treatment for removing organic micropollutants? *Water Science and Technology*, 57, 251–255.

106 AbwVwV Anhang 51 (1989) Allgemeine Rahmen-Abwasserverwaltungsvorschrift über Mindestanforderungen an das Einleiten von Abwasser in Gewässer–Rahmen-AbwasserVwV–vom 8.9.1989, Anhang 51 (Ablagerung von Siedlungsabfällen), GMBl. 40:Nr. 25, S. 527.

107 Kasprzyk-Hordern, B., Ziółek, M. and Nawrocki, J. (2003) Review: catalytic ozonation and methods of enhancing molecular ozone reactions in water treatment. *Applied Catalysis B: Environmental*, 46, 639–669.

108 Sanchez-Polo, M., von Gunten, U. and Rivera-Utrilla, J. (2005) Efficiency of activated carbon to transform ozone into OH radicals: influence of operational parameters. *Water Research*, 39, 3189–3198.

109 AbwVwV Anhang 38 (2000) Allgemeine Rahmen-Abwasserverwaltungsvorschrift über Mindestanforderungen an das Einleiten von Abwasser in Gewässer–Rahmen-AbwasserVwV–vom 01.06.2000, Anhang 38 (Textilherstellung, Textilveredelung).

110 EU Commission (2003) Reference document on best available techniques for the textiles industry, ftp://ftp.jrc.es/pub/eippcb/doc/txt_bref_0703.pdf (accessed November 2008).

111 Libra, J.A. and Sosath, F. (2003) Combination of biological and chemical processes for the treatment of textile wastewater containing reactive dyes. *Journal of Chemical Technology and Biotechnology*, **78**: 1149–1156.

112 Hemmi, M., Krull, R. and Hempel, D.C. (1999) Sequencing batch reactor technology for the purification of concentrated dyehouse liquors. *The Canadian Journal of Chemical Engineering*, **77**, 948–954.

113 Rapp, T. and Wiesmann, U. (2007) Ozonation of C.I. Reactive black 5 and indigo. *Ozone: Science and Engineering*, **29**, 493–502.

114 Lee, J.W., Cha, H.Y., Park, K.Y., Song, K.-G. and Ahn, K.-H. (2005) Operational strategies for an activated sludge process in conjunction with ozone oxidation for zero excess sludge production during winter seasons. *Water Research*, **39**, 1199–1204.

115 Chu, L., Yan, S., Xing, X.-H., Sun, X. and Jurcik, B. (2009) Progress and perspectives of sludge ozonation as a powerful pretreatment method for minimization of excess sludge production. *Water Research*, DOI: 10.1016/j.watres.2009.02.012.

116 Zenz, D.R., Law, K.P., Bouchard, A.B., Lanyon, R., Sobanski, J., Kunetz, T., Haas, C., Schwab, K. and Marinas, B. (2007) Study of effluent disinfection for urban rivers in Chicago. *Water Practice*, **1** (3), 1–17.

117 Masschelein, W.J. (1994) Towards one century application of ozone in water treatment: scope–limitations and perspectives. Proceedings of the International Ozone Symposium "Application of Ozone in Water and Wastewater Treatment" May 26–27, Warsaw, Poland (ed. A.K. Bin), pp. 11–36.

118 Siemers, C. (1995) Betriebserfahrungen mit der Deponiesickerwasserkläranlage Braunschweig, in *2. Fachtagung naßoxidative Abwasserbehandlung*, CUTEC Schriftenreihe No. 20 (ed. A. Vogelpohl).

Part II Ozone Applied

4
Experimental Design

Much experimental work has been carried out on ozonation and AOPs in drinking-, waste- and process-water treatment. And since there is still much to be learned about their mechanisms, and many possibilities of utilizing their oxidizing potential, many experiments will be carried out in the future. Not only researchers but also designers, manufacturers and users of chemical oxidation systems will continue to do bench-scale testing because ozonation and AOPs are so system dependent. Most full-scale applications have to be tried out bench-scale for each system considered. This means that there is a need for not only fundamental information about the mechanisms of ozonation, but also information on how to set up experiments so that they produce results that can be interpreted and used to design larger systems.

To achieve good results experiments should be designed correctly. This seems self-evident, but especially in the case of ozone, the complexity of the system is often underestimated. The goal of this chapter is to provide an overview of the process of designing experiments, concentrating on practical aspects and showing what information is necessary to design larger-scale systems. It offers a framework for the additional information in this book, which will help in understanding this complex topic.

First, the basis for experimental design is laid with a general description of the steps for designing experiments (Section 4.1). This general approach has to be adapted to chemical oxidation, which requires familiarity with the components of an ozonation system. Knowledge of which parameters influence the process, their relative importance and how they affect the process is essential to designing good experiments. This is discussed in Section 4.2. Since the core element of the bench-scale as well as full-scale system is the reactor, familiarity with reactor design in general is also necessary. A short review of reactor design concepts is presented in Section 4.3 to provide a framework for designing bench-scale experiments that provide the basis for larger scale reactors. The section is meant as a quick introduction to the concepts focused on in later chapters of this book.

After the basics, this chapter continues with checklists to illustrate the activity associated with each design step (Section 4.4). The checklists can be returned to when the practical work is in the planning phase. Similarly, the final Section 4.5

Ozonation of Water and Waste Water. 2nd Ed. Ch. Gottschalk, J.A. Libra, and A. Saupe
Copyright © 2010 WILEY-VCH Verlag GmbH & Co. KGaA, Weinheim
ISBN: 978-3-527-31962-6

will be a valuable help as a reference with its tabularized information on ozone throughout an experimenter's career.

Good experimental design can help produce good results with a minimum of effort. There are of course always surprises and unexpected circumstances in the life of an experimenter. However, with good preparation, these can be minimized and perhaps even turned to good use.

In order to choose an appropriate reactor design, it is important to understand what parameters affect the oxidation reactions and how. This section provides the reader with a concise overview to begin with and a guide to the later sections in the book that treat the topics in-depth.

4.1
Experimental Design Process

Good experimental work involves many preliminary steps before the real lab work can begin and then many steps after the lab work has been carried out. The experiments need to be well-designed to produce good results. The time invested in initial planning pays off in less time lost in unnecessary lab work. A good starting point when developing any method is the general problem-solving approach of: defining the problem and the requirements of a suitable solution, identifying possible solutions, ranking them according to appropriateness, designing and implementing the solution, followed by evaluating the effectiveness of the chosen solution and modifying if necessary. This process can be adapted for experimental design and grouped roughly into six steps (Figure 4.1). Perhaps "steps" is a misnomer since they are not meant to be carried out consecutively. An iterative process of defining the goals and system, selecting experimental methods and procedure, carrying out initial experiments, evaluating the data and assessing the results must be used. Since each step influences the others, it is common to work through the loop with preliminary experiments, checking results and then refining methods, before carrying out the majority of the experiments.

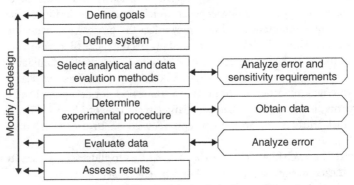

Figure 4.1 Main steps in experimental design including quality-control program.

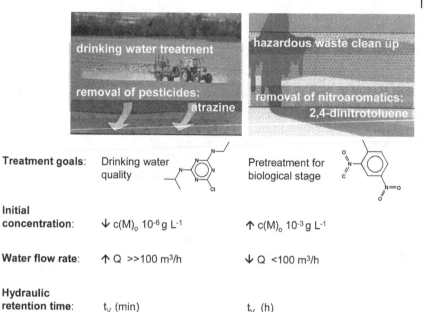

Figure 4.2 Examples of applications and their boundary conditions.

An important task in the initial design process is to consult the wealth of literature available on experimental investigations into ozone applications. However, a common error to be avoided at this stage is to think that ozonation procedures and results gathered on one type of water are easily extrapolatable to another. To illustrate the differences between ozone applications, we will look at the experimental design requirements for organic-compound removal in a drinking-water application and a hazardous-waste application. The general treatment goals and boundary conditions for the two cases can be found in Figure 4.2.

While some differences are obvious, other differences are more subtle. For example, the influent concentrations of the target compounds often differ by at least three orders of magnitude ($\mu g\,l^{-1}$ vs. $mg\,l^{-1}$), the amount of water to be treated differs and different regulations for effluent concentrations apply according to water use. It is apparent that the treatment goals and the size of the treatment plants will differ greatly and that this must be taken into consideration in the steps *define goals and system*. The strict national drinking-water standards that apply to micropollutant removal often require that the concentrations of individual compounds be determined. Permit levels for discharge of treated hazardous waste to natural waters may be set locally using lumped parameters (dissolved organic carbon [DOC] removal, toxicity reduction, concentrations of compound groups, etc.). This difference will, at the very least, affect the step *selection of analytical methods* and the development of the quality-control program.

In order to illustrate the further differences, we can look at the experimental results from the semibatch ozonation of the target compound M for these two

4 Experimental Design

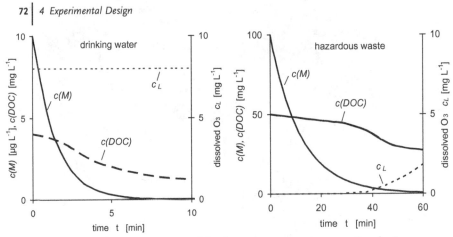

Figure 4.3 Qualitative comparison of the change in concentrations over time for the semibatch ozonation of a target compound M at relatively low concentration in a drinking water and high concentration in a hazardous waste clean-up.

cases. In Figure 4.3 we see that the concentration $c(M)$ decreases over time in both cases. However, the relative concentrations of the reactants differ extremely, both in absolute quantity and in behavior over time.

In drinking-water ozonation, the dissolved ozone concentration c_L remains constant over time and its value is almost a thousand times higher than the target compound concentration $c(M)$. Ozone is in constant supply. The reaction rate is only limited by the reaction kinetics and is called reaction limited. During ozonation in the highly contaminated groundwater, no dissolved ozone can be measured until the target compound is almost eliminated and/or the other ozone-consuming compounds in the complex matrix are oxidized. The transfer rate of ozone into the water limits the reaction rate, that is, the removal rate is mass-transfer limited.

Thus, in the step *define system*, these differences must be considered. The equipment has to be designed to provide the amount of ozone required for reaction, taking into account dose, mass-transfer rate, etc. The appropriate reactor conditions and mode of operation must be chosen based on the required reactions rates and efficiency. In the next step, *select analytical and data-evaluation methods*, we already pointed out that the analytical methods in the two cases will have to be adapted to allow reliable measurements in the different matrices at widely different concentration levels. In addition, the mathematical models used to evaluate and describe the results will differ for the two cases (reaction limited vs. mass-transfer limited). In particular, the different requirements in each system for the *sensitivity and error analysis* must be considered at this stage so that an adequate experimental setup can be developed. The *experimental procedures* and quality control will be tailored to each case to obtain good results. In *data evaluation* the parameters that can be used to answer the questions specific to each application need to be calculated. The errors associated with the parameters will be analyzed to see if the values are accurate enough to answer the questions.

And finally, in *assess results* the experimental results will be assessed to see if the experimental goals were achieved, and whether they are consistent with theoretical considerations and results from comparable ozone applications. Additionally, the ozonation results can be compared to other treatment methods to evaluate cost efficiency. These comparisons are usually specific to the application.

So, after pointing out the differences between applications, we come back to the premise of the book: the basic principles are the same for all applications, but some aspects dominate or are negligible according to the application. In the following section, the design steps are explained such that they can be followed for most applications. As we go into more detail for each design step, it will become clearer that this is an iterative process that usually requires going through the design loop several times. Perhaps the first iteration asks qualitative questions, to be answered with initial experiments and evaluation. These are then used to refine the goals and methods for further experiments to answer more quantitative questions.

4.2
Experimental Design Steps

4.2.1
Define Goals

The process of experimental design begins with determining the purpose of the experiments. Which questions must be answered by the experiments? Which measurements have to be made to answer the questions? It is also the time to consider what results are expected. As pointed out above, more than one iteration of the design loop may be necessary to develop these questions. At first, the goals of the ozonation experiments may be qualitative: Can the target compound be oxidized by ozone? Under which conditions? Is more than one treatment process required to achieve the targeted removal, that is, is a pre- or post-treatment step advantageous? And proceed to more quantitative questions to determine design criteria and operating conditions: How much removal is possible? Which oxidant dose and reaction time are needed? Which by-products are formed? Similar detailed questions concerning requirements for pre- and post-treatment processes can be formulated.

A very important point to decide at the beginning of the experimental process is what degree of treatment must be achieved and how it will be measured? The type of application usually determines the treatment goal as pointed out above for the two examples in Figure 4.2. These goals should then be met in all experiments, since it is extremely difficult and sometimes even impossible to extrapolate experimental results with chemical oxidants. This is illustrated in Figure 4.4 with experimental results from the semibatch ozonation of a target compound ($c(M)_o = 50\,mg\,l^{-1}$ DOC) in waste water with the goal of mineralization.

In these experiments with a water containing aromatic compounds, the DOC removal efficiency $\eta(M)$ was measured over the course of ozonation at three

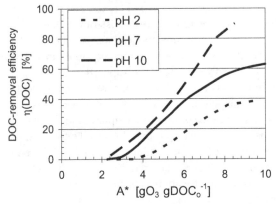

Figure 4.4 Change in DOC-removal efficiency $\eta(M)$ as the specific ozone absorption A^* is increased in waste-water treatment (pH control during treatment).

different pH values. Figure 4.4 shows that the DOC removal efficiency $\eta(M)$ increases initially as the specific ozone absorption A^* (the ratio of the amount of ozone absorbed in the reactor to the amount of the target compound initially present) increases for the three pH values. However, we see that pH has a very large effect on the removal efficiency in this case. The pH affects the relative importance between direct and indirect reactions. High degrees of mineralization can only be achieved at the higher pH values, where the indirect reaction dominates. If the treatment goal is changed from a DOC removal efficiency of 40% to 80% and data was only collected for 40% removal efficiency, we see that linear extrapolation of the results would cause extreme error. At low pH, 80% mineralization is impossible and at neutral pH, the amount of ozone needed would be severely underestimated.

If an oxidation process is combined with one or more additional processes, then the treatment goals in each process must be optimized considering the overall treatment efficiency. A common basis for comparison is necessary in order to compare the results of the process combinations. It is helpful in this situation to calculate relative values (e.g., removal rate, or specific ozone ratios) of each stage based on the initial concentration to the whole treatment train. Then the values are additive over the treatment train. This is discussed in more detail in *data evaluation*.

4.2.2
Define System

The next step is to define the system, which includes the composition of the water to be treated, the chemical oxidant, the reaction system and the surroundings. In order to know exactly what has to be defined, it is important to understand which parameters affect the chemical oxidation reaction and how. This, of course, is a major undertaking and the topic of the whole second half of the book. However, to get started, we can consider the most important influences on the reaction – the water system (Table 4.1). This includes the concentration of the target compound $c(M)$, the reactivity of the oxidant (ozone or $OH°$) with the target compound and with the other constituents in the water, as well as the amount of water to be treated.

The relative reactivity of the target compound and the other water constituents with the oxidant can be estimated from their rate constants k. The rate constants for the direct and indirect reactions with a large number of compounds can be found in the literature [1–5]. These rate constants can be used to estimate whether sufficient reactivity can be expected and under which conditions the direct or indirect reaction would be expected to dominate.

Since many water constituents can influence the chemical oxidation reaction, the composition of the water to be treated should be evaluated carefully. For example, the presence of scavengers such as carbonate, or ozone-consuming compounds such as reduced metal species, natural organic matter or other organics can drastically affect the required ozone dose. Especially in surface waters, seasonal fluctuations in turbidity and DOC due to algal blooms can be expected. Therefore, these constituents (along with expected daily and seasonal variations) should be identified and described as completely as possible, qualitatively and quantitatively. This is especially important if the water to be used in the experiments is not the original water, but rather a synthetic water spiked with the target compound. The effect of the compounds expected in the real water should then be experimentally evaluated.

Lumped parameters are often used to quantify the constituents in real waters: IC – inorganic carbon for carbonates, DOC – for all dissolved organic carbon, and UV extinction at 254 nm for aromatic compounds relative to the DOC concentration ($SUVA_{254}$) or 436 nm for humic-like substances ($SUVA_{436}$) as well as concentrations of individual reduced species of metals – Fe, Mn. If suspended solids are present, their composition can be differentiated between organic and inorganic fractions. In addition, the presence of compounds that affect surface tension σ or ionic strength μ is also of interest because of their affect on the mass-transfer rate $m(O_3)$, the rate of ozone transferred into the water. Here, the conductivity of the water in $\mu S\, cm^{-1}$ and/or the total dissolved solids TDS can be used for indicators of ionic strength.

The total amount of oxidizable compounds can be measured as COD – the chemical oxygen demand, or the biodegradable compounds can be assessed as BOD – the biochemical oxygen demand, in which the cumulative oxygen consumption of bacteria over a certain period of time is measured, often 5 d (BOD_5). Concurrent measurement of oxygen consumption and CO_2 evolution can provide not only a BOD value, but also an estimate of biomineralization possible.

Other methods for assessing biodegradability are based on measuring changes over time in the optical density (assimilable organic carbon – AOC), DOC (biodegradable DOC – BDOC). The AOC analysis was developed as an index of regrowth potential in drinking water. Since it is proportional to the difference in the optical

4 Experimental Design

Table 4.1 Main parameters to characterize the system "water".

Parameters	Symbols	Units	Discussed in
Type of water			
Natural waters: groundwater,			
Surface water; treated waters: tap water,			
Deionized water, etc.			
Target compound			
Initial concentration of M	$c(M)_0$	µg or mg l^{-1}	
Goal–effluent concentration of M	$c(M)$	µg or mg l^{-1}	
Reaction rate constants			
Direct	k_D		
Indirect	k_R		
Relevant water parameters			Chapter 1
That affect the reaction rate			
pH			
Concentrations of:			
Initiators, scavengers, promoters	$c(I)_i, c(S)_i, c(P)_i$	Mg l^{-1}	Chapters 2, 7 and 8
Reduced chemical species	Fe, Mn		
Lumped parameters			
Suspended solids (organic, inorganic)	SS, VSS		
Total dissolved solids	TDS		
Dissolved inorganic carbon	DIC		Chapter 2
Organic carbon	DOC		
	COD		
	BOD$_5$		
	SUVA$_{254}$, SUVA$_{436}$		
That affect treatment design			
Biodegradability of organic carbon	BDOC		Section 9.3
	AOC		
Toxicity	EC$_{50}$		Chapter 1 and Section 9.3
That affect mass transfer			Chapters 6 and 7
Ionic strength	μ	µS cm^{-1}	
Surface tension	σ	N m^{-1}	
Flow rates			
Gas and liquid flow rates, (daily average, min and max)	Q_G, Q_L	l s^{-1}	

density of test organisms that can grow in the solution over time, it is not an absolute measure of carbon concentration. In contrast, BDOC measures biodegradation using DOC, and, therefore, mineralization. In addition, simple and fast toxicity tests (such as inhibition of bioluminescence) can be used to evaluate treatment effectiveness.

Based on this information, considerations of pre- and post-treatment processes can also be included in the experimental design. For example, if high amounts of organic or inorganic ozone-consuming compounds other than the target compound are present, a pretreatment step should be evaluated. A high ratio of COD/DOC can indicate that oxidizable compounds are present. If this is combined with a relatively low ratio of COD/BOD or high BDOC, indicating biodegradable compounds, a biological pretreatment step may be advantageous. This can be applied analogously to a decision about post-treatment steps.

Another important aspect for defining the water system is to determine the liquid flow rate to be treated and expected variations. The average daily, peak and minimum hourly flow rates need to be defined. In experiments the flow rate may not be an independent parameter, but rather dependent on the required sample volume, which in turn depends on sensitivity and error requirements. In full-scale design, though, the flow rates usually can be defined as part of the boundary conditions.

After the water to be treated has been adequately defined, the reaction system is next. The experimental setup used in almost all ozonation experiments and treatment systems will consist of at least the components found in Figure 4.5: an ozone generator, an ambient-air monitor for safety, a reactor system with flow

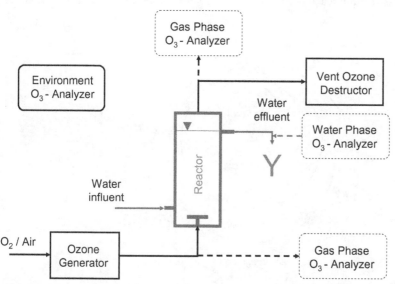

Figure 4.5 Basic components of any experimental ozonation setup (— required, --- optional equipment).

measurements, a water to be ozonated containing the target compound(s), and an ozone destructor for the vent gas. Gas and liquid analyzers for process measurement are optional, but necessary to determine important ozone parameters, such as efficiency and absorption. Each setup will be individual, determined by the experimenter's goals and resources, but the key components of the ozone system should be present.

Two possibilities exist: (i) an operational experimental setup is available or (ii) one has to be designed. Certainly the first option is preferable, however, often even existing setups have to be modified and/or redesigned for the current application. Therefore, the following discussion deals with how to design the experimental setup.

In order to start to design the setup, the requirements derived in the previous steps (definition of goals, water composition, and other boundary conditions) as well as those found in the subsequent steps of evaluation and assessment must be taken into consideration. For instance, we see from the description of components in Figure 4.5 that we already have to know which measurements are required for data evaluation, and which analytical methods and instruments will be used to obtain them before we can design the reactor. Indeed, in bench-scale applications, the equipment size is often determined by how much water is necessary for sampling and analysis. Conversely, how often a sample has to be taken and where is determined by the type of reactor used and its mode of operation. Clearly, a major difficulty in design is deciding where to start.

The traditional chemical-engineering approach, the onion model of process design, is to start with the reactor as the core of the system and work outward (Figure 4.6) to design a plant that fulfils the goals within the given boundary conditions. This approach is also valid for ozonation since the reactor hydrodynamics,

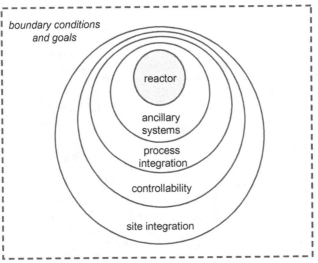

Figure 4.6 The onion model of process design (modified from [8]).

mass transfer and kinetics determine the efficiency of the oxidation processes. The structure of this part of the book follows this approach to some extent. The next section of this chapter (Section 4.3) focuses on reactor-design basics and how the reactor type influences the size. This is expanded in Chapter 5 with more practical information on typical reactors used and ancillary systems (ozone generators and destructors, feed-gas preparation, instrumentation) as well as with a discussion of analytical methods that are available and how to choose them, while Chapters 6 and 7 deal with what influences the mass transfer and kinetics in the reaction system. The mathematical description of the system, the modeling of the reactor, is the topic of Chapter 8. The next layers of the onion model, designing the processes before and after the reactor: the pre- and post-treatment processes, as well as their integration into the whole treatment train, are considered for certain systems in Chapter 9. For the last layers, controllability and integration within the site the reader is referred to further literature sources [5–7].

In full-scale systems, the type and amount of water to be treated and the treatment goals are usually given by the application, sometimes even the type of reactor is specified, so that the design process is fairly straightforward. Everything can be designed around the system requirements following the layer principle. At the bench-scale, a high level of flexibility in the types of water and reactor, and mode of operation is usually desirable, so the design process is more iterative. Preliminary experiments based on previous experience and/or on information from the literature are probably necessary at a smaller scale (e.g., gas-wash flasks) before the bench-scale reactor system can be designed. In general, the experimental system should be designed to allow flexibility in the range of possible operating conditions. This will take the uncertainty into account as well as allow the operating conditions to be varied to find the optimal conditions for each application, especially if a full-scale system is to be designed based on these results.

The reaction system shown in Figure 4.5 consists of the reactor and the feed streams containing the reactants, as well as ancillary equipment such as the ozone generator, feed-gas preparation, instrumentation, etc. The parameters describing the water system and compounds were described in Table 4.1, while some of the main components and parameters of the reaction system that have to be determined are summarized in Table 4.2. Since the reactor and associated equipment are discussed in detail in Section 4.3 and Chapter 5, they will only be briefly presented here.

In the first step, the reactor type and operating mode should be decided upon. Often it is desirable to design the system to accommodate various operating modes to increase flexibility. Further parameters that we need to know for designing the reactor in addition to those from the definition of the water (Table 4.1) are: the required effluent concentration from the goal definition (or desired operating concentration in case a safety margin is used), the expected seasonal and daily variations, as well as the required reaction time to achieve that concentration t_R. This depends on the reaction rates r, how fast ozone reacts with the target compounds and other constituents in the water, which in turn are a function of the reactant concentrations in the reactor. We also need to describe how fast ozone

Table 4.2 Main parameters to characterize the system "reactor".

Parameters	Symbols	Units	discussed in
Reactor			Section 4.3, Chapter 5
Reactor type: stirred tank, bubble column, etc.			
Mode of operation: batch, continuous flow			
Degree of mixing: mixed, plug-flow			
Reactor geometry and volume (gas and liquid)	V_G, V_L	l	
Ozone dosing points (single or multiple)			
Operating parameters			
Flow rates and ratios			
Design and average gas and liquid flow rates	Q_G, Q_L	$l\,s^{-1}$	
G/L–ratio of gas to liquid flow rates	Q_G/Q_L	–	Chapters 5 and 6
Concentrations			
Ozone			
Liquid phase	c_{Lo}	$mg\,l^{-1}$	Chapter 5
Gas phase	c_{Go}, c_{Ge}, c_G	$mg\,l^{-1}$	
Target compound M (expected variations)			
Influent	$c(M)_o$	μg or $mg\,l^{-1}$	
Effluent–(limit and operating point)	$c(M)_e$	μg or $mg\,l^{-1}$	
Reactor conditions			
Reaction time	t_R	s, min or h	Section 4.3
Hydraulic retention time	t_H	s, min or h	
Mass transfer coefficient	$k_L a$	s^{-1}	Chapter 6
Stirrer speed	n	s^{-1}	
Temperature and pressure	T, P	°C or K, Pa	
Reaction rates			
Of ozone in the gas and liquid phase	r_G, r_L	$mg\,l^{-1}\,s^{-1}$	Chapter 7
Of target compound in the liquid phase	$r(M)$	$mg\,l^{-1}\,s^{-1}$	
Ancillary equipment			Chapter 5
pH control			
Ozone gas generator (type and capacity)			
Ozone destructor			
Feed gas type and preparation system			
Analyzer system (gas/liquid/ambient)			
Sample points			

should be transferred into the water, the mass-transfer coefficient $k_L a$. The reactor volumes V and flow rates Q of both phases must then be chosen to accommodate the required hydraulic retention time t_H or for semibatch, the reaction time t_R, as well as the required sample sizes for the analytical measurements.

In addition, the ancillary equipment must be designed. Ozone can be produced in a variety of ways from oxygen or water. These are described in detail in Chapter 5. For illustration purposes, an electrical discharge generator, a common generator type, is depicted in Figure 4.5. The feed gas can be air, oxygen-enriched air or high-purity oxygen. The unreacted high-purity oxygen leaving the reactor or the vent ozone destructor can also be recycled to the feed. The feed-gas preparation system, the capacities of the ozone generator and destructor, as well as the analyzer system (instruments and sampling points) must be determined. A pH-control system should be considered.

So we see from Table 4.2 that, in order to define the reaction system, we require quite a bit of information that usually will be gained from experimental results. Hence, the previous discussion about an iterative method of design. However, this section has not yet provided the experimenter with a method for setting values for all these parameters. This will be remedied in Section 4.3.3 after the reactor-design basics are presented in Section 4.3. But first, the further steps in experimental design are discussed in the next sections.

4.2.3
Select Analytical Methods and Methods of Data Evaluation

The step choosing analytical and data-evaluation methods is very closely connected with the previous steps, defining the goals and system. Often, the definition of the treatment goals has to be adjusted to reflect what is measurable. For instance, assessing mineralization of target compounds may be difficult in the presence of background organic compounds. In both of the above cases, the relatively high DOC concentration of the matrix compared to the DOC of the target compound prohibits using DOC alone to determine the mineralization of the target compound. Therefore, more analytical effort, such as measuring the appearance/disappearance of metabolites, is necessary.

In order to ensure that enough information can be gathered to determine the conditions under which the treatment goals can be reached, the experimental setup has to provide adequate online measurements and sample points. As pointed out above, the size of the bench-scale reaction system is often determined by the amount needed for sampling. This in turn is set by the choice of analytical methods, which have to be sensitive enough to measure the required information with enough certainty. Information on analytical methods for ozone can be found in Section 5.4.

It is crucial that the effects to be investigated are large enough to be considered statistically significant and are reproducible. To evaluate whether the effects will be large enough, an error and sensitivity analysis of the methods must be made. Often existing setups will have to be modified to produce measurable effects. With the results of such an analysis, the methods can be fine tuned, so that the experimental goals can be correctly defined and then achieved.

In order to carry out a sensitivity analysis, the errors in the various measurements as well as the data-evaluation methods must be known. The basic tool for evaluating experimental data and for interpreting the results is always the mass balance, whether we are only interested in the degree of removal in a bench-scale experiment with a gas-wash bottle or in designing a full-scale treatment system. A mass balance (also called a material balance) is a mathematical description of the law of conservation of mass that for nonnuclear reactions can be described verbally as *what goes in has to come out (Transport)* or *be stored (Accumulation)* or *reacted (Reaction)* so that:

ACCUMULATION = TRANSPORT + REACTION

The mathematical form of the mass-balance equation to be used is determined by the type of reactor and its hydrodynamics. This is discussed in more detail in Section 4.3 as well as in Chapter 5. To illustrate the method we will look at the system we already saw in Figure 4.5, which consists of an ozone source, a water to be ozonated with one or more compounds of interest M, and a reactor. If both gas and water continuously flow into the reactor and are completely mixed, it can be represented as a continuous-flow stirred-tank reactor (CFSTR) as shown in Figure 4.7.

With the parameters from Tables 4.1 and 4.2 we can set up the mass balances on the reaction system. Here, we will use the mass balance for the CFSTR to illustrate the relationships between the parameters. In a CFSTR we can assume that the gas and liquid phases are ideally mixed and the concentration of the

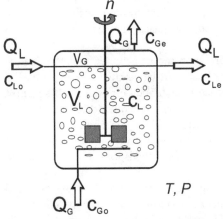

Figure 4.7 Operating parameters necessary for ozone mass balance(s) on a continuous-flow stirred-tank reactor (for operation in semibatch mode: $Q_L = 0$).

effluent is the same as that in the reactor ($c_L = c_{Le}$, $c_G = c_{Ge}$), which allows us to use a simple integral mass balance.

Further simplification comes from assuming that compound M is nonvolatile and is not stripped, so that only the mass transfer for ozone must be considered. Also, most oxidation reactions in drinking- and waste-water treatment involve irreversible reactions in the liquid phase at constant density and temperature. Therefore, we do not have to consider changes in the reaction volume or liquid volumetric flow rate Q_L. If the concentrations change over time, however, the nonsteady-state mass balances must be used.

4.2.3.1 Ozone

For the absorption of any gas, for example, ozone, in a CFSTR, the nonsteady-state mass balances are as follows:

liquid phase:

$$V_L \cdot \frac{dc_L}{dt} = Q_L(c_{Lo} - c_L) + k_L a \cdot V_L \left(c_L^* - c_L\right) - r_L \cdot V_L \tag{4.1}$$

gas phase:

$$V_G \cdot \frac{dc_G}{dt} = Q_G(c_{Go} - c_G) - k_L a \cdot V_L \left(c_L^* - c_L\right) - r_G \cdot V_G \tag{4.2}$$

Equations 4.1 and 4.2 can be combined for the total material balance on the reactor:

$$V_L \cdot \frac{dc_L}{dt} + V_G \cdot \frac{dc_G}{dt} = Q_L(c_{Lo} - c_L) + Q_G(c_{Go} - c_G) - r_L \cdot V_L - r_G \cdot V_G \tag{4.3}$$

This can be simplified for steady state, where there is no change over time dc_G/dt and $dc_L/dt = 0$, so that:

$$0 = Q_L(c_{Lo} - c_L) - r_L \cdot V_L - Q_G(c_{Go} - c_G) - r_G \cdot V_G \tag{4.4}$$

or for a semibatch with no reaction in the gas phase:

$$V_L \cdot \frac{dc_L}{dt} + V_G \cdot \frac{dc_G}{dt} = Q_G(c_{Go} - c_G) - r_L \cdot V_L \tag{4.5}$$

4.2.3.2 Target Compound M

In writing the mass balance for the target compound M, only the liquid-phase mass balance need be considered if the compound is nonvolatile. Mass transfer does not take place, so that:

$$V_L \cdot \frac{dc(M)}{dt} = Q_L(c(M)_o - c(M)) - r(M) \cdot V_L \qquad (4.6)$$

Many of the system parameters necessary to evaluate the data and discuss the results are based on these mass balances. For example, we can use them to evaluate the ozone consumption $D(O_3)$, which is the specific mass of ozone consumed in the reaction per volume of water treated. It is calculated from the ozone transferred from the gas to the liquid phase minus the ozone that leaves the reactor in the liquid phase. At steady state, Equation 4.4 can be rearranged to obtain $D(O_3)$ based on the condition that there is no ozone in the liquid influent $c_{Lo} = 0$ and there is no reaction with M in the gas phase $r_G = 0$.

$$D(O_3) = \frac{r_L \cdot V_L}{Q_L} = \frac{Q_G}{Q_L} \cdot (c_{Go} - c_{Ge}) - c_{Le} \qquad (4.7)$$

It is important to check, though, that the gas system has been properly designed so that r_G due to the loss of ozone by decomposition in the gas phase is negligible.

In semibatch systems, the values of the parameters change as the reaction proceeds (Figure 4.8). Then, differential equations can be evaluated at any time of interest t, or integrated to find the sum for the reaction time. For example, $D(O_3)$

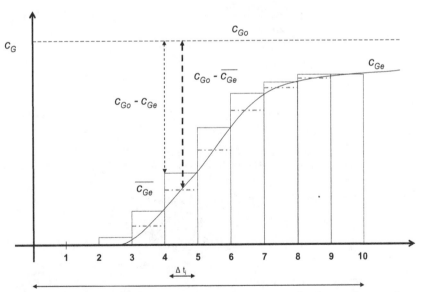

Figure 4.8 Changes in the effluent ozone gas concentration in semibatch ozonation over time and example time intervals Δt_i used for the approximation of c_{Ge} in Table 4.3.

can be evaluated for semibatch systems by rearranging Equation 4.5 and assuming the change in the stored ozone in the gas phase is negligible ($V_G \, dc_G/dt = 0$):

$$D(O_3) = \int_0^t \frac{Q_G}{V_L}(c_{Go} - c_{Ge})dt - \int_{c_{L(t=0)}}^{c_{L(t=t)}} c_L \qquad (4.8)$$

Equations for calculating commonly found system parameters are found in Table 4.3 for nonsteady-state semibatch systems as well as for continuous-flow systems at steady state. The equations are listed in Table 4.3 in integral form for semibatch as well as the equations that can be used to approximate them. Many of the parameters are interrelated. Some of these interrelationships are shown for the continuous-flow system.

4.2.4
Determine Experimental Procedure

The next step is to determine the experimental procedure for the experiments. This involves developing a plan for experiments that will answer the questions formulated in the step *define goals*. This starts with preliminary experiments to characterize the hydrodynamics and mass transfer in the reaction system for the range of operating conditions planned. Knowledge about the type of mixing in the reactor is necessary to determine where samples should be taken. Information on the range of mass-transfer rates achievable in the reactor can be used in the planning of the individual experiments to ensure that the reaction rate is not mass-transfer limited if kinetic data is to be gathered.

An experimental plan for each experiment is needed, which includes an objective, the expected results, a schedule with operating conditions, sampling points and intervals, equipment and analytical needs, data collection and handling, and quality control. The plan must take the error and sensitivity analysis from the previous step into consideration as well as any time constraints, that is, for CFSTR – it takes approximately 3–5 times the hydraulic retention time t_H to reach steady state or for semibatch – some equipment requires a warm-up phase to reach constant operation (e.g., ozone generators often take an hour to reach constant output). Adequate provisions must be made to show reproducibility and statistically significant results.

When setting up the experimental plan, an important strategy is to carry out the first experiments at the extreme ranges of the variables being investigated [9]. If there is no effect at the minimum and maximum values of one variable, an influence at an intermediate value is not to be expected. In this way the important influences can be singled out and investigated further. In addition, many developments have been made in the use of statistics to aid in experimental design over the last decades. Their most important contribution is that they provide a mathematical framework for changing pertinent factors simultaneously, in order to produce statistically significant results in a small number of experiments. The

Table 4.3 Overview of important parameters in ozonation experiments and their equations ($c(O_3)$, $c(M)$ also possible as mol l^{-1}).

Parameter	Symbol	Unit	Semibatch system	Continuous-flow system	Eqn. No.
Specific mass					
Introduced ozone dose or applied ozone dose or ozone input	I	mg l^{-1}	$I = \int_0^t \frac{Q_G}{V_L} \cdot c_{Go} dt$ $I = \sum_{i=1}^n \frac{Q_G \cdot c_{Go}(t_i)}{V_L(t_i)} \cdot \Delta t_i$	$I = \frac{Q_G}{Q_L} \cdot c_{Go}$	T 4-1
Absorbed ozone dose or transferred ozone dose or ozone absorption[a]	A	mg l^{-1}	$A = \int_0^t \frac{Q_G}{V_L}(c_{Go} - c_{Ge}) dt$ $A = \sum_{i=1}^n \left[\frac{Q_G}{V_L(t_i)}(c_{Go}(t_i) - \overline{c_{Ge}}(\Delta t_i)) \cdot \Delta t_i \right]$		T 4-2
Consumed ozone dose or ozone consumption[a]	$D(O_3)$	mg l^{-1}	$D(O_3) = \int_0^t \frac{Q_G}{V_L}(c_{Go} - c_{Ge}) dt - \int_{c_{L(t=0)}}^{c_{L(t=t)}} c_L$ $= A - \int_{c_{L(t=0)}}^{c_{L(t=t)}} c_L$ $D(O_3) = \sum_{i=1}^n \left[\frac{Q_G}{V_L(t_i)}(c_{Go}(t_i) - \overline{c_{Ge}}(\Delta t_i)) \cdot \Delta t_i - \overline{c_L}(\Delta t_i) \right]$	$D(O_3) = \frac{Q_G}{Q_L}(c_{Go} - c_{Ge}) - c_{Le}$ $= A - c_{Le}$	T 4-3
Specific mass flow rate					
Ozone feed rate or ozone dose rate	$F(O_3)$	mg l^{-1} s^{-1}	$F(O_3) = \frac{Q_G c_{Go}}{V_L} = \frac{I}{t_R}$	$F(O_3) = \frac{Q_G c_{Go}}{V_L} = \frac{I}{t_H}$	T 4-4
Reaction rates					
Ozone absorption rate (instantaneous or average)	$r_A(O_3)$	mg l^{-1} s^{-1}	$r_A(O_3) = \frac{dA}{dt}$ $\overline{r_A(O_3)} = \frac{A}{t_R}$	$r_A(O_3) = \frac{Q_G(c_{Go} - c_{Ge})}{V_L} = \frac{A}{t_H}$	T 4-5

Quantity	Symbol	Units	Formula	Eq.
Ozone consumption rate (instantaneous or average)	$r(O_3) = r_L$	$mg\,l^{-1}\,s^{-1}$	$r(O_3) = \dfrac{dD(O_3)}{dt}$ $\overline{r(O_3)} = \dfrac{D(O_3)}{t_R}$	$r(O_3) = \left(\dfrac{Q_G(c_{Go}-c_{Ge})}{Q_L} - c_{Le}\right)\cdot\dfrac{1}{t_H} = \dfrac{D(O_3)}{t_H}$ $= r_A(O_3) - \dfrac{c_{Le}}{t_H}$	T 4-6 T 4-7
Pollutant removal rate	$r(M)$	$mg\,l^{-1}\,s^{-1}$	$r(M) = \dfrac{c(M)_o - c(M)_e}{t_e - t_o} = \dfrac{c(M)_o - c(M)_e}{t_R}$	$r(M) = \dfrac{c(M)_o - c(M)_e}{t_H}$	T 4-8
Specific doses					
ct		$mg\,l^{-1}\,s$	$c \cdot t = c_L(t) \cdot t_R$	$c \cdot t = c_L(t) \cdot t_H$	T 4-9
Ratios					
Specific ozone dose or input	I^*	$gO_3\,g^{-1}M$	$I^* = \dfrac{I}{c(M)_o} = \dfrac{F(O_3) \cdot t_R}{c(M)_o}$	$I^* = \dfrac{I}{c(M)_o} = \dfrac{F(O_3) \cdot t_H}{c(M)_o}$	T 4-10
Specific ozone absorption	A^*	$gO_3\,g^{-1}M$	$A^* = \dfrac{A}{c(M)_o} = \dfrac{\overline{r_A(O_3)} \cdot t_R}{c(M)_o}$	$A^* = \dfrac{A}{c(M)_o} = \dfrac{\overline{r_A(O_3)} \cdot t_H}{c(M)_o}$	T 4-11
Ozone-yield coefficient	$Y(O_3/M)$	$g\,O_3\,g^{-1}M$	$Y(O_3/M) = \dfrac{r(O_3)}{r(M)} = \dfrac{D(O_3)}{c(M)_o - c(M)_e}$	$Y(O_3/M) = \dfrac{r(O_3)}{r(M)} = \dfrac{D(O_3)}{c(M)_o - c(M)_e}$	
Efficiencies					
Ozone-transfer efficiency	$\eta(O_3)$	%	$\eta(O_3) = \dfrac{c_{Go} - c_{Ge}}{c_{Go}} = \dfrac{A}{I} = \dfrac{A^*}{I^*}$	$\eta(O_3) = \dfrac{c_{Go} - c_{Ge}}{c_{Go}} = \dfrac{A}{I} = \dfrac{A^*}{I^*}$	T 4-12
Degree of pollutant removal	$\eta(M)$	%	$\eta(M) = \dfrac{c(M)_o - c(M)_e}{c(M)_o}$	$\eta(M) = \dfrac{c(M)_o - c(M)_e}{c(M)_o}$	T 4-13
Mass transfer					
Mass-transfer rate	$m(O_3)$	$mg\,l^{-1}\,s$	$m(O_3) = k_L a (c_L^* - c_L)$		T 4-14
Mass-transfer enhancement factor	E	–	$E = \dfrac{r(O_3)}{k_L a (c_L^* - c_L)}$	$E = \dfrac{r(O_3)}{k_L a (c_L^* - c_L)}$	T 4-15
Additional calculations for c_{Ge} or c_L (averaged over reaction time for one time interval Δt_i; see also Figure 4.8)	$\Delta t_i = t_i - t_{i-1}$	–	$\overline{c_{Ge}(\Delta t_i)} = \dfrac{(c_{Ge}(t_{i-1}) + c_{Ge}(t_i))}{2}$; $\overline{c_L(\Delta t_i)} = \dfrac{(c_L(t_{i-1}) + c_L(t_i))}{2}$	$E = \dfrac{r(O_3)}{k_L a (c_L^* - c_L)}$; $t_R = n \cdot \Delta t_i$; $t_R = t_e - t_o$	

description of such methods, though, is beyond the scope of this book and the reader is referred to the literature [10–13].

4.2.5
Evaluate Data

The parameters introduced in the previous section are used to evaluate the data. The complete list in Table 4.3 seems like a lot of information and the question arises if all of these parameters are necessary to report, and if not all, which ones. The answer is that a nonredundant set of information that adequately describes the system and results must be measured and reported. They can be grouped according to the ones that define the system, describe the experimental procedure, and are used to assess the results (Table 4.4).

Besides the operating parameters, information on pollutant removal is indispensable, that is, initial pollutant concentration $c(M)_o$, degree of pollutant removal $\eta(M)$. If various treatment schemes are to be compared for example, one stage ozonation to a two stage biological/ozonation process, it is advantageous to calculate the degree of removal achieved in each stage as part of the overall removal $\eta(M)_\Sigma$ so that the degrees of removal are additive, for example:

$$\eta(M)_\Sigma = \frac{c(M)_o - c(M)_i}{c(M)_o} + \frac{c(M)_i - c(M)_e}{c(M)_o} \tag{4.9}$$

Table 4.4 Parameters for the evaluation of an ozonation experiment.

Required information	Define system		Experimental procedure	Assess results	
–	Reactor	Water	Ozonation (input)[a]	Amount of ozone used[a]	Pollutant removal[a]
Generally	V_L, H, D $k_L a$, E n_{STR} T, P	$c(M)_o$ pH $c(TIC)^b$	Q_G, Q_L t_R or t_H $c_{Go}\left(c_L^*\right)$ $F(O_3)$ or I^*	$\eta(O_3)$ $r(O_3)$, $r_A(O_3)$ c_L A^{*c} $Y(O_3/M)^c$ ct-value^d	$\eta(M)$ $r(M)$
	Reactor type	Buffer type	pH-control		
For kinetic assessment	–	–	–	$k_d(O_3)^d$	$k_D(M)$, $k_R(M)$

a For each stage and pollutant as well as the whole system.
b Sum of concentrations of HCO_3^- and CO_3^{2-}.
c More important in experiments on waste-water ozonation.
d Important in drinking-water ozonation.

Additional important parameters are those for ozone, that is, ozone consumption rate $r(O_3)$, specific ozone input I^*, specific ozone absorption A^*. Combined parameters allow a quick comparison of results, that is, ozone yield coefficient $Y(O_3/M)$. Only a few of the calculated parameters in Table 4.4 are interrelated and, therefore, redundant. For example if the specific input I^* and specific ozone absorption A^* are given, then the ozone-transfer efficiency $\eta(O_3)$ can be calculated from the two values. If it is desired to model the process or estimate reactor volumes using kinetic parameters, the kinetic coefficients k_D and k_R must also be measured if not available in the literature.

Unfortunately, the most common shortcoming in published literature is not too much, but rather too little information given. Often the system and the results are not adequately described, so that comparison of results cannot be made. This will be addressed in more detail in the section on designing treatment systems after the important aspects of designing reactors are presented below.

4.2.6
Assess Results

Last but not least in this iterative process is the step *assess the results*. This means to check if the desired goals have been reached and compare the experimental results with others. An example of experimental values for the two different applications discussed in Figure 4.2 above can be found in Tables 4.5 and 4.6. They give an interesting comparison of the difference in values found in drinking- and waste-water applications.

Both experiments were conducted in semibatch mode in the same reactor, a completely mixed stirred-tank reactor (STR), with almost identical initial liquid volumes V_L of 7.5 and 7.0 l. Also the same gas-flow rates Q_G were applied resulting in an identical mass-transfer coefficient $k_L a$ for pure water. And in both experiments the waters were ozonated at pH 7 so that elimination of the target compounds can occur by direct reactions or by indirect hydroxyl radical reactions.

Some differences in the operating parameters existed such as a higher influent gas concentrations c_{Go} in the hazardous waste clean-up application resulting in an almost 2.5 times higher ozone feed rate. But the major difference was in the composition of the two waters. The DOC of the waste water was almost 15 times higher than that of the drinking water, which contained $3.6 \, \text{mg} \, l^{-1}$ DOC from natural organic matter (NOM). Moreover, the initial concentration of the target compound $c(M)_o$ in the waste water was as much as 10 000 times higher, approx. $100 \, \text{mg} \, l^{-1}$, compared to $10 \, \mu\text{g} \, l^{-1}$ in the drinking water (Table 4.5).

The primary goal in the drinking-water application was to reach the legally required target concentration of less than $0.1 \, \mu\text{g} \, l^{-1}$, which was equivalent to a degree of removal $\eta(M)$ of more than 99% of the initial atrazine concentration used in this experiment. Similarly, the primary goal in the hazardous waste clean-up was an almost complete removal of the target compound 2,4-dinitrotoluene, with $\eta(M) = 98\%$. In both experiments the ozonation was continued at least until the desired target compound concentration or degree of removal was reached.

Table 4.5 Main parameters to characterize the system "water".

Application			Drinking water	Hazardous waste clean-up
Reference			[14]	[15]
Parameters	Symbol	Unit		
System "water"				
Type of water	–	–	Model drinking water = Berlin tap water	Model groundwater = buffered deionized water
Target compound (M)				
Name of M	–	–	Atrazine	2,4 DNT
Molar mass	MW	$g\,mol^{-1}$	215.69	182.14
Initial concentration of M	$c(M)_o$	$mg\,l^{-1}$	1.0×10^{-2}	1.14×10^2
Primary goal				
Effluent concentration of M	$c(M)_e$	$mg\,l^{-1}$ ($\mu g\,l^{-1}$)	$<1.0 \times 10^{-4}$ (<0.1)	–
Degree of removal of M	$\eta(M)$	%	>99[a]	98
Secondary goal				
Degree of mineralization	$\eta(DOC)$	%	–	85
Reaction rate constants				
Direct	k_D	$l\,mol^{-1}\,s^{-1}$	6.35	n.a.
Indirect	k_R	$l\,mol^{-1}\,s^{-1}$	1.8×10^{10}	n.a.
Relevant water parameters				
pH-value	pH	–	pH 7	pH 7
Lumped parameters				
Organic carbon	$c(DOC)_o$	$mg\,l^{-1}$	3.6	52.5
Ozone decomposition rate constant	$k_d\,(O_3)$	$l\,mol^{-1}\,s^{-1}$	0.035	n.a.

a primary goal defined by $c(M)_e$ and not by degree of removal of M; $\eta(M)$ derived from indicated concentration difference.
n.a., not available.

The reaction time required to reach this goal for both compounds was on a similar order of magnitude: atrazine transformation took 84 min, while 2,4-DNT took 130 min. However, the reaction rates of ozone were nearly 13 times lower and the target compound removal rate $r(M)$ was as much as 10 000 times lower in drinking-water treatment (Table 4.7). Therefore, a much higher specific ozone dose I^* was required, almost 3000 times higher than for the waste-water

Table 4.6 Main parameters to characterize the system "reactor".

Application			Drinking water	Hazardous waste clean-up
Reference			[14]	[15]
Parameters	Symbol	Unit		
System "reactor"				
Reactor type	–	–	STR	STR
Mode of operation	–	–	Semibatch	Semibatch
Mixing (gas and liquid phase)	–	–	Both phases completely mixed	Both phases completely mixed
Reactor liquid volume	V_L	L	7.5	7.0
Range of operating parameters				
Temperature and pressure	$T; p$	°C; Pa	20; 1200	20; 1200
Design gas-flow rate	Q_G	l s^{-1}	8.33 × 10^{-3}	8.33 × 10^{-3}
		l h^{-1}	30	30
Ozone concentration in the gas phase (influent)	c_{Go}	mg l^{-1}	10	23
Ozone dose or feed rate	$F(O_3)$	mg l^{-1} s^{-1}	11.1 × 10^{-3}	27.4 × 10^{-3}
Stirrer speed	n_{STR}	s^{-1}	25	25
Mass-transfer coefficient	$k_L a$	h^{-1}	12	12
Theoretical ozone equilibrium concentration at T, p (see Section 6.1.3, Figure 6.5)	c_L^*	mg l^{-1}	3.7	8.4
Measured ozone concentration in the liquid phase	c_L	mg l^{-1}	2.8	Initially not detectable, rising to approx. 4.3
Reaction time (*to reach primary goal*)	t_R	min	84	130
Ancillary equipment				
pH control	–	–	None	Phosphate buffer
Ozone gas generator (Fischer type M 501, Germany)	$m(O_3)_o$	g O$_3$ h^{-1}	Max. 5.0	Max. 5.0
Ozone destructor	–	–	Activated carbon	Activated carbon
Analyzer system	–	–	Gas/liquid/ambient	Gas/liquid/ambient
Sample point	–	–	Reactor bottom, nearby stirrer	Reactor bottom, nearby stirrer

Table 4.7 Illustration of data evaluation using experimental values from semibatch treatability studies for the drinking water and hazardous-waste applications in Figure 4.2.

Application			Drinking water	Hazardous waste clean-up
Target compound (M)			Atrazine	2,4–DNT
Reference			[14]	[15]
Reaction time to reach primary treatment goal	t_R	min	84	130
Efficiencies				
Pollutant removal efficiency	$\eta(M)$	%	>99	98
Ozone transfer efficiency	$\eta(O_3)$	%	15	88
Reaction rates				
Reaction rate of ozone in the gas phase (ozone absorption rate)	$r_A(O_3)$	$mg\,l^{-1}\,s^{-1}$	1.8×10^{-3}	24×10^{-3}
Reaction rate of ozone in the liquid phase (ozone consumption rate)	$r(O_3)$	$mg\,l^{-1}\,s^{-1}$	1.8×10^{-3}	23×10^{-3}
Reaction rate of target compound in the liquid phase	$r(M)$	$mg\,l^{-1}\,s^{-1}$	2.0×10^{-6}	14×10^{-3}
Ratios				
Ozone yield coefficient	$Y(O_3/M)$	$g\,O_3\,g^{-1}\,M$	934	1.7
Specific ozone dose or input	I^*	$g\,O_3\,g^{-1}\,M$	5580	1.9
		$g\,O_3\,g^{-1}\,DOC$		4.1
Specific ozone absorption	A^*	$g\,O_3\,g^{-1}\,M$	924	1.7
		$g\,O_3\,g^{-1}\,DOC$		3.6
Kinetic data				
Observed kinetic rate coefficient	k' (obs.)	s^{-1}	0.92×10^{-3}	n.a.
		min^{-1}	0.055	n.a.

application. The values for specific ozone absorption A^* and ozone yield coefficient $Y(O_3/M)$ were also dramatically higher by a factor of almost a thousand. In addition, the ozone-transfer efficiency $\eta(O_3)$ was very low in the drinking-water application, while the dissolved ozone concentration was comparatively high (compare c_L to c_L^* in Table 4.6).

These differences are mainly due to the very low initial concentration of the target compound in drinking-water treatment and the different oxidation pathways involved. Since the direct reaction rate constant of atrazine is low, much ozone decomposes to produce hydroxyl radicals, which, however, are consumed to a large extent by scavengers like inorganic carbon and NOM in the drinking water.

Nevertheless, the production of hydroxyl radicals is necessary since atrazine oxidation proceeds mainly via the indirect pathway. Comparison of the reaction rate constants in Table 4.5 shows a much higher value for the indirect reaction ($k_D = 6.35 \, l\,mol^{-1}\,s^{-1}$ vs. $k_R = 1.8 \times 10^{10} \, l\,mol^{-1}\,s^{-1}$).

The high values of I^*, A^* and $Y(O_3/M)$ underscore the high loss of ozone to decay in the drinking water. These specific values, however, are not of great importance in evaluating micropollutant removal in drinking-water treatment. Instead, the dominating question in drinking-water treatment is which treatment alternatives can be used to reach the required concentration of the target compound and/or what can be done to reduce the necessary treatment time without applying even more ozone. For example, evaluation of an alternative treatment with the AOP combination of ozone and hydrogen peroxide (O_3/H_2O_2) reduced the treatment time for reaching the treatment goal to only 16 minutes [14]. Gottschalk measured a 6-fold higher removal rate coefficient ($k'_{obs} = 5.5 \times 10^{-3} \, s^{-1}$ vs. $0.92 \times 10^{-3} \, s^{-1}$) at a 60% lower ozone feed rate ($F(O_3) = 4.4 \times 10^{-3} \, mg\,l^{-1}\,s^{-1}$) and a molar feed ratio H_2O_2 to ozone of 0.88. In general, the results compared well with those of other researchers (e.g., [16, 17]).

In contrast, I^* or A^* are of great importance in waste-water treatment to give an overall indication of the amount of ozone that is necessary to reach the treatment goal under the specific reaction conditions used in the experiment. When the treatment goal is a certain degree of mineralization, the values are given as a ratio of ozone to the lumped parameter DOC or COD. Their values indicate the (global) effectiveness of the whole treatment process. This takes into account the effects of complex oxidation reactions in the experiment, for example, direct and indirect reactions of ozone with the target compound 2,4-DNT, secondary reactions leading to partial mineralization, and oxidation of nitrite to nitrate, as well as (partial) mass-transfer limitations.

They can be used to compare experiments under different treatment conditions and/or in different reactors when the water systems are similar. For example, Saupe [15] compared the specific amount of ozone necessary to treat various contaminants in a model water in a two-stage chemical-biological process (CBP). Specific ozone doses in the range of $I^* = 5.0$ to $10\,g \, O_3 \, g^{-1}$ DOC were necessary to reach a comparatively high degree of overall mineralization in both stages, for example, a DOC removal of $\eta(DOC) \geq 85\%$. And at the same time such doses allowed for an almost complete removal of the target compounds, for example, $\eta(M) \geq 95\%$. Further comparisons using I^* or A^* for two-stage treatment processes can be found in Section 9.3.

However, it is important to keep in mind that ozonation reactions are substrate specific and can be very dependent on the water composition, so that it is almost impossible to predict the specific ozone dose or absorption based on literature values or theoretical considerations. Therefore, minimum values have to be assessed individually for every water type in experiments under optimized reaction conditions. Nevertheless, comparisons with the results of other authors have underlined the value of this method to assess the results.

4.3
Reactor Design

This section will provide the framework required to design bench-scale experiments to provide results that can be used to extrapolate to larger scales. Since the reactor is the core element of both bench-scale as well as full-scale systems, familiarity with reactor design in general is necessary to design ozonation experiments. The first section briefly introduces most of the reaction-engineering concepts used in later chapters of this book. It is meant to be a quick refresher only, with an emphasis on simpler ozonation applications. More detailed explanations should be sought in the abundant general chemical reaction engineering literature [9, 18] or in the ozone-specific literature [6, 19]. In order to make a choice between the reactor types for the reactor design, we need to know the advantages and disadvantages of the various types. The second section offers a comparison between two common reactor types. In the last section a short example is given of how experimental information can be used to size a system.

4.3.1
Reactor Types

To design a reactor, we have to determine the reactor type to be used. Then we can calculate the size needed. Reactor types are commonly differentiated according to their mode of operation (discontinuous–batch reactor; continuous–flow reactor; half-continuous–semibatch reactor) and mixing (mixed or plug-flow) (Tables 4.8 and 4.9). The type of reactor influences how the reactant concentrations vary within the reactor and how long the material is held in the reactor, that is, the residence time of the material. There is often a distribution of residence times in flow reactors, so that we use the mean residence time t_H to characterize the reactor. In systems with solids retention, for example in biological reactors, it is common to differentiate between the residence times of the liquids and solids. They are often called hydraulic and solids retention times, respectively. In disinfection applications the mean residence time is called the contact time.

4.3.1.1 Operating Mode
In batch mode, the reactants are placed in a container and then allowed to react. To describe the number of phases present in a reactor, the terms *homogeneous* (one phase) and *heterogeneous* (at least two phases) are used.

When ozone is one of the reactants, it has to be introduced into the reactor:

1) In a *homogeneous* reaction, it is dissolved in a miscible liquid that is placed in the reactor.
2) In a *heterogeneous* reaction, it is transferred from the gas phase into the liquid phase in the reactor or from a nonmiscible liquid phase.

If the amount of ozone necessary for the reaction is larger than can be dissolved into the liquid volume or transferred from the gas volume of a closed container,

Table 4.8 Mode of reactor operation and its effect on reaction rates.

Mode of operation		Effect on reaction rate	Comments
Batch (discontinuous)		High reaction rates.	Storage and process control necessary. Downtime for filling and emptying.
Semibatch (half-continuous)		High reaction rates	Continuous or intermittent flow in one phase. Storage and process control necessary. Downtime reduced due to continuous addition of at least one reactant.
Continuous-flow		More difficult to achieve high reaction rates.	Reduced process control and storage requirements. No downtime. Two-phase flow can be cocurrent or countercurrent.

ozone has to be continuously introduced, either by being produced in the liquid phase in the reactor or by a constant flow of a gas or liquid containing ozone into the reactor. The common mode of operation in this case is to have an ozone-containing gas flow continuously into the reactor while the liquid remains in the reactor. This is often referred to as batch in the literature, but is called the semibatch mode in this book to characterize the operating modes of both phases.

In fermentation reactions the liquid phase is often fed continuously or intermittently to the batch reactor to supply substrate. This semibatch mode is called fed-batch. Another variation of the batch-operating mode found mainly in biological wastewater treatment is the sequencing batch mode. This commonly refers to a modification of the activated sludge process in which a sequence of biological reactions can be carried out in the same reactor by modifying the reaction conditions, often to achieve biological nutrient removal. The downtime for the cycling is integrated into the process.

The continuous-flow mode also has several variations depending on the direction of the flow of the phases relative to each other: countercurrent, cocurrent, and alternating cocurrent/countercurrent.

4.3.1.2 Mixing

Continuous-flow reactors are distinguished according to their hydraulic flow pattern. The two extremes of ideal flow reactors are (Table 4.9):

Table 4.9 Type of reactor mixing.

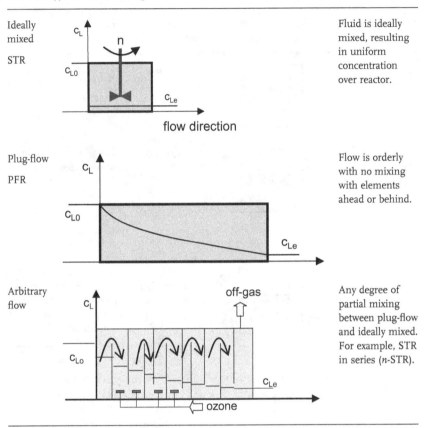

- The continuous (flow) stirred-tank reactor (CSTR or often CFSTR) – in which the liquid as well as the gas phase are ideally mixed. This is sometimes referred to as completely mixed flow (CMF). The concentration of material in the effluent is the same as that in the reactor. The residence time of material in a CSTR varies: some material leaves immediately upon entering and some never leaves. The mean residence time is used to describe the average time the material remains in the reactor.

- The plug-flow reactor (PFR), often called a tube reactor or tubular reactor – in which mixing over the length of the reactor (axial mixing) does not occur. Radially – across the diameter of the tube – the fluid is ideally mixed. The changes in the concentration of the material axially are due only to reaction. All material leaving the reactor has the same residence time and it is equivalent to the mean residence time.

4.3 Reactor Design

In reality, a reactor often has an arbitrary flow pattern between these two limits. Such a flow pattern can usually be described by comparing its residence time distribution to that of a number n of ideally mixed reactors connected in series (n-CFSTR). The higher the number of reactors connected in series, the closer the flow pattern and reaction rates approach that of plug flow.

Another way of describing an arbitrary flow pattern is to use a dispersion model. A dispersion term containing a dispersion coefficient is introduced into the mass balance for plug flow.

4.3.2
Comparison of Reactor Types

In order to make a choice between the reactor types for the reactor design, we need to know the advantages and disadvantages of the various types. These are presented in detail in Chapter 5. Here we will concentrate on comparing the reactor volumes required to achieve the same degree of compound removal $\eta(M)$. To do this we need appropriate design equations for the three types of reactors. These can be obtained from a mass balance on the reactor and are listed in Table 4.10 in terms of the concentrations of the target compound $c(M)$ as well as the degree of removal $\eta(M)$ (also called conversion in reaction engineering). From the equations, we see that the required reactor volume V_L depends on the rate of the reaction $r(M)$ taking place within the reactor and the desired degree of compound removal $\eta(M)$. For ozonation reactions, the reaction rate is generally very dependent on their concentrations. The higher the concentration, the higher the reaction rate. The concentrations in turn are dependent on the type of reactor used to carry out the reaction. The batch reactor and PFR have similar concentration

Table 4.10 Design equations for constant volume and temperature reactions.

Reactor type	Mass balance	Reaction time t_R or hydraulic retention time t_H in terms of	
		Concentration $c(M)$	Degree of removal $\eta(M)$
Batch	$\dfrac{dc(M)}{dt} = -r(M)$	$t_R = \displaystyle\int_{c(M)_0}^{c(M)} \dfrac{dc(M)}{-r(M)}$	$t_R = c(M)_0 \displaystyle\int_0^{\eta(M)} \dfrac{d\eta(M)}{r(M)}$
At steady state			
PFR	$0 = -Q_L \dfrac{dc(M)}{dV_L} - r(M)$	$t_H = \dfrac{V_L}{Q_L} = \displaystyle\int_{c(M)_0}^{c(M)} \dfrac{dc(M)}{-r(M)}$	$t_H = c(M)_0 \displaystyle\int_0^{\eta(M)} \dfrac{d\eta(M)}{r(M)}$
CFSTR	$0 = Q_L(c(M)_0 - c(M)) - r(M) \cdot V_L$	$t_H = \dfrac{V_L}{Q_L} = \dfrac{(c(M)_0 - c(M))}{r(M)}$	$t_H = \dfrac{c(M)_0 \cdot \eta(M)}{r(M)}$

profiles. The concentration decreases over time in the batch reactor and over the length in a PFR. In contrast, in CFSTR the reactor concentration is the same as that in the effluent.

If we compare the equations for batch and plug-flow reactors in Table 4.10, we can see that both have the same equation for the reaction time. Thus, both require the same reaction times to achieve the same degree of treatment for a certain reaction rate $r(M)$. However, batch reactors require additional time in their operation cycle. The downtimes for filling, emptying, cleaning, etc. must be added to the reaction time to find the total cycle time t_Σ. The volume is then calculated from the total reaction time t_T, the volume of water to be treated V_L over a certain period, and the amount of time the reactor will be operated t_T during that period.

$$V_L = \frac{V_T}{t_T} \cdot t_\Sigma \tag{4.10}$$

The volume of a batch reactor is, therefore, larger than a PFR. Looking at the equations for PFR and CFSTR, we cannot see offhand which requires a smaller volume. However, we can see that if we know the hydraulic retention time t_H required to achieve the removal efficiency $\eta(M)$, the size for both reactors can easily be calculated from the influent flow rate. Otherwise, the reaction rate $r(M)$ must be known. How to describe the reaction rate with the rate law is discussed in depth later in Chapter 7. Nonetheless, in order to compare the volumes for the various reactor types at this point, we will introduce just a few basic concepts.

For instance, we need to briefly consider what influences the reaction rate. The reaction rate is usually somehow dependent on the concentration of each compound. This dependency is described by the power n associated with the concentration in the reaction rate equation. This power n is called the order of the reaction with respect to the considered reactant. Normally chemical oxidation reactions are dependent on both the concentration of the compound to be removed M and the concentration of the oxidant, ozone or $OH°$. Since the total order of the reaction is the sum of the orders for each reactant, chemical oxidation reactions are usually second-order reactions.

If the concentration of the reactant does not influence the reaction rate ($n = 0$, i.e., zero-order kinetics) the volume for both reactor types is the same. However, for higher orders ($n = 1$ or 2, i.e., first- or second-order kinetics) the effect of concentration is very strong. Figure 4.9 compares the reactor volumes for the two reactor types CFSTR to PFR. For first-order reactions, four times more volume is required in a CFSTR than a PFR to achieve 90% removal. For higher degrees of removal, the ratio increases exponentially. This is due to the very low effluent concentrations required. Since in CFSTR the reactor concentration is the same as the effluent concentration, the reaction rate is very low, requiring much more volume than the PFR or batch reactor in which the reaction rate starts out fast due to high initial concentrations, slowing down as the concentration reaches the desired effluent concentration.

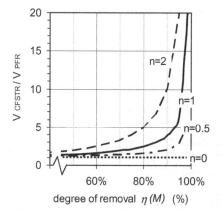

Figure 4.9 Comparison of continuous-flow stirred-tank and plug-flow reactor volumes required for a given degree of compound removal as a function of reaction order.

Quite often, treatment plants operate as CFSTR even though this requires a larger reactor volume than batch or plug-flow operation. The continuous mode is usually chosen due to ease of operation control, especially if the production of the water/waste water is continuous. The high degree of mixing is not necessarily by choice, but rather due to the extreme difficulty in attaining plug flow with two-phase flow. If the ozone consumption rate is high, ozone must be continuously added to the reactor or added at multiple points, disturbing the plug flow. However, since there is much to be gained from plug-flow operation, many reactor variations (very long tube reactors, use of baffles to achieve n-CFSTR, intermittent injection of ozone gas, etc.) have been advanced to improve the reaction kinetics.

When choosing the reactor type and configuration, it is important to note that the choice affects not only the required reactor volume, but also the ozone consumption and oxidation products. If compound oxidation is achieved mainly via the direct reaction, then the ozone consumption is also usually lower in PFR or batch than CFSTR. Less ozone is lost to decomposition when the target compound concentration is high. Furthermore, the oxidation products and thus the degree of removal achievable can differ in the reactor types. In PFR or batch reactors, sequential reactions can occur, where the easily oxidizable compounds are transformed before the more retractable ones, while in CFSTR with the continuous addition of target compounds, simultaneous competitive reactions take place. Thus, care must be taken to design the experimental setup if the results are to be used to design a larger scale.

The majority of lab experiments on determining the treatability of a water/waste water are made in semibatch reactors. The rates of disappearance of the target compound and ozone consumption can be measured over time and kinetic parameters can be measured. Chapter 7 is devoted to the reaction kinetics and methods for determining the required rate constants can be found there. Although batch data can often be used for sizing a continuous-flow stirred-tank reactor in many chemical reactions, this may not apply in ozonation experiments because of the unspecific oxidation reactions.

4.3.3
Design of Chemical Oxidation Reactors

If treatability is shown, the next step is to evaluate the feasibility of a pilot or full-scale treatment scheme. Looking back at the onion model in Figure 4.6, we see that this involves, at the core, designing a reactor system for the larger scale and then the system around it. Both the technical and economic aspects must be evaluated. Usually various alternatives are compared.

The procedure for the conceptual design of a whole treatment system with economic analysis is beyond the scope of this book. Information on design can be found in the literature: for plant design in general [8], for ozone drinking-water plants [5, 6] and examples of design [20]. However, in order to be able to design experiments and an experimental setup, it is essential to know what information is needed in the full-scale design process so that this can be gained from experiments. Therefore, how the information gathered in the experiments can be used for sizing a system is briefly sketched here according to the layers in the onion model. The procedure is summarized in Table 4.11. Table 4.11 can be used as a rough guide to draw attention to some of the important questions to be considered when going from smaller to larger scales, for example, when going from initial experiments in gas-wash flasks to a lab-scale setup, or from lab-scale to pilot scale. It is, however, by no means complete. The emphasis is laid on a few quantitative aspects of sizing the equipment. For information on important considerations for scaling up equipment the reader is referred to the literature [21].

To begin with, the goals and boundary conditions for the reaction system must be determined. The first two steps of the experimental design process provide the necessary information: *define treatment goals* and *define system* (here the water system). In the step *define treatment goals*, values for control parameters are set, for example, for the effluent concentration. In the step *define system*, the system is divided into two parts: a description of the water to be treated (Table 4.1) and the reaction system (reactor and ancillary equipment) that will be used to treat it (Table 4.2). The parameters listed in Table 4.1 describing the quantity and quality of the influent water are usually known so that the table can be filled in, while the values in Table 4.2 for the reaction system are what we seek.

Most of the parameters in Table 4.2 are chosen based on the experimental setup used. However, the reactor volume and the ozone generator size must be determined for the larger scale. In addition, values for I, $F(O_3)$ from Table 4.4 in data evaluation are needed. For the initial design of the experimental setup, values from preliminary experiments or from the literature are used. For full-scale, lab-scale or pilot-plant results are used. The reactor hydrodynamics should be similar to those in which the ozone consumption, etc. were determined. Experimental results from batch tests cannot be used for continuous systems and vice versa (see also Figure 4.9).

Table 4.11 Possible method to design the reaction system using the layers of the onion model and information from Tables 4.1, 4.2 and 4.4.

Goals and boundary conditions	Determine using the design steps *define treatment goals* and *define system*
Reaction system	
Volume of reactor V_L	Evaluate for daily and seasonal fluctuations, that is, average daily, peak-hour and minimum hourly influent flow rates and concentrations
• Semibatch	→ use total cycle time t_Σ evaluated from t_R + downtimes (Equation 4.10) $$V_L = \frac{V_T}{t_T} \cdot t_\Sigma$$
• PFR	→ use $t_H = t_R$ from semibatch experiments $V_L = Q_L \cdot t_H$
• CFSTR	→ use t_H from CFSTR experiments $V_L = Q_L \cdot t_H$ → or calculate V_L from design equation in Table 4.7 $$V_L = \frac{Q_L \cdot (c(M)_o - c(M))}{r(M)}$$ with an appropriate rate equation for $r(M)$, using kinetic coefficients measured in a semibatch reactor
Ozone feed rate $F(O_3)$	If a different initial concentration of M is used than in the experiments, $F(O_3)$ must be adjusted. → use the specific ozone input I^* and (initial) mass of compound to be treated
• Semibatch	$$F(O_3) = \frac{I^* \cdot c(M)_o}{t_R}$$
• Continuous flow	$$F(O_3) = \frac{I^* \cdot c(M)_o}{t_H} = \frac{I^* \cdot c(M)_o \cdot Q_L}{V_L}$$
Ancillary systems	
Ozone gas generator Production rate $m_{PR}(O_3)$	Evaluate for standby capacity, in addition to flow rate fluctuations. → from ozone feed rate $F(O_3)$ and reactor volume V_L $m_{PR}(O_3) = F(O_3) \cdot V_L$
Gas flow rate Q_G	→ from ozone feed rate $F(O_3)$ and ozone gas concentration c_{Go} $$Q_G = \frac{F(O_3) \cdot V_L}{c_{Go}}$$ → if the gas-flow rate has to be adjusted to a larger scale it must be checked that the mass-transfer rate $k_L a$ remains adequate

Table 4.11 Continued

Goals and boundary conditions	Determine using the design steps *define treatment goals* and *define system*
Ozone destructor	Design to destroy maximum ozone production rate $m_{PR}(O_3)_{max}$, that is, with no reaction
Process integration	Determine whether pre- or post-treatment is required to meet effluent requirements. Evaluate:
Pre- or post-treatment	→ solids removal before ozonation → pH adjustment before or after the ozonation reactor → biological treatment stage to reduce ozone consumption
Controllability	Choose instrumentation for analysis and control
Process control	→ determine which parameter(s) indicate when the treatment goal is achieved: for example, dissolved ozone concentration c_L or a lumped parameter for aromatic compounds $SUVA_{254}$ → consider manual or automatic control schemes
pH control	→ determine whether pH control is necessary during ozonation
Storage	→ determine storage requirements for semibatch operation based on cycle times (from reaction time t_R + downtimes)
Site integration	
Energy consumption	→ estimate energy requirements for: • oxygen production (if pure oxygen used); • gas compression; • ozone production rate; • additional for AOPs, for example, UV radiation intensity.
Chemicals	→ consider chemicals needed for pH control during oxidation or for pH adjustment before discharge → additional for AOPs: H_2O_2, catalysts

4.3.3.1 Reaction System

The required reactor size can then be calculated using the information in Table 4.1, the experimental data in Tables 4.2 and 4.4 and an appropriate safety factor. The operational variations in flow rate and concentration must also be taken into account for the design of the reactor as well as to determine the necessary equipment redundancy. This procedure is shown in Table 4.11 and briefly described here.

The reactor volume V_L for the semibatch reactor mode is calculated considering the given total volume of water V_T to be treated in a daily period t_T and the total cycle time t_Σ estimated from an experimentally determined t_R (the reaction time to achieve the required removal efficiency) and the downtime for filling, emptying, cleaning. For a CFSTR the volume is calculated from the experimentally determined hydraulic retention time t_H.

The necessary ozone feed rate $F(O_3)$ is the rate measured in the experiments if the initial influent concentration of M is the same. The required ozone gas-flow rate and concentration can be calculated for the larger reactor volume using Equation T 1-4 in Table 4.4. If a different initial concentration of M is used than in the experiments, $F(O_3)$ must be adjusted. This can be done using the specific ozone input I^* that was measured in the experiments and the reaction time.

4.3.3.2 Ancillary Systems

Using the values for $F(O_3)$ and the reactor volume, the production rate of the ozone gas generator $m_{PR}(O_3)$ can be calculated. In choosing the number and size of the ozone gas generators for operation and backup, it is important to check whether the capital costs of generators show significant economies of scale. The unit cost per kg of ozone may decrease as the production rate of the generator increases [6]. The required gas-flow rate Q_G can then be calculated from $F(O_3)$ and the initial ozone gas concentration c_{Go}. In scaling up the gas-flow rate, it is important to keep in mind the connection between the gas-flow rate and the mass-transfer rate. The dependency differs according to reactor type and is discussed in Chapter 6.

Generally, the ozone destructor has to be designed based on the maximum production capacity of the ozone generator $m_{PR}(O_3)_{max}$. This simply is a safety measure in case the reactor would be empty and no ozone was consumed by reaction in the reactor.

4.3.3.3 Process Integration

The next layer of the onion is the aspect of process integration. Here, questions dealing with pre- and post-treatment are looked at such as:

- Are treatment stages necessary before the oxidation step to increase removal efficiency or reduce operating costs? For example, easily oxidizable compounds can be removed before ozonation in a pretreatment step – dissolved compounds can be oxidized biologically (organics) or with air (inorganics such as Fe or Mn) and/or solids can be filtered or settled in order to prevent clogging and/or unnecessarily high ozone consumption.

- Is another treatment process required after the ozonation stage to achieve the final treatment goals? For instance, solids can be produced by the coagulating effects of ozone that have to be settled or filtered in a separate stage. In addition, nonbiodegradable organic compounds can be converted to biodegradable ones in the course of oxidation that can be removed more effectively in a subsequent biological stage.

The use of process combinations with biological treatment is treated in detail in Section 9.3.

4.3.3.4 Controllability

In this step the instrumentation for analysis and control must be chosen. In order to do this various questions must be answered in the experiments such as – How does the pH affect the treatment process? Is pH adjustment before or after treatment necessary? Does the pH have to be controlled during ozonation? Another important question is: Which parameter(s) indicate when the treatment goal is achieved? For example, in batch treatment the rise in the dissolved ozone concentration c_L over time may correlate with the degree of treatment (see Figure 4.3), so that it can be used to trigger the end of a batch cycle. Or perhaps a lumped parameter for aromatic compounds such as $SUVA_{254}$ can be used to indicate the degree of compound transformation. Further considerations are whether to use manual or automatic control schemes.

Associated with considerations for the process control are the storage requirements. Is storage capacity required? For semibatch operation storage requirements are determined based on cycle times (from reaction time t_R + downtimes).

4.3.3.5 Site Integration

There are many aspects to be considered in site integration. Only two are mentioned here: energy consumption and chemicals. When determining the energy requirements for ozone generation, both the energy for the feed-gas production (oxygen or compressed air) and preparation as well as the energy for the ozone generator must be considered.

Chemicals for pH control during oxidation or for pH adjustment before discharge may be needed. In the case of AOPs, provisions for a supply of H_2O_2 or catalysts have to be made and the costs estimated.

4.4
Checklists for Experimental Design

To help the reader identify the activity associated with each step of the experimental design process on the more intuitive level, checklists are presented in Table 4.12. They are not comprehensive, but can help the beginning experimenter get started. We have placed them at the beginning of Part B to give an overview of the information necessary for successful experimentation. Each reader will spend more or less time initially reading through the checklists. They can be returned to when the practical work is in the planning phase. It is important to use them critically, checking which points apply to the situation at hand and to modify and add appropriate points that have been missed.

4.4.1
Checklists for Each Experimental Design Step

1. **Define Goals**

   – Experimental goals could be one or more of the following:
   - determination of process feasibility;
   - determination of minimal use of ozone for required pollutant removal efficiency;
   - determination of reaction kinetics;
   - determination of scale-up procedure;
   - determination of best process or combination of processes;
   - etc.

2. **Define System**

   Water or Waste water

   – Define the composition of the (waste) water. Which individual compounds: organic as well as inorganic are present? Can the type of compound at least be determined? How can it be quantified (lumped parameter, individual analysis)?

   – Consider using a synthetic (waste) water to test certain hypotheses.

   – Check the recent literature on biodegradation in order to determine the conditions under which the compound(s) in question are biodegradable.

   – Make a theoretical analysis of the most probable behavior of the individual compounds as well as the complete water matrix during ozonation.

   – Determine the most probable oxidation products, and how to measure them.

   – Consider that ions contained in the raw water and/or occurring from oxidized substituted organics might act as promoters, inhibitors or scavengers in the radical reaction cycle process (see Chapter 2) or influence the mass transfer of the system (see Section 6.3).

   Oxidant

   – Select the most appropriate type of chemical agent, checking whether there are similar/competitive oxidation processes available (e.g., application of AOPs etc.) which might be more efficient or economical.

   – Consider combined treatment, for example, oxidative and biological processes.

- Respect technical constraints of each treatment step, as well as of combined processes. In general, the feasibility of combined processes (in terms of operating costs) depends on the performance and effectiveness of the oxidation process.
- If choosing ozone as the chemical agent, consider the interdependencies between ozone production by the two most common types of generators and system parameters, for example, mass transfer (Section 5.2).

Reaction System (Chemical and Biological)

- Choose between the two operating principles, batch or continuous flow, and types of reactors.
- Consider possible advantages and disadvantages of each operating principle before setting up the equipment, that is, volume of influent water required, pumping and storage requirements, length of experimental run.
- Build and operate the reactor system considering all safety precautions.

Total System

- Evaluate the reactor hydrodynamics by determining the retention time distribution (see [18–22]).
- Evaluate the mass transfer by determining the mass-transfer coefficient for a range of operating conditions using the same water that will be oxidized later if possible.
- Determine the optimal operating conditions with respect to the experimental goals.
- Characterize the reactor at these operating conditions.

3. **Select Analytical Methods**

   Oxidants

   - see Section 5.5

   Pollutants and Water Matrix

   - Assess TOC and TIC if possible; consider that TIC values indicate the amount of scavengers.
   - DOC is the best parameter for total pollutant removal in (waste) water application, DOC characterizes mineralization uniquely in chemical and biological processes.
   - Measure $SUVA_{254}$ as an indicator of the sum of aromatic compounds or $SUVA_{436}$ for humic-like substances.
   - Avoid COD because of possible interferences of some inorganics (NO_2^-, H_2O_2 etc.), very different oxidation state of individual substances, and since it does not measure mineralization.

- Use analytical procedures adapted to the individual pollutants (e.g., GC, HPLC, LC, MS, etc.).
- Stop the further oxidation of compounds after sampling by adding Na_2SO_3.
- Ions contained in raw water and/or occurring as oxidation products (e.g., HCO_3^-, CO_3^{2-}, Cl^-, NO_2^-, NO_3^-, SO_4^{2-}, PO_4^{3-}) are best measured with ion chromatography (IC).
- pH: Consider a continuous control of pH since it might change during oxidation due to the production of organic acids.
- Analyze analytical errors and their effect on the results, for example, a sensitivity analysis with the Gaussian error propagation method.

Toxicity

- see Chapter 1 and Section 9.3.3

4. **Determine Experimental Procedure**

    - Decide what information is necessary to achieve the experimental goal(s).
    - Generally, data are necessary that allow the reaction system to be balanced. If possible an online determination is of advantage.
    - Decide on and implement a quality-control program to assure reproducibility and minimal error.
    - Determination in advance:
        - information about the inorganic and organic matrices of the water
    - Determination online:
        - ozone gas concentration at the inlet;
        - ozone gas concentration at the outlet;
        - liquid concentration of the oxidants and pollutants;
        - information about the inorganic and organic (e.g., $SUVA_{254}$, $SUVA_{436}$) matrices of the water;
        - pH-value;
        - temperature.
    - Determination during/after experiment:
        - liquid concentration of the oxidants (if not possible online) and pollutants;
        - information about the inorganic and organic matrices of the water.

    in case of AOP:

    - $F(H_2O_2)/F(O_3)$: specific hydrogen peroxide dose rate;

- O_3/UV: detailed information about the UV-intensity, wavelength spectrum, illuminance and penetration.
  – In general, every experiment should at least be performed twice in order to check whether the same results can be achieved during two independent treatments.

5. **Evaluate Data**
   – Assess the treatment results for each individual process step as well as for the whole system.
   – Use DOC as the main parameter if mineralization is required.
   – Calculate parameters, for example, ozone consumption rate $r(O_3)$, ozone yield coefficient $Y(O_3/M)$ (see Table 2.2).
   – Make sure the set of parameters reported adequately characterizes the system and results so that they can be of value to other experimenters.

6. **Assess Results**
   – Compare results with those found in the literature.
   – Compare results to experimental goals.
   – If the goals have been reached, modeling or scaling-up of the results can be undertaken.
   – If the experimental goals have not been reached, new experiments must be planned and carried out.

   Iterative Process!

4.5
Ozone Data Sheet

gas: blue colored
water: purple blue in concentration higher than $20\,\mathrm{mg\,l^{-1}}$

This section contains useful information on physical properties and conversion factors for ozone (Table 4.12–4.14), as well as some common conversion formulas.

Table 4.12 Physical properties of ozone.

Property	Value	Unit	Reference
Density	2.144	$g\,l^{-1}$	[23]
Diffusion coefficient	1.26×10^{-9} (20 °C)	$m^2\,s^{-1}$ (measured)	[24]
	1.76×10^{-9} (20 °C)	$m^2\,s^{-1}$ (measured)	[25]
	1.75×10^{-9} (20 °C)	$m^2\,s^{-1}$ (calculated)	[26]
	1.82×10^{-9} (20 °C)	$m^2\,s^{-1}$ (calculated)	[27]
Extinction coefficient	3300 (λ = 254 nm)	$l\,mol^{-1}$	–
	3150 (λ = 258 nm)		[28]
Boiling point (at 101.3 kPa)	−111.3	°C	[23]
Melting point	−193.0	°C	[28]
Molar weight (MW)	48	$g\,mol^{-1}$	
Redox potential, E_0^H (in aq. solution for pH = 0 and O_2/O_3 gas)	+2.07 (25 °C)	V	[28]
Vaporization heat	316.3 (101.3 kPa at boiling point)	kJ/kg^{-1}	[23]
Viscosity (dynamic)	0.0042 (−195 °C)	Pa s	[23]
	0.00155 (−183 °C)	Pa s	
Temperature	solubility ratio s	Henry's Law constant $H_C = s^{-1}$	[29]

(°C)	(K)	(–) (dimensionless)	(–) (dimensionless)
5	278.15	0.45	2.20
10	283.15	0.37	2.71
15	288.15	0.35	2.86
20	293.15	0.30	3.30
25	298.15	0.27	3.68
30	303.15	0.24	4.15

4 Experimental Design

Table 4.13 Physical properties of ozone and other gases.

Gas	Property	Value	Unit
O_3	density	2.14	$g\,l^{-1}$
O_2	density	1.43	$g\,l^{-1}$
Air	density	1.29	$g\,l^{-1}$
Water	density	1000	$g\,l^{-1}$

Table 4.14 Conversion table for ozone gas-phase concentrations.

Ozone concentrations

c_G (wt.)[a]	c_G (vol.)	$y(O_3)$	$c_G^{\,b}$	$c_G^{\,c}$
Weight %	Volume %	Mole fraction in gas	$g\,m^{-3}$	$g\,m^{-3}$
1	0.7	0.007	14.1	13.9
2	1.3	0.013	28.4	28.0
3	2.0	0.020	42.7	42.1
4	2.7	0.027	57.2	56.4
5	3.4	0.033	71.6	70.7
6	4.1	0.040	86.2	85.1
7	4.8	0.047	100.9	99.6
8	5.5	0.054	115.8	114.3
9	6.2	0.061	130.7	129.0
10	6.9	0.068	145.7	143.8
11	7.6	0.075	160.9	158.8
12	8.3	0.082	176.1	173.8
13	9.1	0.089	191.4	189.0
14	9.8	0.097	206.8	204.2
15	10.5	0.104	222.4	219.6
16	11.3	0.111	238.1	235.1
17	12.0	0.119	253.9	250.6
18	12.8	0.126	269.8	266.3
19	13.5	0.133	285.8	282.1
20	14.3	0.141	301.9	298.0

a   For an ozone/oxygen gas mixture at STP: 1 ppm = 2.14 mg m^{-3} = 1 cm^3 m^{-3}; 20 °C, 101.3 kPa.
b   According to ideal gas law at: STP: T = 0 °C, P = 1.013 × 10^5 Pa.
c   According to ideal gas law at: NTP: T = 0 °C, P = 1.0 × 10^5 Pa.

Conversion formula

ideal gas law:

$$p(O_3) = y(O_3) \times P_{abs} = c_G \frac{\Re T}{MW(O_3)} \rightarrow c_G = \frac{y(O_3) \cdot P_{abs} MW(O_3)}{\Re T}$$

with:

$\Re$ = ideal gas constant (8.3145 J mol^{-1} K^{-1} = 8.3145 kPa l mol^{-1} K^{-1})

MW(O$_3$) = 48 g mol^{-1}

conversion c_G to c_G (vol.) as vol.%:

$$c_G(\text{vol.}) = \frac{c_G V_n}{MW(O_3) \, 1000} \cdot 100 = y(O_3) \cdot 100$$

with:

V_n = molar volume (22.4 l mol^{-1})

$y(O_3)$ = ozone mole fraction in the gas

calculation of c_G (wt.) as wt.% for O$_3$ in O$_2$:

$$c_G(\text{wt.}) = \frac{y(O_3) \cdot 48}{[(1 - y(O_3)) \cdot 32 + y(O_3) \cdot 48]} \cdot 100$$

calculation of $y(O_3)$ from c_G (wt.) as wt.% for O$_3$ in O$_2$:

$$y(O_3) = \frac{c_G(\text{wt.})/(\rho(O_3) \cdot 100)}{[(1 - c_G(\text{wt.})/100)/\rho(O_2) + c_G(\text{wt.})/(\rho(O_3) \cdot 100)]}$$

References

1 Hoigné, J. and Bader, H. (1983) Rate constants of reactions of ozone with organic and inorganic compounds in water – I. Non dissociating organic compounds. *Water Research*, **17**, 173–183.
2 Hoigné, J. and Bader, H. (1983) Rate constants of reactions of ozone with organic and inorganic compounds in water – II. Dissociating organic compounds. *Water Research*, **17**, 185–194.
3 Hoigné, J., Bader, H., Haag, W.R. and Staehelin, J. (1985) Rate constants of reactions of ozone with organic and inorganic compounds in water – III. Inorganic compounds and radicals. *Water Research*, **19**, 173–183.
4 Madden, K.P. NDRL Kinetics Database, Radiation Chemistry Data Center of the Notre Dame Radiation Laboratory, http://allen.rad.nd.edu (accessed December 2008).
5 DVGW (1999) *Technical Rule – Code of Practice W625: Anlagen zur Erzeugung und Dosierung von Ozon (Plants for the Production and Dosage of Ozone)*, DVGW German Technical and Scientific Association for Gas and Water, Bonn.

6. Langlais, B., Reckhow, D.A. and Brink, D.R. (eds) (1991), *Ozone in Water Treatment – Application and Engineering*, Cooperative Research Report, American Water Works Association Research Foundation: Company Générale des Eaux and Lewis Publisher, Chelsea, MI, ISBN: 0-87371-477-1.
7. Rakness, K.L. (2005) *Ozone in Drinking Water Treatment: Process Design, Operation and Optimization*, American Water Works Association, ISBN 1583213791.
8. Koolen, J.L.A. (2001) *Design of Simple and Robust Process Plants*, Wiley-VCH Verlag GmbH, Weinheim.
9. Fogler, H.S. (2006) *Elements of Chemical Reaction Engineering*, 4th edn, Prentice-Hall, New Jersey.
10. Box, G.E.P., Hunter, J.S. and Hunter, W.G. (2005) *Statistics for Experimenters: Design, Innovation, and Discovery*, 2nd edn, John Wiley & Sons, Inc., New Jersey.
11. NIST/SEMATECH e-Handbook of Statistical Methods, http://www.itl.nist.gov/div898/handbook/ (accessed December 2008).
12. Haaland, P.D. (1989) *Experimental Designs in Biotechnology*, Marcel Dekker Inc., New York, Basel.
13. Montgomery, D.C. (2005) *Design and Analysis of Experiments*, 6th edn, John Wiley & Sons, Inc., New Jersey.
14. Gottschalk, C. (1997) *Oxidation organischer Mikroverunreinigungen in natürlichen und synthetischen Wässern mit Ozon und Ozon/Wasserstoffperoxid*, Shaker Verlag, Aachen.
15. Saupe, A. (1997) *Sequentielle chemisch-biologische Behandlung von Modellabwässern mit 2,4-Dinitrotoluol, 4-Nitroanilin und 2,6-Dimethylphenol unter Einsatz von Ozon*, VDI-Fortschritt-Berichte Reihe 15 (Umwelttechnik) Nr. 189, VDI-Verlag, Düsseldorf.
16. Beltrán, F.J., García-Araya, J.F. and Acedo, B. (1994) Advanced oxidation of atrazine in water – I ozonation. *Water Research*, 28, 2153–2164.
17. Beltrán, F.J., García-Araya, J.F. and Acedo, B. (1994) Advanced oxidation of atrazine in water – II ozonation combined with ultraviolet radiation. *Water Research*, 28, 2165–2174.
18. Levenspiel, O. (1999) *Chemical Reaction Engineering*, 3rd edn, John Wiley & Sons, Inc., New York.
19. Beltrán, F.J. (2004) *Ozone Reaction Kinetics for Water and Wastewater Systems*, Lewis Publisher, Boca Raton, London, New York, Washington DC.
20. EPA (2005) 815-R-05-013. Technologies and Costs Document for the Final Long Term 2 Enhanced Surface Water Treatment Rule and Final Stage 2 Disinfectants and Disinfection By-products Rule, EPA Office of Water (4606-M), pp. 4-34–4-45, http://www.epa.gov/ogwdw/disinfection/lt2/pdfs/costs_lt2-stage2_technologies.pdf (accessed October 2008).
21. Zlokarnik, M. (2006) *Scale-up in Chemical Engineering*, 2nd edn, Wiley-VCH Verlag GmbH, Weinheim.
22. Levenspiel, O. (1972) *Chemical Reaction Engineering*, 2nd edn, John Wiley & Sons, Inc., New York.
23. Air Liquide (2005) Gas Encyclopaedia. ozone. http://encyclopedia.airliquide.com (accessed March 2009)
24. Matrozov, V., Kachtunov, S. and Stephanov, S. (1978) Experimental determination of the molecular diffusion, *Journal of Applied Chemistry of the USSR*, 49, 1251–1555.
25. Johnson, P.N. and Davis, R.A. (1996) Diffusivity of ozone in water. *Journal of Chemical and Engineering Data*, 41, 1485–1487.
26. Wilke, C.R. and Chang, P. (1955) Correlation of diffusion coefficients in dilute solutions. *American Institute of Chemical Engineering Journal*, 1, 264–270.
27. Reid, R.C., Prausnitz, J.M. and Sherwood, T.K. (1977) *The Properties of Gases and Liquids*, 3rd edn, McGraw-Hill, New York.
28. Hoigné, J. (1998) Chemistry of aqueous ozone and transformation of pollutants by ozonation and advanced oxidation processes, in *The Handbook of Environmental Chemistry Vol. 5 Part C, Quality and Treatment of Drinking Water II* (ed. J. Hrubec), Springer-Verlag, Berlin, Heidelberg.
29. Kosak-Channing, L.F. and Helz, G.R. (1983) Solubility of ozone in aqueous solutions of 0–0.6 M ionic strength at 5–30 °C. *Environmental Science and Technology*, 17, 145–149.

5
Experimental Equipment and Analytical Methods

Reaction mechanisms and experimental observations are not independent of the system in which they are made; therefore, the design of the experimental setup and how the experiment is run affect the outcome. In general, an experimental setup consists of an ozone generator, reactor, flow meters and online analysis of at least the influent and effluent ozone-gas concentrations and ambient-air monitor (Figure 5.1). Each setup should be tailored to the experimental goals and the resources available. The choice of the equipment and procedures should be based on knowledge of how they influence the results. It is important to note that experimental setups and procedures from drinking-water treatment cannot be applied on waste water without appropriate evaluation and vice versa.

This chapter provides some essentials about the individual components and how they affect ozonation. First, the material of the equipment necessary for containing ozone (Section 5.1) is examined, followed by the equipment for producing it (Section 5.2), and bringing the reactants together (Section 5.3). Methods to measure ozone, with their advantages/disadvantages (Section 5.4) and the safety aspects to consider (Section 5.5) are then discussed. This is rounded off with a list of common questions, problems and pitfalls that we have come across over the years (Section 5.6). We hope it will help you get started with your own experiments.

5.1
Materials in Contact with Ozone

Since ozone is a very strong oxidant, all materials in contact with this gas have to be highly corrosion resistant. This has to be considered for all components in the ozone system including the ozone generator as well as all instruments. A short overview of materials appropriate for the various system parts is given in Table 5.1. More detailed information on material requirements can be found in technical guidelines and references (e.g., [7–9]).

The material choice depends on the type of water to be treated, the required ozone doses, scale and required life span of the equipment. The design ozone concentrations and system pressure are important factors. Ozone generation with oxygen can achieve much higher ozone concentrations in the gas and liquid

Ozonation of Water and Waste Water. 2nd Ed. Ch. Gottschalk, J.A. Libra, and A. Saupe
Copyright © 2010 WILEY-VCH Verlag GmbH & Co. KGaA, Weinheim
ISBN: 978-3-527-31962-6

5 Experimental Equipment and Analytical Methods

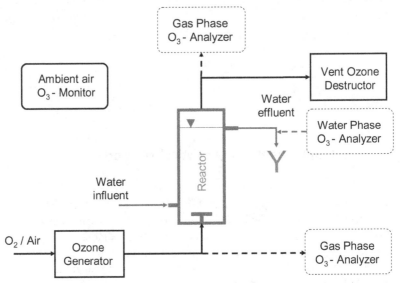

Figure 5.1 Basic components of any experimental ozonation setup (— required, --- optional equipment).

phases, which can be more corrosive for reactors and piping than air-fed ozonation systems. Increasing the system pressure to raise dissolved-ozone concentrations requires not only material with higher structural strength, but more corrosion resistance in the liquid phase.

In addition to safety concerns, materials not able to withstand oxidation by ozone or other reactive species can give off contaminants to the water being ozonated. This is especially to be avoided in systems requiring high-purity water, such as semiconductor applications. Further requirements are that the materials do not cause ozone decomposition or adsorb the compounds to be treated. Table 5.1 lists some of the common components in an ozonation system and some of the preferred materials used for each component.

Generally, the materials used in full-scale and pilot-plant applications are different from those used in lab-scale applications. The investment costs and safety considerations for automated, round-the-clock full-scale applications differ drastically from those at lab-scale. Furthermore, the surface to volume ratio is usually much higher at lab-scale. Materials that accelerate ozone decomposition or interact with compounds through adsorption or leak contaminants should be avoided. Often, materials with shorter lifetimes or fragility are acceptable at lab-scale. Indeed, the flexibility of lab-scale investigations allows new materials to be explored. For instance, the study of reactors with bubble-free ozone contactors made of semipermeable membranes (often polymeric) in laboratory reactors has increased in the last decade. Such reactors are often called membrane reactors. They are not yet applied in commercial drinking- or waste-water ozonation systems.

Table 5.1 Materials resistant to ozone.

System components	Preferred materials	Remarks; Examples of application (Ref.)
Reactors	(Quartz) glass Stainless steel (No. 1.4435 or 1.4404)	Full-scale reactors for treatment of industrial waste waters
	PVC (polyvinylchloride), Concrete	Progressively attacked by ozone Full-scale reactors for drinking-water treatment
Pipes and valves	Glass Stainless steel	Glass prone to breakage Fast corrosion possible with high concentration of salts
	PTFE (polytetrafluoroethylene) PFA (perfluoralkoxy) Kalrez® (Du Pont)	*PTFE, PFA and Kalrez ®*: expensive, scarcely oxidizable, stable over time,
	PVC (polyvinylchloride) PVDF (Polyvinylidenfluoride) PVA (polyvinylalkoxy) (Norprene®)	*PVC, PVDF, PVA*: less expensive, slowly oxidizable, less stable over time
Seals	PTFE (polytetrafluoroethylene) PFA (perfluoralkoxy) Kalrez® (Du Pont) Kynar® (PVDF) VITON®	
Membranes in membrane reactors	Ceramics Fused alumina (capillary membrane) PTFE (polytetrafluoroethylene)	Hydrophobic [1] Novel membrane reactor [2] Resistance higher than PVDF [3]; Mass-transfer studies with nine different membranes [4]
	PVDF (polyvinylidenfluoride)	Resistance lower than PTFE [3]; Hollow-fiber membrane contactor [5]; flat-sheet membrane contactor [6]
Conventional gas contactors	Ceramics Fused alumina PTFE (polytetrafluoroethylene) Stainless steel (No. 1.4435 or 1.4404)	Comparatively expensive

In the following sections, some general aspects of material choice are briefly discussed, while more detailed information is found in the later sections dealing with specific components.

5.1.1
Materials in Pilot- or Full-Scale Applications

Materials in full-scale applications have to be designed for long lifetimes and high safety standards at reasonable costs of investment and maintenance. High pressures exclude the use of materials like glass.

5.1.1.1 Reactors
In full-scale applications reactors are often made from concrete or stainless steel. Large horizontal concrete basins are most often used for low-pressure installations in large-scale drinking water units. Stainless steel is used for reactors in industrial waste-water ozonation. The use of polymeric material such as PVC has seldom been reported in full-scale applications, since it is slowly attacked by ozone. Even with stainless steel, however, corrosion can still be a problem, especially when treating waste waters that often contain high concentrations of salts. Nevertheless, such reactors are best made of stainless steel because of the possibility to operate them at elevated pressures, for example, 200–600 kPa, which can readily be achieved with commercially available ozone generators [10].

The materials used in the whole reaction system have to be evaluated for their appropriateness for the application and their ozone resistance. For instance, a requirement of ozone applications in the semiconductor industry is absolute cleanliness. Therefore, the dissolved ozone that is used as a cleaning agent for silicon wafers (see Section 9.4) has to be produced in superpure quality. To achieve this, not only the influent water and gases must be specially treated, but the generator itself must conform to strict material standards. The water must be superpure, that is, free of organics and particles, and the concentration of dissolved metals (e.g., Na, Al, K, Ca, Ti, Cr, Mn, Fe, Ni, Cu, Zn) must be below 0.01 ppb [11]. This is achieved by using ozone-generator equipment with high-quality material, for example, electropolished stainless steel, highly purified influent gases (oxygen and carbon dioxide grade 4.5 or better) and not doping the gas with nitrogen in order to prevent the production of nitric acid as completely as possible (see Section 5.2.1.1). Some producers of ozone generators provide such equipment (e.g., [11, 12]).

5.1.1.2 Piping
The location of the piping determines the material choice. Before the generator, the piping must be appropriate for the chosen gas (air or oxygen). After the generator, the ozone concentration of the dry gas must be considered. And after the reactor, the material will be exposed to wet ozone-containing off-gas until the ozone is destroyed in the destructor. Very fast corrosion (formation of holes over the course of weeks, especially at improper welds) has been observed in off-gas

piping in full-scale applications, even when made from stainless steel. The problem is most evident when aerosols, for example containing chloride, escape from the reactor into the pipes where they form a very corrosive wet film.

5.1.2
Materials in Lab-Scale Experiments

For the experimenter in the laboratory, not only do materials have to be chosen on the basis of their corrosion resistance, but also for their possible effects on the ozone and carbon balances. Some materials (e.g., silver and other metals) react with ozone, enhancing ozone decomposition. This can be especially detrimental in drinking-water and high-purity-water (semiconductor) ozone applications, where the additional ozone decomposition can prevent a precise balance on the ozone consumption or contaminants can be released into the water. Furthermore, the surfaces of the equipment can adsorb the compound(s) being studied. Therefore, materials in the whole laboratory setup must be carefully chosen for their appropriateness in ozonation experiments on the basis of corrosion resistance, ozone decomposition, and as sources of contaminants or sinks for target compounds. This includes the reactor, ozone generator, all piping for ozone gas and ozonated water, especially the seals, together with the sampling system and instrumentation.

5.1.2.1 Reactors

The most common materials for lab-scale reactors are glass, plexiglass and PVC. Plexiglass and PVC are very inexpensive materials that can be used for lab-scale ozone reactors, however, they are slowly but progressively attacked by ozone. Bubble columns or tube reactors can easily be constructed from glass, plexiglass or PVC tubes. With PVC gas tightness is best achieved by welding, but it can only be operated at ambient pressure ($P_{abs} \approx 100\,kPa$). With a view to system cleanliness in laboratory experiments, use of PVC is only advisable in waste-water treatment, whereas quartz glass is highly appropriate for most laboratory purposes. To avoid adsorption effects, especially in studies on the treatment of trace contaminants, it is advisable to let the liquid phase contact only glass and stainless steel. In stirred reactors for example, the stirrer seals (unless the stirrer is a magnetic bar) and all connectors for piping are best placed at the top of the reactor, so that the liquid does not come into contact with them.

During the last decade, the use of membranes as semipermeable reactors for bubbleless contacting between two phases has increased in laboratory applications. Various configurations have been investigated, for example, porous hollow fibers with liquid phase inside the fibers and gas outside [5] or vice versa [2] as well as flat-sheet (plate) types with single [4] or alternating gas/liquid layers [6]. They are discussed in further detail in Section 5.3.3. In order to be used in such ozone-contacting applications, the membranes need to be ozone resistant and have good mass-transfer characteristics. In addition, the membranes must be hydrophobic, since otherwise the liquid would drain into it and the mass transfer would be drastically decreased [1]. In an early study on hollow-fiber membranes it was confirmed that PTFE and PVDF provide high resistance to ozone [3]. It was pointed out that higher crystallinity of PVDF resulted in better ozone resistance. A comparison of the ozone mass-transfer performance of nine different PTFE and PVDF flat-sheet membranes (e.g., porous and nonporous, different pore size, thickness and pore volume) showed that the factors to be considered when selecting a membrane should be mechanical strength, bubble point and cost [4]. Additionally, the long-term performance is of importance.

5.1.2.2 Piping

As discussed above for full-scale applications, the location of the piping determines the material choice. However, since the surface to volume ratio is usually much higher at lab-scale, it is very important to avoid materials that accelerate ozone decomposition or interact with the reactor contents. Contamination due to oxidation of the piping material or loss of compounds due to adsorption on surfaces can cause unwanted effects for the experimenter.

5.2
Ozone Generation

Since ozone, a three-oxygen-atom modification of molecular oxygen, is an unstable molecule, it has to be generated onsite. It can be produced from air, pure oxygen gas or water by applying some type of energy. If air or pure oxygen is the ozone source, the energy is needed to split the oxygen molecule into two oxygen atoms, which then recombine with the remaining oxygen molecules that have not been split. If water is the ozone source, the energy liberates oxygen atoms from the water molecules that can recombine to molecular oxygen or to ozone molecules. Various energy sources can be used to produce ozone. The methods differ in their working principles and ozone sources and are summarized in Table 5.2.

The first two methods of ozone production, electrical and electrochemical, and the working principles behind them, electrical discharge (ED) and electrolysis (EL), are the only ones of practical importance both in bench- and full-scale applications. These will be discussed in detail in Sections 5.2.1 and 5.2.2. Of the two, the most widespread technology is electrical discharge using air or oxygen as the feed gas. Especially dielectric barrier discharge (DBD) ozone generators (see Section 5.2.1) are most often used in water applications, from small laboratory to large industrial scale [10]. A large variety of manufacturers exist, but only few produce DBDOGs resp. EDOGs of large industrial scale.

On the other hand, ozone production is also possible through the electrolysis of water. This method of ozone production has gained some importance in special fields of application and only a few manufacturers are on the market. From the technical viewpoint ELOGs have advantages in areas where highly purified water (e.g., by distillation, nanofiltration or reverse osmosis) is already being produced for the production process, where it can be produced easily and cost efficiently in

Table 5.2 Overview of types of ozone generation, working principles and fields of application.

Method of ozone generation	Working principle	Ozone source	Field of application
Electrical	Electrical discharge (ED)	Air or O_2	Common standard from laboratory to full-scale
Electrochemical	Electrolysis (EL)	Water (highly purified)	Predominately for pure water applications, laboratory to small industrial scale
Photochemical ($\lambda < 185$ nm)	Irradiation (abstraction of electrons)	O_2 (air), water (drinking water quality or highly purified)	Very seldom, solely experimental
Radiation Chemistry	X-rays, radioactive γ-rays	Water (highly purified)	Very seldom, solely experimental
Thermal	Light arc ionization	Water	Very seldom, solely experimental

the quantity needed for ozone production, or where the introduction of a gas could exert detrimental effects on the product quality, as in wafer-cleaning applications in the semiconductor industry. Another area of application might be pharmaceutical production processes where the use of a highly reactive gas like ozone may be prohibitive because of legislative or safety reasons.

The choice of generator mainly depends on the type of application and the required ozone-production capacity. Important criteria for choosing a generator are the ozone concentration that can be achieved and the specific energy consumption that is measured as energy input (kWh) per unit mass of ozone being produced (kg^{-1} O_3). These, however, are not only influenced by the ozone generator, but also by the type and quality of the gas or water used as the ozone source. The feed gas or water must be adequately prepared for both generator types. This source preparation also requires energy and must be considered in designing the system. Some characteristic operating parameters of electrical discharge and electrolytic ozone generators are summarized for a few examples in Table 5.3. Source preparation systems as well as specific energy consumption differ considerably between the two methods of ozone production. In particular, the specific energy consumption of ELOGs is very high, at least by a factor of ten compared with the EDOGs.

After this short overview a closer look into how these generators work is given in Sections 5.2.1 and 5.2.2, including the underlying chemistry as well as important design and operating parameters.

Table 5.3 Characteristic operating parameters of dielectric barrier discharge and electrolytic ozone generators.

Parameters	Unit	Type of ozone generator		
		Dielectric barrier discharge (DBDOG)		Electrolytic (ELOG)
Reference	–	[11–13]	[13, 14]	[13]
–	–	Lab-scale	Full-scale	Small-scale
Ozone source	–	O_2	O_2 or (seldom) air	High-purity feed-water, conductivity $<20\,\mu S\,cm^{-1}$
Source preparation system	–	Gas compression, cooling, filtering, drying		Ion-exchange, ultra- or nanofiltration, reverse osmosis, distillation
$P_{abs}{}^a$	kPa	250–800	250–800	600^b
T_G or T_L	°C	5–40	5–10	<30
Q_G or Q_L	N m^3 h^{-1}	0.210–1.38 (at STP)	50–230 (O_2 at STP)	0.120–0.360
c_G or c_L (ELOG)	g m^{-3}	85–285	86–148 (O_2)	25
O_3-production capacity	kg O h^{-1}	0.03–0.50	10–20 (O_2)	0.003–0.009
Required power	kW	0.75–4.4	80–150	0.600–1.800
Specific energy consumption	kWh kg^{-1} O_3	8–13 (feed-gas O_2)	7.5–10 (feed-gas O_2)	200
Q_{LC}	m^3 h^{-1}	0.090–0.600	11–22 or air cooling	not reported

a Feedgas inlet pressure; module or system pressure is normally at 100 to 250 kPa.
b System pressure in delivered ozonated water.

5.2.1
Electrical Discharge Ozone Generators (EDOGs)

In electrical-discharge ozone generators (EDOG), ozone is produced using energy from electrons in an electrical field between two electrodes. The electrodes are separated by a space or gap containing a gas. A discharge of electrons from one of the electrodes ionizes the gas. The ionization is limited to a small region around the electrode and produces a collection of electrons, ions, radicals and neutral or excited molecules called a plasma, in the case of an ozone generator, a nonthermal plasma. Although much of the electrical energy is lost to heat, the temperatures are still relatively low compared to thermal plasmas where temperatures reach >5000 K. The ions generated function as the charge carriers to the other electrode. When one of the electrons in this plasma collides with an oxygen molecule, it transfers part of its energy to the oxygen, causing it to dissociate into monoatomic, reactive atoms. These collide with other oxygen molecules. Overall, in a complex reaction mechanism, some of the oxygen atoms form ozone, while others recombine to molecular oxygen.

The discharge can be caused either by high voltage or by an inhomogeneous electric field, such as one produced by a sharply curved surface. A discharge occurs when the potential gradient (the strength of the electric field) is high enough to cause ionization in the surrounding medium. The glow produced by the ionized gas surrounds the surface like a crown—thus the term corona discharge. Such discharges are produced in ozone generators using asymmetric electrodes with point-to-plate or wire-to-cylinder electrode geometries. When high voltage is applied to electrodes placed in parallel with a small gap in-between, a large number of microdischarges spread over the surface is produced. An additional insulating material (a dielectric) is usually placed in the path between the electrodes to prevent shorting between the electrodes.

The terminology used to describe ozone generators in the literature can be confusing. Sometimes, the term corona discharge generator is used in the literature to classify all generators using electrical discharges, while in other sources corona discharge generators are only those with two asymmetric electrodes, and those with parallel electrodes are referred to as dielectric-barrier discharge (DBD) generators. This distinction is employed in this book. Both types can be used, however, the DBD generator is of more practical relevance in commercially available ozone generators for water applications [15].

In dielectric barrier discharge ozone generators (DBDOG), the electrodes are usually two parallel plates or concentric cylinders arranged with a certain distance to each other to form a discharge gap, with a width of 1–2 mm. The working principle of a tubular DBDOG is shown in Figure 5.2. The electrodes are isolated from each other by a dielectric (nonconducting) barrier material and the discharge gap. Various dielectric materials can be used such as glass, quartz, ceramics or polymers. The air or oxygen gas is passed through this gap. When a high-voltage alternating current is established between the two electrodes, a large number of statistically distributed microdischarges cause very fast ionization to occur. The dielectric barrier helps distribute the microdischarges over the entire electrode surface. During these microdischarges, electrons produce oxygen atoms and various ionized species from oxygen molecules that act as the charge carriers necessary for current flow [16]. The process of ozone formation is discussed in detail below.

5.2.1.1 Chemistry
Ozone synthesis by electrical discharge, which is sometimes also called silent discharge, is a complex process. There are six main reactions that occur between

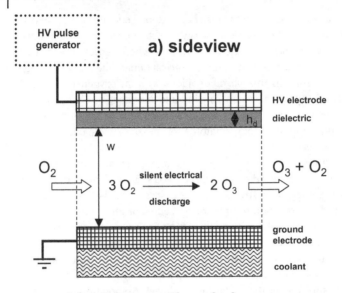

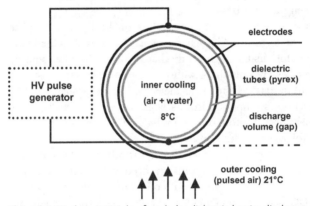

Figure 5.2 Working principle of a tubular dielectric barrier discharge ozone generator (DBDOG).

the electrodes in an oxygen-fed dielectric barrier discharge, including reactions of initiation, that is, generating oxygen atoms and excited oxygen molecules from electrons (5.1 and 5.2), ozone formation (5.3 and 5.4) and ozone decomposition (5.5 and 5.6) [17]:

$$O_2 + e^- \rightarrow O + O + e^- \tag{5.1}$$

$$O_2 + e^- \rightarrow O_2^\circ + e^- \tag{5.2}$$

$$O + O_2 + M \rightarrow O_3 + M \tag{5.3}$$

$$O_2^\circ + O_2 \rightarrow O_3 + O \tag{5.4}$$

$$O + O_3 \rightarrow 2\,O_2 \tag{5.5}$$

$$O + O \rightarrow O_2 \tag{5.6}$$

There are many more reactions involved, but they are well understood [16] and can be quantitatively modeled. For example, Pontiga *et al.* have employed a model comprising ten species and 79 reactions for modeling and simulation of ozone generation in a wire-to-cylinder corona discharge ozone generator [15].

When trying to improve ozone generation, there is always a trade-off between the conversion efficiency of the electrical energy and the ozone yield (expressed as the ratio of the concentrations of ozone formed per oxygen atom produced $c(O_3)/c(O)$). The optimum is a compromise between losing energy to ions and obtaining a reasonable yield. The conversion efficiency itself is a function of the oxygen atom concentration $c(O)/c(O_2)$ and reaches its maximum (of unity) at oxygen atom concentrations below 10^{-3} to 10^{-4} [16].

With respect to the formation of ozone we see from Equation 5.3 that ozone is mainly formed as a result of the reaction between oxygen molecules and oxygen atoms in the presence of a "third collision partner" (M). When oxygen is used as feed gas O, O_2 and O_3 themselves are the collision partners [18]. From the chemical viewpoint ozone generation in air is a much more complex process than in pure oxygen [16]. Intense research has shown that nitrogen (N_2) also acts as a third collision partner. It provides, for example, additional reaction paths for the formation of oxygen atoms (see 5.7), and thus of ozone,

$$N + O_2 \rightarrow NO + O \tag{5.7}$$

In addition, carbon dioxide and argon can also act as collision partners. This finding has gained important practical relevance in the operation of so-called high-performance dielectric barrier discharge ozone generators, that is, the ones producing high ozone-gas concentrations using oxygen as the feed gas, especially from a liquid-oxygen supply. They require the addition of a small amount of nitrogen as a third collision partner in the feed gas to ensure optimum performance and constant, long-term ozone generation [19]. This procedure is commonly called "feed-gas doping". Because highly corrosive nitric acids can be formed from the nitric oxides in the presence of water vapor, carbon dioxide (CO_2) or Argon (Ar) are sometimes used as a dopant [20]. This is especially relevant in ozone applications in the semiconductor industry, where the corrosive potential of nitric acids must be avoided.

5.2.1.2 Engineering and Operation

Many factors play a role in determining the efficiency of a dielectric barrier discharge ozone generator. Some important factors are the power density, the electrode geometry and its surface area, the dielectric material, the type (air or oxygen) and quality (dryness and cleanliness) of the feed gas as well as its pressure inside the gap. Since ozone decomposition (see 5.5 and 5.6) is enhanced by higher temperatures (e.g., $T_G > 50\,°C$) the adequate cooling of the electrodes is also essential for efficient ozone formation (e.g., [18]).

Two parameters are commonly used to judge the overall energy efficiency of generators – the ozone productivity (g O_3 kWh^{-1}) or its inverse – the specific energy consumption (kWh g^{-1} O_3) (see values in Table 5.2). For information on the dependency of the ozone productivity on the various parameters of influence in a dielectric barrier discharge ozone generator the reader is referred to the study of Haverkamp et al. [21].

Many generator variations have been tried over the years. In particular, the electrode geometry, dielectric material and cooling systems have been varied considerably. The first system that came into operation for full-scale water treatment at the beginning of the twentieth century, was the plate-type system (developed by Marius Paul Otto). During the 1930s the tubular-type generator, often called Van der Made- or Welsbach-type, was developed and has been most frequently used in large-scale ozone-generating systems since the 1960s [10]. Today, ozone generators contain several thousand parallel DBD electrodes with metallized glass as the dielectric. Also, several manufacturers employ ceramics for the dielectric, which is highly efficient in heat transfer [22] and allows for fewer dielectric tubes than in the glass systems as well as a more energy-efficient ozone production [13].

Ozone generators for semiconductor applications requiring ultraclean ozone must be especially designed for that application [11]. Concentrations of various metals (Al, Ti, Cr, Fe, Mn, Ni, Cu, Zn) as well as other inorganic compounds (Na, K, Ca) must not exceed very low concentrations in the ozone gas (<5 ng m^{-3}) or in the ozonated water (<50 ng m^{-3}) being fed to the point of use. Therefore, all components of the ozone-generation system must be carefully chosen to avoid trace-level contamination.

Lab-scale electrical-discharge ozone generators often work at high-frequency (kHz-range) generated from conventional line voltage of 230 or 400 V (Europe) and 120 or 208 V (USA) at 50 or 60 Hz. They are usually operated at ambient pressure (P_{abs} = 100 kPa). In these systems only very few electrodes are employed. In full-scale (industrial) ozone generators high-power alternating current (medium-frequency: 200–650 Hz; high-voltage: 8500–10 000 V) is applied to several thousands of electrodes contained in one housing. These generators can also be operated at elevated pressures of P_{abs} up to 600 kPa.

Unfortunately, only part of the electrical energy supplied to the generator is used for the formation of ozone, the rest is transformed into heat. In electrical-discharge ozone generators only about 25% of the energy consumed is effectively utilized. Since ozone decomposes fast at elevated temperatures, an efficient cooling system has to be installed. Smaller generators use air cooling, while larger ones use water cooling for the ozone-containing gas being kept at T_G = 5–10 °C (Table 5.2). The concept of double-side air cooling of the electrodes (see Figure 5.2b) helps to increase the ozone concentrations as well as the energy efficiency, and is mostly applied in small-scale ozone generators, for example, for laboratory applications.

Work continues on improving the electrical efficiency of EDOGs. For instance, the benefits of advanced cell geometry, optimized dielectric material and an improved cooling system using air plus nebulized water on the energy efficiency of ozone production were demonstrated in a laboratory study, in which a novel surface discharge configuration was compared to a conventional tubular (annular) configuration [18].

5.2.1.3 Type of Feed Gas and its Preparation

The type of feed gas used, air or oxygen, determines the achievable ozone-gas concentration and the gas-preparation requirements. The higher the oxygen content, the higher the ozone concentration possible. Ambient air contains O_2 in about 21 vol.% (at STP) and is thus a cheap and ubiquitous resource for ozone production. Until today its main use and advantage is in applications where large mass flows are required at comparatively low ozone-gas concentrations, for example, in drinking water ozonation systems.

The major disadvantages of using ambient air as feed gas for ozone generation are:

- air is extremely energy inefficient,
- a high-quality air preparation system is required, which is capable of achieving a dew point of about $-60\,°C$ (213 K at 1 bar abs, corresponding to a max. absolute humidity of $6.7 \times 10^{-6}\,kg\,H_2O\,kg^{-1}$ air) and normally consists of the following sequence of installations: filter, compressor, water and/or refrigerant cooler with separator, absorption dryer and filter [23] and
- production of highly corrosive nitrogen oxides: NO, N_2O, NO_2, N_2O_5 ([19, 24]).

Although drying the air to such an extent also consumes a considerable amount of energy, corrosion in the generator and piping to the reactor can be effectively prevented this way.

It is because of the aforementioned disadvantages that the use of air as feed gas is more and more replaced by the use of pure oxygen, which provides the advantage of containing almost five times more oxygen than ambient air. In industrial applications it is either delivered in tanks as liquid oxygen (LOX) or produced onsite. Tank delivery is especially suitable in larger applications if additional investment costs are to be avoided [19]. Otherwise it has to be produced onsite from ambient air, usually with pressure swing adsorption (PSA) or vacuum pressure swing adsorption (VPSA) [23], which also is an energy-consuming process. In both cases, some air or nitrogen gas is needed as a make-up gas for efficient ozone production. The make-up gas supplies nitrogen that acts as a "third collision partner" in the ozone formation reaction (as discussed in the chemistry section, see 5.3). Normally, less effort is necessary in lab-scale ozone applications. Oxygen is supplied from cylinders and the gas has only to be dried to a dew-point of $-40\,°C$.

If ozone is used in the semiconductor industry, high-purity oxygen (grade 4.5 or better) is used as feed gas, along with small amounts of nitrogen or carbon dioxide gas as the doping gas (each of grade 5 or better). The water-vapor content in the feed gas must also be kept extremely low to hold corrosive nitric oxides or acids at the lowest possible level.

5.2.1.4 Ozone Concentration, Production Capacity and Specific Energy Consumption

Theoretically, almost fivefold higher ozone concentrations in the gas can be achieved with pure oxygen compared to air. State-of-the-art full-scale EDOGs achieve c_G = 3–5 wt.% from air and 10–15 wt.% from oxygen. Lab-scale systems can produce up to 23 wt.% ozone (353 g O_3 m^{-3} at STP) from pure oxygen. Electrical-discharge ozone generators are manufactured for a wide range of ozone production capacities. Small ones produce some ten grams of ozone per hour while the biggest ones have a production rate up to 200 kg O_3 h^{-1} from pure oxygen at comparatively high concentrations (c_G = 10 to 15 wt.%).

When oxygen is used as feed gas, the specific energy consumption in small and medium-scale systems often ranges between 6–8 kWh kg^{-1} O_3 (e.g., [9, 25]). Large-scale oxygen-fed EDOGs typically have a specific energy consumption of about 10 kWh kg^{-1} O_3 [13]. Although research on new generators continues and some laboratory research studies of novel methods to achieve a very low specific energy consumption have been reported (e.g., 2.5 kWh kg^{-1} O_3 by [26]) improvements in energy efficiency are still needed.

In contrast, the specific energy consumption is about twice as high when ozone is being produced from air instead of pure oxygen [23]. This is due to the decreased ozone production rate with air at similar power consumption levels. When air is used in the same ozone generator (same power supply), the ozone production rate (kg O_3 h^{-1}) decreases by 50% or more. Furthermore, when considering the energy consumption of the overall system, it is important to include the energy for feed-gas production and its preparation.

5.2.1.5 Use of EDOGs in Laboratory Experiments

Even if pure oxygen from pressurized cylinders is used in lab-scale applications, gas purification is recommended, especially for experiments in which ozone is used to eliminate trace organic compounds, for example, preparation of water for semiconductor cleaning applications, drinking- or ground-water treatment. It is advisable to install the following gas-purification system in front of the generator: gas drying by means of an absorptive material, for example, silica gel, a molecular sieve (0.4 mm) and a microfilter (4–7 μm) to remove any particles from the gas [27]. In semiconductor applications a particle filter (0.003 μm) is also necessary after the ozone generator.

When air is used as the feed gas in lab-scale ozonation equipment, a high-quality air-drying and oil-removal system similar to full-scale applications has to be installed following the compressor. Otherwise the generator may be destroyed from moisture, dust, oil, hydrocarbons and hydrogen.

Planning experiments in the lab, it is important to recognize the basic operating characteristics of electrical-discharge ozone generators. At constant power, the ozone-gas concentration (c_G) decreases with increasing gas flow rate (Q_G) and this dependency is not linear. The deviation from linearity between the two parameters is largest at lower gas flow rates (Figure 5.3). Different generators will show different curve types. Figure 5.3 gives an example of an electrical-discharge ozone

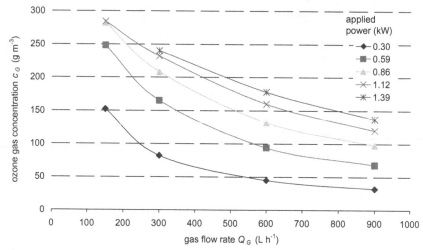

Figure 5.3 Characteristic dependency of the ozone-gas concentration on the gas flow rate in an electric discharge laboratory ozone generator (From ASTeX Sorbios [28]).

generator with a nominal ozone-production capacity of $0.090\,\mathrm{kg\,O_3\,h^{-1}}$ (at $Q_G = 0.600\,\mathrm{m^3\,h^{-1}}$ and $c_G = 0.150\,\mathrm{kg\,O_3\,m^{-3}}$; [28].

The implications of this inverse relationship between Q_G and c_G can be illustrated using an example of mass transfer in a bubble column. To increase the $k_L a$-value, the gas flow rate must be increased. Due to the operating characteristics of EDOGs the increase in Q_G will decrease c_G. If instead c_G is held constant as Q_G is increased by raising the power or the voltage applied, this will result in the ozone mass flow rate also being raised. This in turn will considerably influence the oxidation rate more than the effect of the higher $k_L a$-value. Therefore, especially when the reaction is mass-transfer limited (see Section 6.2) which often occurs in wastewater ozonation, investigations of single effects must be well planned. Careful evaluation of whether the higher $k_L a$-value matches with other goals of the experiment, for example, the intention to use a high ozone-gas concentration or achieve a high ozone-transfer efficiency has to be made.

Another characteristic of EDOGs of importance for experimental planning is that there is a certain minimum gas flow rate Q_G at which the highest ozone concentration is produced. Below these Q_G-values the ozone-production capacity becomes unstable. The highest mass flow rates, on the other hand, can only be achieved at high power input as well as high gas flow rates.

5.2.2
Electrolytic Ozone Generators (ELOGs)

In the electrolytic ozone generator (ELOG) ozone is produced *in-situ* from the electrolysis of high-purity water (Figure 5.4). Therefore, no mass transfer from the

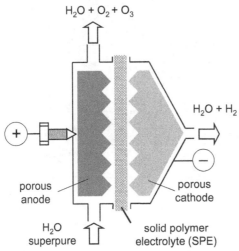

Figure 5.4 Working principle of an electrolytic ozone generator (ELOG) (after Fischer, [29]).

gas to the liquid phase is involved and only very fine gas bubbles are formed so that the gas is immediately dissolved in the water. And since the absence of gaseous ozone means that another chemical is avoided and all its related safety precautions, the use of ELOGs is in compliance with FDA regulations and therefore of advantage in the production of foodstuffs and pharmaceuticals.

In the electrolytic cell, water is split into molecular hydrogen H_2 as well as oxygen (O_2) and ozone (O_3) by the action of electrons supplied by the catalytic properties of the electrode material. Ozone is produced according to the overall reaction:

$$3 H_2O \rightarrow O_3 + 6 e^- + 6 H^+ \quad (5.8)$$

The anode is made of a porous, water-penetrable and current-conductive carrier material coated with an active layer, mostly PbO_2, but recently a novel diamond anode was developed [30]. At the site where the active catalytic layer and the electrolyte membrane touch, ozone is produced at currents of up to 50 A (DC) [30]. At the theoretical optimum at a current efficiency of 100%, 298.5 mg of ozone can be produced per Ah. However, in practical applications current efficiency is usually below 25%. The highest current efficiency obtained with a novel diamond anode and SPE electrolyte membrane (Diachem®-SPE-electrodes) was 24% [30].

Water, O_2 and O_3 leave the cell on the anode side, whereas H_2 is produced at the cathode side of the electrolytic cell in which a solid-polymer electrolyte membrane (SPE) separates the anode and the cathode side.

Since all materials of the cell have to be electrochemically very stable and have to provide a high conductivity, the cell consists of refined metals or metal oxides at their highest oxidation level. Also, the feed water has to be of high purity, because it has to pass the porous anode and cathode materials without clogging or causing chemical damage. Therefore, ions and other impurities that are contained in normal drinking water have to be removed by ion-exchange, ultra- or nanofiltration, reverse osmosis or distillation.

Electrolytic ozone generators are supplied by a limited number of producers (e.g., [13, 30]). The ozone-production capacity of one cell is between 0.05 and $4\,g\,O_3\,h^{-1}$, but several cells can be combined in one generator. The principle dependency of the ozone-production capacity (of one cell) on the voltage, the current and temperature is shown in Figure 5.5.

Again, cell temperature strongly influences the ozone production and efficient cooling is necessary. Cooling is mostly achieved by maintaining a high water flow rate through the cell (e.g., $Q_L = 50\text{--}2250\,l\,h^{-1}$; [13, 30]), however, systems with air cooling have also been developed. Insight into the operating characteristics and parameters of influence is given in [30]. The specific energy consumption is very high, at least $200\,kWh\,kg^{-1}\,O_3$ (about 20 times the value of the EDOGs) [30]. Although in a recent study on an electrochemical reactor with planar perforated $\beta\text{-}PbO_2$-coated electrodes a value of $60\,kWh\,kg^{-1}\,O_3$ was achieved [31], these very high values make it evident that this type of ozone generator cannot be economically used in (large-scale) drinking- and waste-water treatment systems. Further practical drawbacks are operation at relative high pressures as well as high maintenance.

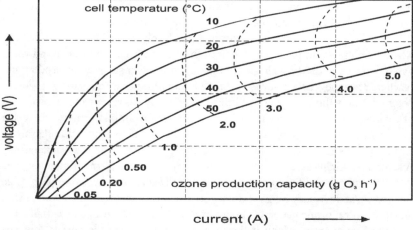

Figure 5.5 Dependency of the ozone-production capacity in an electrolytic ozone generator (ELOG) on the applied voltage and current (with cell temperature as a parameter) (from: Fischer [29]).

5.2.2.1 Use of ELOGs in Laboratory Experiments

The temperature of the liquid in the system should be kept below $T_L = 30\,°C$ because at pH 7 the half-life time of dissolved ozone falls below 12 min at higher temperatures. This requires very high flow rates of purified feed water to be directed through the cell, unless the cell is air cooled or the feed water is cooled before entering the ELOG.

The fact that dissolved ozone is produced brings a very important advantage to subsequent ozone applications: mass transfer from the gas to liquid phase is not required. Efficient mixing (e.g., with static mixers) of the ozone-rich pure water stream with the (waste-)water stream to be treated, though, is required. During this *in-situ* ozone production, the liquid-ozone concentration (c_L) can easily reach the solubility level (c_L^*), depending on the pressure (P) and temperature (T) in the cell. Oversaturation of the feed water will immediately occur, when the pressure drops. Due to this potential degassing ozone gas destruction in the reactor off-gas is also required for this system.

5.3
Reactors Used for Ozonation

This section provides an overview of the types of reactors commonly used for ozonation in lab-scale studies. Some basics on the hydrodynamic behavior and mass-transfer characteristics of reactors that were already presented in Section 4.3 are briefly reviewed in Section 5.3.1. Then, practical aspects and examples of reactors are examined. The discussion is divided according to the number of phases present in the reactor. When ozone is generated as a gas by an electrical-discharge generator, gas and liquid phases have to be brought into contact in the reactor. In Section 5.3.2 we look at directly gassed reactors with more than one phase in the reactor (*heterogeneous systems*), where absorption of ozone from the gas phase is accompanied by a simultaneous reaction in the liquid phase.

In Section 5.3.3 we examine reactors that can be considered *homogeneous systems*, where only the liquid phase is present in the reactor. This can be achieved when gaseous ozone is used by physically separating the two steps of absorption and reaction. If ozone gas is absorbed into the liquid phase in a contactor or absorber before it enters the reactor, we call this an indirectly gassed system. This can be of interest for reaction systems with low ozone consumption rates, where the mass-transfer rate is faster than the reaction rate and time required to dissolve a sufficient amount of ozone into the liquid phase is shorter than the reaction time. A second possibility also exists that is growing in importance – membrane reactors in which the gas and liquid phases are physically separated by a semipermeable membrane and virtually no gas bubbles are produced. In addition, systems using an electrolytical ozone generator to produce ozone *in situ* can also be classified as homogenous systems as long as the gases remain dissolved. We refer to this type as a *nongassed system*.

The chapter continues with a look at gas diffusers appropriate for introducing gaseous ozone into heterogeneous systems (Section 5.3.4), followed by a discussion of the choice of the operating mode (Section 5.3.5) and some important aspects of experimental procedure (Section 5.3.6). Special consideration is given to the practical aspects of design for laboratory experiment, rather than full-scale applications.

The type and dimensions of the reactor should match the purpose of the investigation. If the experimental goal is to investigate reactions, not develop a new reactor, the use of a reactor type for which correlations and experience already exist is preferred. When designing the reaction system, it is important to consider the size relationships between the components in the entire experimental setup. In experiments investigating reaction kinetics, the ozone generator must supply adequate quantities of ozone to the reactor. Therefore, the volume of the reactor and the concentration of ozone-consuming compounds have to match the required capacity of the ozone-generation system. Especially for waters containing high concentrations of (highly reactive) compounds, for example, in waste-water ozonation experiments, operating conditions where ozone becomes depleted in the off-gas should be avoided. This would limit the reaction rate. As a result of these considerations, waste-water investigations often have smaller liquid volumes (e.g., $V_L = 1–5\,l$) than drinking-water applications for similar generator sizes.

5.3.1
Overview of Hydrodynamic Behavior and Mass Transfer

A multitude of reactors have been developed over the years and each reactor type has its characteristic hydrodynamic behavior. Knowledge of the hydrodynamic behavior as well as its mass-transfer characteristics is important for designing a system as well as evaluating experimental results. These concepts were introduced briefly in Section 4.3, and will be expanded upon here. Reactors can be operated according to two modes: batch where the reactants are placed into the reactor for a defined length of time and then removed, or continuous-flow where at least one phase is continuously introduced into the reactor. Ozone reactors are often operated semibatch where the liquid is filled into the reactor batchwise but the gas is continuously introduced into the reactor.

From the viewpoint of hydrodynamics or mixing, there are two types of ideal continuous-flow reactors: the ideally mixed reactor represented by the continuous-flow stirred-tank reactor (CFSTR), in which the liquid as well as the gas phase are ideally mixed versus the plug-flow reactor (PFR), in which axial mixing (over the length of the reactor) ideally does not occur (Figure 5.6). For both types of reactors the reaction rates between ozone and the target substances can be obtained from mass balances on the reaction partners. The appropriate design equations are discussed in Section 4.3.2 or in Levenspiel [32] or Beltrán [33].

The degree of mixing can significantly influence the reaction rates and required reactor size. For example, plug-flow conditions provide higher reaction rates than

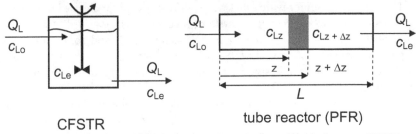

Figure 5.6 Schematic sketch of ideal reactors: continuous flow stirred-tank reactor (CFSTR) and plug-flow or tube reactor (PFR).

those in CFSTR, consequently PFR require smaller reactor volumes. Therefore, it is important to know the degree of mixing to be expected from the various reactor configurations. The degree of mixing/dispersion is often characterized by the dimensionless Bodenstein number, which is defined as follows:

$$\mathrm{Bo} = \frac{v_\mathrm{L} L}{D_\mathrm{ax}} \tag{5.9}$$

with

v_L superficial liquid velocity (m s^{-1})
L length of the reactor (m)
D_ax axial coefficient of dispersion (s m^{-2}).

Small values of Bo indicate ideal mixing, while large numbers indicate plug-flow behavior.

Tracer studies should be performed to verify the hydrodynamic behavior of the reactor before starting ozonation experiments. Again the appropriate design equations can be found in Levenspiel [32] or Beltrán [33]. Table 5.4 summarizes characteristic features of the most important reactor types. Most often real reactors only approach the ideal conditions.

Bubble columns (BCs) and stirred-tank reactors (STRs) are the most frequently used types of reactors in laboratory ozonation experiments. While STRs can generally be assumed to behave like perfectly mixed reactors with respect to the liquid phase, this applies for BCs only when the ratio of height (H) to diameter (D) is small ($H/D \leq 10$). Almost complete mixing of the gas phase can also be assumed in small lab-scale STRs, whereas plug-flow or less mixed behavior are often found for the gas phase in bubble columns ([37–39]) and packed towers [40].

In order to achieve plug flow in the liquid phase, reactors with a high height to diameter ratio H/D such as tube reactors are required. Tube reactors though are not commonly used for ozonation of large flow rates such as in drinking-water applications, since the introduction of a second phase in a tube reactor complicates

Table 5.4 Characteristics of gas–liquid contacting systems.

Type of reactor or mass-transfer system	Reference	Type of hydrodynamic behavior	$k_L a$ (s^{-1})	Specific power consumption (kW m^{-3})
Bubble columns (BCs)	[34]	Mixed flow for liquid Plug flow for gas	0.005–0.01	0.01–0.1
Packed Tower	[34]	Plug flow for liquid Mixed flow for gas	0.005–0.02	0.01–0.2
Plate Tower	[34]	Mixed flow for liquid Mixed flow for gas	0.01–0.05	0.01–0.2
Stirred-Tank Reactors (STRs)	[35]	Completely mixed liquid Mixed flow for gas	0.02–0.05	0.1–1
Submerged Impinging Zone Reactor (IZR)[a]	[35]	Mixed flow for liquid Mixed flow for gas	0.07–0.7	0.2–2
Jet Loop Reactor	[35]	Mixed flow for liquid Mixed flow for gas	0.1–0.6	5–50
Mixing devices				
Static Mixers	[34]	Plug flow for liquid Plug flow for gas	0.01–2	10–500
Semipermeable membrane reactors				
Flat plate[b]	[4]	Plug flow for liquid Plug flow or mixed flow for gas	0.005	n.a.
Hollow fiber[c]	[36]	Plug flow for liquid Plug flow or mixed flow for gas	0.01–0.09	n.a.

a [35]: $Q_G/V_L = 80\,\text{Nm}^3\,\text{h}^{-1}\,\text{m}^{-3}$.
b [4]: Re = 2000; pore size 0.07–6 µm; membrane thickness 0.076–0.254 mm.
c [36]: Re = 2000; $d = 1.8\,\text{mm}$, $a = 20\,\text{cm}^{-1}$.

the mixing patterns. In addition, it may be hard to keep an adequate supply of ozone in the system if the tube reactor is indirectly gassed. A common reactor configuration though that approaches the higher reaction rates of the PFR is the CFSTR in series (n-CFSTR) shown in Figure 5.7, where the liquid volume is divided into smaller stages and the gas can be continuously introduced into more than one stage.

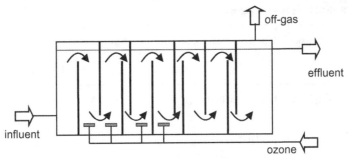

Figure 5.7 Schematic sketch of a conventional ozonation contactor with eight chambers (four are gassed) for drinking-water treatment.

With the continuing improvement in semipermeable membranes, their use in reactors in a variety of applications needing mass transfer and separation has increased. In ozone applications the membranes are usually operated such that the gas and liquid phase remain physically separated and ozone diffuses through a hydrophobic membrane into the liquid phase. The liquid flow through the long channels of the membrane plates or hollow fibers approximates plug flow. The extent of mixing in the gas phase depends on the geometry of the configuration. The $k_L a$ values for the membranes listed in Table 5.4 are in the low range compared to the other devices. Unfortunately, values for the specific power consumption are not available from the studies cited in Table 5.4.

Generally, knowledge of the true reactor hydrodynamics is required to optimize the performance of gas–liquid reaction systems. This can be gained by measuring the hydraulic retention time distribution of the system. After the hydrodynamics of the system have been determined, process modeling can be used to aid optimization. In particular, the use of computational fluid dynamics (CFD) has greatly improved the possibilities and ease of modeling nonideal systems (e.g., [41], see also Section 8.2.3).

CFD has often been used in recent years to assess and optimize the hydrodynamics of full-scale ozonation reactors for drinking-water treatment. For instance, modeling helped determine how mass transfer could be improved in an existing plant, resulting in the doubling of the ct-value as well as the inactivation efficiency [42]. It was also used to analyze the current conditions in ozonation reactors at the Amsterdam water-treatment works in order to improve the disinfection efficiency of the system [43]. The analysis showed that under current conditions there was a strong deviation from plug flow and that a redesign of the existing contactor basin, creating additional chambers, should be considered. Improving the plug flow instead of increasing the ozone dose would improve the disinfection efficiency.

5.3.2
Directly Gassed Reactors

Water or waste-water ozonation—regardless of the scale of equipment—is mostly performed in *directly gassed* systems, where the ozone-containing gas is produced by an electrical-discharge ozone generator and is introduced into the reactor through some type of gas diffuser. Since two phases, the gas and the liquid, are present, they are also called *heterogeneous* systems.

5.3.2.1 Bubble Columns and Similar Reactors

Bubble columns and airlift reactors are frequently used in lab-scale ozonation experiments. Reactors with a liquid phase volume of V_L = 2–10 l, and a height-to-diameter-ratio of H/D = 5–10 have proven to be useful. The ozone/oxygen (or ozone/air) gas mixture is supplied through a diffuser. Fine-pore diffusers (porosity 3, 10–40 µm hole diameter) made of ceramic, fused alumina or stainless steel porous plates are often used. PTFE membranes either porous or with punched holes are an alternative for the ozone gas-to-water transfer [44]. Section 5.3.4 provides more information on gas contactors or gas diffusers.

Most columns are operated in a concurrent mode, both liquid and gas flow upwards. A countercurrent mode of operation, up-flow gas and down-flow liquid, has seldom been reported for lab-scale studies, but can easily be approximated by means of applying an internal recycle-flow of the liquid, pumping it from the bottom to the top of the reactor. The advantage is an increased level of the dissolved-ozone concentration c_L in the reactor (effluent), which is especially important when the reaction rate is low and low concentrations of target compounds are required.

The mass-transfer rate in bubble columns is determined by the energy introduced by the gas flow rate and the internal recirculation. Moderate $k_L a$-values in the range of 0.005–0.01 s^{-1} can be achieved in simple bubble columns ([34]; Table 5.4). Since mixing in bubble columns is only due to the power dissipated by the gas flow very small gas flow rates might cause incomplete mixing in such reactors. Sometimes, modifications are made to bubble columns such as the downflow mode of operation already mentioned (e.g., a U-tube reactor was studied for hydrodynamics and ozone mass transfer [45]). Especially of interest are modifications to enhance mass transfer such as impinging-jet reactors with $k_L a$-values in the range of 0.01–0.1 s^{-1} [46] or 0.07–0.7 s^{-1} ([35, 47]), which can be used in pilot- or full-scale applications.

A very simple type of a bubble column, which has not yet been mentioned, is a gas-wash bottle. This very small-scale system (V_L = 0.2–1.0 l) may be used for basic studies, in which general effects (e.g., influence of pH and/or buffer solutions, specific ozone dose) are to be assessed. Its use is not recommended for detailed studies, because the mass-transfer coefficient is often low and its dependency on the gas flow rate is unknown or difficult to measure. Often, there is no possibility to insert sensors or establish a reliable measuring system for exact balancing of the ozone consumption. However, different ozonation conditions can easily be tested in them. The effect of the ozonation time or the ozone-gas concentration can be examined by operating the flasks for a certain period of time, preferably without withdrawal of solution during the ozonation, and analyzing the solution at the end. A variation of the gas flow rate is not recommended.

5.3.2.2 Stirred-Tank Reactors

Stirred-tank reactors (STR) are the most frequently used reactors in lab-scale ozonation, partially due to the ease in modeling completely mixed phases as well as the two independent ways to influence the mass transfer (stirrer and gas flow rate), but they are seldom used in full-scale applications. From the viewpoint of mass transfer, the main advantage of STRs is that the stirrer speed can be varied, and thus also the ozone mass-transfer coefficient, independently of the gas flow rate. There are various modifications with regard to the types of gas diffusers or the construction of the stirrers possible. Normally, lab-scale reactors are equipped with coarse diffusers, such as a ring pipe with holes of 0.1–1.0 mm diameter and the stirrer energy is relied on to create smaller bubbles. The $k_L a$-values are in the range of 0.02 to 0.05 s^{-1} (see Table 5.4), which are considerably higher than those of bubble columns.

In order to give a general idea of the dimensions of a STR, important features of three STRs that have been successfully operated in various fields of water or waste-water applications are summarized in Table 5.5. Besides the differences in volumes of the reactors used by Gottschalk [27] or Sotelo et al. [48] and Beltrán et al. [49], they also differ in the type of the stirrers, the gas spargers and the baffles. A special PTFE plate with holes in its upper side, rotating with high speed above a PTFE tube serving as the gas sparger, as well as nonvertical, swirled baffles were used by Gottschalk for drinking-water ozonation [27]. In contrast, a well-known, "classical" setup is the six-blade Rushton-type stirrer, a fine porous diffuser and four vertical baffles as were used by Sotelo et al. [48] and Beltrán et al. [49]. A completely different type of STR is represented by the agitated cell that was used by Beltrán and Alvarez [50].

The agitated cell reactor is a reactor specially designed to measure kinetic parameters (Figure 5.8, [51]). It consists of two chambers, one for the liquid phase and another one for the gas phase, and each chamber is equipped with a stirrer so that they can be independently mixed. In this reactor the mass-transfer area can be varied independently of the gas flow rate by installing a porous plate with a defined number of holes to create a defined contact area gas–liquid between the two chambers. The value of k_L can then be determined from the measurement of $k_L a$.

Beltrán and coworkers have successfully used the agitated cell as well as "classical" STRs working in the semibatch mode to determine the reaction rate constants of fast direct reactions of ozone with various water pollutants, for example, phenol, resorchinol, phloroglucinol, azo-dyes and PAHs (Table 5.5; see also [33]).

Table 5.5 Examples of STRs successfully used in lab-scale ozonation studies.

Reference	Gottschalk [27]	Sotelo et al. [48]	Beltrán et al. [49]	Beltrán and Alvarez [50]
Water/waste water	Drinking water	Synthetic waste water	Synthetic waste water	Synthetic waste waters
Pollutants/model contaminants	Atrazine (micropollutants)	Resorchinol and Phloroglucinol	PAHs	Phenol and 4 Azo Dyes
Type of stirred-tank reactor	"Modified" setup	"Classical" setup	"Classical" setup	Agitated cell[a]
Volume and dimensions	$V_L = 8 l$ $H = 260$ mm $H/D = 1.3$	$V_L = 0.5$ H = n. r. H/D = n. r.	$V_L = 4 l$ H = n. r. H/D = n. r.	$V_L = 0.3 l$ $D = 75$ mm $H/D \approx 0.9$
Material	Quartz glass, top and bottom stainless steel, coated with PTFE layer	Quartz glass	Quartz glass	glass
Stirrer	PTFE plate ($h = 15$ mm, $d = 65$ mm) with six 5 by 5 mm grooves at its bottom	Six-blade Rushton-type, stainless steel	Six-blade Rushton-type, stainless steel	n.r.
Gas sparger	8 mm o.d. PTFE-tube with 1.0 mm holes on top	2 mm i.d. bubbler	Fine-pore diffuser 16–40 μm	None; defined holes in interface plate
Baffles	3 baffles swirled, nonvertical, PTFE	4 baffles, vertical, stainless steel	4 baffles, vertical, stainless steel	n.r.
Operating conditions				
T (°C)	20 ± 1	1–20	10, 20	20 (phenol), 15 (dyes)
Q_G (l h^{-1})	30 (at STP)	110–70	25	60
n_{STR} (min^{-1})	1500	100–700	1000	75
$k_L a$ (h^{-1})[b]	12	2.8–13.3	126	0.8
$p(O_3)_o$ (Pa)	6–130	12–871	116–1015	30–2220

a For constructive details see [51].
b At Q_G, T; n.r. = not reported.

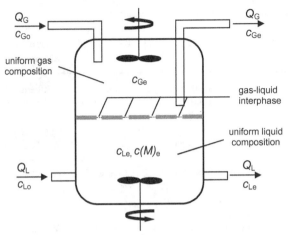

Figure 5.8 Schematic sketch of an agitated cell (from Levenspiel and Godfrey [51]).

5.3.3
Indirectly and Nongassed Reactors

In contrast to the two-phase *heterogeneous* reaction systems discussed above, *homogeneous* reaction systems contain only one phase. In water and waste-water applications it is the liquid phase. The challenge of such a system is how to provide dissolved ozone to the reactor and to assure that it is available in sufficient quantities. Various methods are possible. Ozone can be transferred from the gas into the liquid phase in a separate vessel prior to the water entering the reactor, it can be produced *in situ* using an electrolytic ozone generator, or it can be provided continuously through a semipermeable membrane. The benefit of a homogeneous reaction system is that it is easier to achieve plug flow, which allows for higher reactions rates than those reachable in fully mixed phases [32].

5.3.3.1 Tube Reactors
In order to achieve plug-flow, ozonation tube reactors are mostly operated as indirectly gassed or nongassed *homogeneous* systems since the direct gassing in a tube reactor can cause unwanted axial mixing. Ozone is either absorbed into the water in a separate vessel located before the reactor or it is produced *in situ* using an electrolytic ozone generator. The two-stage absorber/reactor system only makes sense when the time required for absorption is significantly less than that required for reaction. The liquid stream to the absorber can be either a slip-stream (e.g., absorption into a recycle stream to the reactor) or full-stream process. Such indirectly gassed and nongassed plug-flow systems are more often found at lab-scale. At full-scale, plug flow is often approximated with a CFSTR in series in which the first stages are directly gassed (see Figure 5.7).

Tube reactors can easily be constructed using conventional piping, for example, made from stainless steel, PTFE or PVC. Space-saving constructions such as coiled flexible piping are possible. In such an *indirectly gassed* tube reactor Sunder and Hempel studied the ozone oxidation of tri- and perchloroethene contained in a model ground water system [52]. The ozone gas was first absorbed in a highly efficient ozone absorber, the so-called Aquatector®, which was operated in the slip-stream mode. The tube reactor had the following dimensions, $l_R = 14.8\,\text{m}$, $d_R = 18\,\text{mm}$ and was operated at $Q_L = 220\,\text{l}\,\text{h}^{-1}$ and a hydraulic retention time of $t_H = 62\,\text{s}$. The Reynolds and Bodenstein numbers (Re = 4300 and Bo = 600) in the reactor indicate that ideal *plug flow* was achieved. This system is comparable to a conventional water pipe, where variations in the hydraulic retention time can easily be realized by the tube length. Another application of the two-stage absorber/tube-reactor system was a lab-scale study on ozone disinfection of *bacillus subtilis* [53]. Here, the absorber was a static mixer in front of a bubble column.

Alternatively, instead of using gaseous ozone, ozone can be produced *in situ* by an electrochemical ozone generator. In such a *nongassed* system, ozone can remain dissolved in the liquid phase until it reacts in the system. This may be especially advantageous in ultrapure water applications in the pharmaceutical or semiconductor industries [30].

5.3.3.2 Membrane Reactors

Another possible configuration gaining in popularity over the last decade is the use of semipermeable membranes to transfer gaseous ozone to the liquid without mixing the two phases. In a membrane reactor the gas and water are physically separated by the membrane and virtually no gas bubbles are produced in the liquid. The membranes are either plates or hollow fibers and are usually arranged in parallel in one housing or module. The hollow fibers can be operated with the gas outside and the liquid inside or vice versa. Figure 5.9 shows a schematic sketch of such a membrane reactor. In this reactor the gas molecules pass through the fiber membranes and dissolve in the liquid that flows inside the hollow fibers. No bubbles occur in the liquid phase when the total pressure of the gas is lower than the pressure of the liquid.

Important parameters that characterize the membranes as well as the contactor modules are shown for both plate- and hollow-fiber-type membranes in Table 5.6.

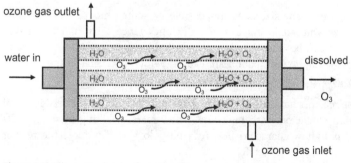

Figure 5.9 Schematic sketch of a hollow-fiber membrane reactor.

Table 5.6 Characteristic features and parameters of membranes and reactors composed thereof used for ozone mass transfer in water and waste-water treatment.

Plate-type (flat-sheet) membranes ([4, 6])	Hollow-fiber-type membranes ([3, 5])
Membrane features and parameters	
Membrane material	Membrane material; Skin (e.g., on shell side)
Pore size (µm)	Pore size skin (nm), Pore size lumen (µm)
Thickness (µm)	Outer (o.d.)/Inner diameter (i.d.) (mm)
Porosity, ε or Pore volume (%)	
Tortuosity, χ	
Specific contact area (m^2/membrane)	Specific contact area (m^2/hollow fiber)
Design of contactor module	
Number of sheets or plates in housing	Number of hollow fibers in housing
Channel or flow path length (m)	Fiber or contact length (m)
Total membrane (cross-sectional) area (m^2)	Total membrane area (outer surface) (m^2)
Hydraulic diameter (m)	Housing diameter (m)
Contactor volume (m^3)	Contactor volume (m^3)

During the last decade various types of membrane reactors have been used for ozonation ([1, 2, 4, 5, 36]). Most of the applications have been in lab-scale.

The following advantages of membrane reactors compared to the use of directly gassed reactors are often claimed (e.g., [4, 6]).

1. The *specific surface area* of membrane contactors is much higher than in conventional contactors and may result in a higher volumetric mass-transfer coefficient $k_L a$ (this is, however, not always the case; compare for example, the values for STRs in Table 5.4).

2. *Bubble-free operation* prevents foaming.

3. The *effluent gas* mixture (O$_2$/O$_3$) can be more easily *recycled* back to the ozone generator because of the relatively low moisture content.

4. *Scale-up* to almost any size is easily possible by adding modules of the same type and size without losing transfer efficiency since a constant ozone-gas concentration can be established at the gas/liquid interface. Also no changes in the specific energy dissipation are caused, which is crucial in the scale-up of directly gassed reactors.

5. Membrane reactors have a *small footprint*: the membranes can be used in small cross-section contactors with high linear flow rates, resulting in small units with good plug-flow characteristics. And plug-flow itself reduces the reactor volume compared to STR or BC.

Concerning bubble-free operation it has to be pointed out, that the total gas pressure has to be held lower than the pressure in the liquid ($P_G < P_L$). Otherwise the membrane acts as a bubble diffuser. Also, failure-free pressure regulation is necessary since the membranes are susceptible to pressure drops. Furthermore, it is important to consider that the maximum absolute pressure for PTFE and other synthetic membranes is below 300 kPa.

An important application for membrane reactors could be in the study of reaction kinetics. Jansen et al. pointed out that the hollow-fiber membrane contactor may be a useful tool for the experimental assessment of reaction kinetics, since it has an exact surface area, the hydrodynamics in the fiber can easily be assessed and the gas/liquid absorption process can be operated at steady state [5].

Nevertheless, some disadvantages also have to be considered:

1. The *ozone-transfer efficiency* $\eta(O_3)$ is normally well below 100%; it is often in the range of 30–40%. Recycling of the off-gas is possible but costly and not easy to handle.

2. Comparatively *high investment costs* result from rather short membrane lifetimes. For example, in semiconductor applications membrane replacement can be necessary every one to four years.

3. The *specific energy consumption*, mainly caused by pumping to achieve high water flow rates, and thus the operating costs may be comparatively high.

5.3.4
Types of Gas Contactors

Most types of mass-transfer systems currently available can be used to transfer ozone from the gas into water, as long as they are made of a material resistant to ozone that does not cause excessive ozone decomposition. An efficient gas contactor must produce a large mass-transfer area while consuming little energy. Most systems use the liquid as the continuous phase, while the gas phase is discontinuous and present as bubbles, although devices in which the phases are reversed, for example, spray towers, have also been used. Furthermore, the use of semipermeable membranes to transfer gases in which both phases are continuous on either side of the membrane has grown over recent years. Since an important use of mass-transfer systems in water applications is to transfer air into water, these devices are often called aerators or aeration systems. This section will only briefly present common contactors on the lab-scale. For more detailed information, the reader is referred to Chapter 6, as well as the wealth of literature on aeration (e.g., [54, 55]).

Mass-transfer systems are selected based on their ability to provide the required mass-transfer rate as well as on their energy efficiency, the ratio of the mass transferred to the energy required for the transfer. In general, the smaller the bubble size, the higher the mass-transfer rate that can be achieved. Smaller bubbles unfortunately usually require higher energy inputs. Other important criteria are their operational characteristics, that is, their reliability, service life and maintenance. Often there is a trade-off between reliability and energy efficiency.

The various types of contactors can be grouped according to how the energy is delivered into the system: mechanical devices, for example, surface aerators that disperse water into the gas phase, diffusers that disperse gas into the water, and combined systems such as injectors and turbines that usually involve dispersing both phases. The latter two types are more common in ozonation.

The choice of the type of gas contactor is also related to the type of reactor or absorber that will be used. Fine pore diffusers or injectors and static mixers that utilize the high shear created by forcing liquid and gas through constrictions are often installed in bubble columns, while turbines that use the mechanical energy of the stirrer to make fine bubbles from a coarse diffuser are often used in STR. The potential of the water to cause blockage or clogging also is a determining factor in contactor choice. If precipitates, such as carbonates, aluminum or ferrous oxides, manganese oxides, calcium oxalate or organic polymers are expected, fine-pore diffusers should be avoided.

Mass-transfer systems are commonly characterized by the diameter of the bubbles produced, for example, micro (d_B = 0.01–0.2 mm), fine ($d_B \approx$ 1–3 mm) or coarse (d_B > 3 mm) bubbles [54], and by the size of the pores in diffusers that also can range from fine to coarse. It is important to note that the size of bubble produced is not only dependent on the pore size of the diffuser or energy input into the system but is also dependent on the water and its constituents. In systems relying on a diffuser to disperse the gas, an increase in mass transfer is only possible by increasing the gas flow rate. However, two phenomena limit the benefits of such an increase. High gas flow rates can lead to higher bubble coalescence so that only small increases in the specific surface area are seen. In addition, an increase in the gas flow rate to an electrical-discharge ozone generator results in a decrease in ozone-gas concentrations because of their typical operating characteristics (see Figure 5.3). This reduces the driving force for mass transfer, the concentration gradient. Thus, a higher energy input via an increased gas flow rate may be counteracted by higher bubble coalescence and a lower driving force.

In lab-scale stirred-tank reactors that rely on the stirrer to produce fine bubbles, ring pipes with 0.1–1.0 mm i.d. holes that emit coarse bubbles are common. Although fine porous plate diffusers (d_P = 10–50 μm) have also often been used in STRs (e.g., [49]), they are most often used in bubble columns (e.g., [38, 56]). Coarse (d_P = 50–100 μm) porous disks are the most frequently applied diffusers in large-scale ozonation contactors for drinking-water treatment systems (Figure 5.7; [10]). They are seldom used in industrial waste-water treatment applications since the potential for blockage is high. Other gas-contactor types are used less often, but some examples are provided below in order to highlight the diversity and possibilities available for mass-transfer systems.

Many modifications have been made on standard reactors to optimize the mass transfer. For example, specialized reactor types like the impinging-jet bubble column [46] or the submerged impinging zone reactor (IZR) ([35, 47],) have been constructed for very high mass-transfer rates (Table 5.4). They often involve two-phase gas/water injector nozzles, which can achieve high mass-transfer rates. However, this requires high gas flow rates and liquid recirculation. Therefore, such systems are seldom applied in lab-scale ozonation experiments; instead they are used in pilot- or full-scale bubble-column applications requiring high mass-transfer rates [14].

A lab-scale tube reactor made from PTFE into which ozone gas was introduced by an *injection nozzle* made of glass was used by Hemmi et al. [57]. The system was successfully applied for the ozonation of dyehouse liquors with the purpose of color removal and partial DOC oxidation. Another example of a lab-scale system using an injector nozzle as the gas contactor and a tube reactor for ozonation of small concentrations of tri- and perchloroethene is given in [52]. The injector nozzle coupled with the highly efficient Aquatector® ozone-absorption unit was installed in front of the tube reactor. Both the gas and liquid were partially recycled in this system. In demineralized water 90% of the ozone produced was absorbed and dissolved-ozone concentrations ranged up to $100\,\mu mol\,l^{-1}$ (c_L = $5\,mg\,l^{-1}$, T = $20\,°C$).

Static mixers can also be used for gas–liquid mixing. High mass-transfer rates can be achieved at large liquid flow rates and with a high specific power consumption (Table 5.4). A high pressure drop develops in the system meaning that a lot of energy is needed to "push" the liquid through the mixer. Martin et al. modeled the transfer characteristic of static mixers and compared the efficiency of three commercially available systems [34]. Static mixers are small and easy to handle. From the viewpoint of full-scale application further operational advantages are: inline setting up in a pipe, compact dimensioning, no mobile part(s) and very little maintenance requirement (if no abrasion occurs!). An example of a lab-scale application of a static mixer is given in a study on drinking-water ozone disinfection [53]. The complete reaction system consisted of a small static mixer (length of 75 mm, volume of 15 ml) being installed in front of a small bubble column, which was used to accomplish ozone mass transfer, and a subsequent tube reactor for disinfection, which was hydraulically uncoupled from the bubble column.

As was already pointed out in the last section *membranes* have been increasingly used for bubble-free ozone mass transfer from the gas to the liquid over the last decade. The achievable ozone concentration in the liquid is a function of the gas concentration, the pressure, the ratio of the liquid to gas flow rates as well as the ozone consumption rate in the liquid. For example the LIQUOZON™ is such a kind of a membrane contactor [11]. The system is especially attractive for producing ozonated water for wafer cleaning in semiconductor applications, where the ultrapure water on the one hand does not contain substances being prone to clogging and on the other hand the ozonated water has to be absolutely free of gas bubbles. Dissolved-ozone concentrations up to $95\,mg\,l^{-1}$ at a pressure of up to P_{gauge} = $250\,kPa$ and flow rates between 0.5 and $20\,l\,min^{-1}$ are feasible with this system.

5.3.5
Mode of Operation

Reactors can be operated either in a batch or continuous-flow mode. The combination, batch with respect to the liquid and continuous flow with respect to the gas, is called *semibatch*. Often, this fine distinction is ignored and it is commonly referred to as batch. Continuous-flow reactors can be either plug flow or ideally mixed. Batch reactors are usually completely mixed in respect to the liquid phase. Large differences in reaction rates and products can occur depending on the mode of operation and hydrodynamic behavior of the reactor. A general introduction to the differences was given in Section 4.3 and so is only summarized here; for more detailed information on reaction engineering the reader is referred to the literature (e.g., [32, 33, 51, 58]).

In the oxidation of compounds with ozone, the concentrations of both ozone and the target compound M usually influence the reaction rate. In general, the higher the concentration, the higher the reaction rate. To achieve high degrees of target compound removal, low concentrations of M must be reached in the effluent. In continuous-flow completely-mixed reactor systems, the concentrations in the reactor and in the effluent are virtually the same (Figure 5.6); as a consequence, the reaction rate is relatively low. From the reaction-engineering viewpoint, batch processes (even in ideally mixed reactors) behave similarly to those in a continuous plug-flow reactor PFR, the concentration is reduced to the effluent value over time (batch) or the length of the reactor (PFR). Therefore, both batch and PFR can achieve higher reaction rates than are possible in the completely mixed system of a CFSTR, This behavior has for example been shown in a study on ozone disinfection [53]. In a study on the removal of hazardous organic compounds in three model waste waters, the savings due to the higher reaction rates with a batch operating mode as compared to a continuous flow mixed system were quantified ([56, 59]). Reductions of 50 to 60% in the specific ozone doses I^* were observed for the same degree of pollutant removal (η_S = 98%) in batch treatment.

The majority of lab-scale ozonation experiments reported in the literature have been performed in one-stage semibatch *heterogeneous* systems, with liquid-phase reactor volumes in the range V_L = 1–10 l. Most full-scale applications, however, are operated in continuous flow for both phases. Therefore, the purpose of the investigation must be clear in order to choose which of the two modes should be used in the experiments. The advantages of continuous-flow operation, for example, reduced process control and storage requirements, have to be weighed against possible disadvantages, for example, reduced reaction rates and ozone efficiency.

Multistage CFSTRs in series (n-CFSTR) are often used in full-scale applications. However, they are not often used in lab-scale experiments, even though the improvement in ozone efficiency due to lower ozone consumption can be dramatic. From a practical point of view, the setup and handling of multistage systems is more complicated than running a one-stage system.

The choice of reactor type, number of stages and mode of operation must be made in view of the experimental goals. Some important aspects to consider when conducting batch or continuous-flow lab-scale experiments on drinking- or wastewater ozonation are summarized in the following section.

5.3.6
Experimental Procedure

5.3.6.1 Batch Experiments

Batch, or more precisely, semibatch experiments in a *heterogeneous* system are comparatively quick and easy to perform. In their simplest form, they can be used to assess the gross effect or the tendency of influence that a certain parameter of interest, for example, pH, concentration, ozone dose rate, has on the treatment result.

An advantage of batch operation is that normally only a small amount of solution is required. Enough solution, though, must be available to perform all necessary analyses, as well as to show reproducibility. This is especially important to verify for combined chemical/biological processes, where more treatment as well as more analyses are carried out.

It is very important to apply the same oxidation conditions in each batch, and vary only one parameter at a time, unless statistical experimental design is being applied (e.g., [53]). Since ozone reactions depend much on concentrations and the pH, be especially careful to assess the liquid-ozone concentration c_L as well as the pH value as a function of the reaction time. In drinking-water studies, especially when micropollutants are treated, the dissolved-ozone concentration may easily be held constant throughout the whole time of experiment. In semibatch kinetic studies (in the slow kinetic regime; see Section 6.3.2) care has to be taken to establish the steady-state ozone concentration in the liquid before the micropollutants are injected into the water [27]. In contrast, in waste-water ozonation both parameters are often observed to vary within the reaction time. The liquid-ozone concentration shows an increase, whereas the pH decreases due to the formation of acidic reaction products (organic acids).

It is not recommended to withdraw large amounts of liquid from the ozone reactor during the treatment, since the hydrodynamic conditions as well as the ratio of mass of ozone per mass of pollutant remaining may change and thus influence the gas–liquid mass transfer or the oxidation rates. The mass transfer might also be influenced by the oxidation, for example, when surface active agents are oxidized, however, this can normally not be avoided by changes in the experimental procedure (see also Section 6.2.3).

During waste-water ozonation, the specific ozone dose I^* or the specific ozone absorption A^* (see Table 4.3 and Figure 4.8; best recorded and computed online) are recommended as the measure for analyzing process performance. Since in waste-water ozonation, the ozone off-gas concentration often increases during the batch treatment, the specific ozone absorption will not vary linearly with the treatment time. Thus, in waste-water ozonation a recommended procedure for conducting a series of batch ozonations with minimum parameter variation is: fill the reactor to the desired level, ozonate, withdraw the whole solution, fill again to the same level and repeat the oxidation with a new set of ozonation conditions (preferably varying the oxidation time or the specific ozone absorption) in the next run. If the goal is to minimize the specific ozone absorption A^* or ozone consumption $D(O_3)^*$ semibatch ozonation experiments are the best and easiest method. Keep in mind that in a continuous process, the same results can only be achieved in a plug-flow reactor but not in a CFSTR (see Section 4.3.2).

5.3.6.2 Continuous-Flow Experiments

The continuous mode of operation requires a comparatively high effort in the laboratory and often the experiments cannot be performed as quickly as in the batch mode.

Quite a lot of maintenance has to be dedicated to the system to insure stable liquid and gas flow rates. Carbonates can plug up or block a fine-pore diffuser for ozone over time, causing the pressure in front of the reactor to increase, the gas-flow rate to sink and the ozone concentration in the influent gas to increase (if an EDOG is used). The experimenter will be severely unsatisfied by this situation although the total ozone mass flow into the system will be unchanged and the oxidation process might not be affected negatively, if the reaction depends (more) on the total ozone dosage than on the ozone concentration. However, the mass-transfer and reaction rates might differ.

The size of the system and the desired range of hydraulic retention times determine how much feed water is required. Large reactors or short hydraulic retention times require that a comparatively large amount of feed water has to be prepared and stored in a way that guarantees the concentration of the influent remains stable over the necessary period of time. Cool and dark places are often required, but cooling may become problematic if very large volumes are involved. Considering the amount of liquid to be stored, it has to be kept in mind that it takes at least three times the hydraulic retention time before steady state is achieved even in completely mixed reactors [32].

5.3.6.3 Process Combinations

When processes are combined, such as an ozonation stage with a biological one, the requirements to reach the experimental goals in each stage must be fulfilled. For example, a sequential chemical-biological system with two comparatively large reactors with liquid volumes of 2 and 7.5 l was operated in series with the specific intent to track a possible adaption of the biomass in the biological stage [56]. Unfortunately, since it is normally unknown how long it will take for an adaptation to occur (if ever), a lot of water has to be treated in such a case. This should be considered when planning the sizes of the experimental setup.

Furthermore, if a multistage sequential chemical-biological system is set up, care has to be taken, that ozone – either liquid or gaseous – cannot enter the biological stage, since it would kill the biomass due to oxidation. Either the ozone concentration in the off-gas can be controlled at zero, which would most likely mean that c_L is also near zero [38]) or the gas phase has to be kept separate from the biological stage, for example, with a gas trap [56].

In order to allow for more flexibility in the operation of a sequential chemical-biological system the two stages do not have to be directly coupled. A storage tank or drain in between allows changes in experimental parameters to be made more independently. A procedure considering this is outlined in Section 9.3.3.

5.4 Ozone Measurement

The following section gives an overview of the different methods available to measure ozone in the gas and liquid phases. For quick reference the methods including detection limits are summarized in Table 5.8 so that the reader can choose an analytical method that fits his or her system at a glance. All important information for example, interference, detection limit, as well as the original reference with the detailed description of the method, necessary for its application can be found in this table. The methods are described in ascending order of their purchase costs.

5.4.1 Methods

5.4.1.1 Iodometric Method (Gas and Liquid)

This wet analytical method can be used for the determination of the ozone concentration in the gas and/or liquid phase. The measurement takes place in the liquid phase, though, so that to measure a process gas containing ozone, the gas must first be bubbled through a flask containing potassium iodide KI. To measure a liquid-ozone concentration, a water sample is mixed with a KI solution. The iodide I^- is oxidized by ozone. The reaction product iodine I_2 is titrated immediately with sodium thiosulfate $Na_2S_2O_3$ to a pale yellow color. With a starch indicator the endpoint of titration can be intensified (deep blue). The ozone concentration can be calculated by the consumption of $Na_2S_2O_3$.

$$KI + O_3 + H_2O \rightarrow I_2 + O_2 + KOH \tag{5.10}$$

$$3 I_2 + 6 S_2O_3^{2-} \rightarrow 6 I^- + 3 S_4O_6^{2-} \tag{5.11}$$

advantages: Purchase price for this method is very low.
disadvantages: Iodide is oxidized by substances with an electrochemical potential E_0 higher than 0.54 eV. This means nearly no selectivity. (e.g., Cl_2, Br^-, H_2O_2, Mn-components, organic peroxides).
The measurement is time consuming.

5.4.1.2 UV Absorption (Gas and Liquid)

This photometric method utilizes the absorption maximum of ozone at 254 nm. This is very close to the wavelength of the mercury resonance line at 253.7 nm, so the source of radiation is generally a low-pressure mercury lamp. The decrease in the UV intensity at $\lambda = 254$ nm due to absorption in the sample is proportional to the concentration of ozone based on the Lambert–Beer law of absorption:

$$I_l = I_o 10^{\varepsilon c(M) l} \tag{5.12}$$

I_l Intensity passing through the absorption cell containing the sample
I_o Intensity passing through the absorption cell containing the reference
$c(M)$ concentration in M
l internal width of the absorption cell in cm
ε molar extinction coefficient in $M^{-1} cm^{-1}$
$\varepsilon_{254 nm}$ about $3000 \, (M \, cm)^{-1}$, depending on the literature source

Temperature and pressure influence the density of the gas sample, which changes the number of ozone molecules in the absorption cell. This effect is addressed by directly measuring temperature and pressure and including their actual values in the calculation:

$$c(M) = -\frac{1}{\varepsilon l} \frac{T}{273 \, K} \frac{14.695 \, psi}{P} \ln \frac{I}{I_o} \tag{5.13}$$

T sample temperature in K
P sample pressure in psi

This method can be used for both gas and liquid phase samples. The law of Lambert–Beer is valid between $0.1 < \ln I_l/I_o < 1$. So that the concentration range for this method depends on the width of the absorption cell. Systems exist that can measure up to $200 \, mg \, l^{-1}$ in the liquid phase, and up to $400 \, mg \, l^{-1}$ in the gas phase.

Most organic substances and some inorganics absorb at 254 nm and can potentially cause interference with this method especially in the liquid phase. However, this can be exploited using a differential measurement of the sample with and without ozone to determine the ozone as well as the organic concentration. After measuring absorption in the sample, ozone is destroyed with $Na_2S_2O_3$. In distilled water and tap water it is possible to measure the liquid-ozone concentration and the unsaturated organic substances by using $Na_2S_2O_3$. The first absorption value at $\lambda = 254$ nm represents the sum of the organic substances and the liquid-ozone concentration. After ozone destruction, the residual absorption is correlated to the organic matter [60]. This sensitivity and concentration range of this method is determined by the relative concentrations and absorption coefficients of ozone and the organic substances.

advantages: The method is very easy and simple. Continuous measurement is possible.

disadvantages: Aromatic pollutants in water absorb UV-radiation at $\lambda = 254$ nm and can interfere with the measurement. Additionally, bubbles in ozonated water can severely disturb the measurement. This can be avoided by reducing pressure drops in the sampling system that could lead to the degassing of ozone or other dissolved gases [61].

5.4.1.3 Visible-Light Absorption (Gas and Liquid)

Instead of ultraviolet light the yellow-red Chappuis bands (500–700 nm) can be used as light source. Again the Lambert–Beer equation is the basis for the calculation of the ozone concentration.

Since the specific absorption of ozone at 254 nm is about 2000 times greater than that at the yellow-red Chappuis bands, the measurement sensitivity is reduced compared to the UV method.

This lower sensitivity can be improved by increasing the path length by using mirrors and by adding a second nonabsorbing visible wavelength for example, blue light LED. By directing the two bands of light along the same path, the reference signal can be used to correct for intensity loss in the first band of light due to factors others than the absorption by ozone [62].

advantages: The method is very easy and simple. Continuous measurement is possible and less expensive than UV absorption
disadvantages: Lower sensitivity than UV absorption

5.4.1.4 Indigo Method (Liquid)

The concentration of aqueous ozone can be determined using the decolorization reaction of indigo trisulfate with ozone. The ozone concentration is proportional to the loss of color and can be measured photometrically (λ = 600 nm). The reaction is stoichiometric and extremely fast. The indigo molecule contains only one C = C double bond that can be expected to react with ozone directly and with a very high reaction rate (see Figure 5.10).

One mole ozone decolorizes one mole of aqueous indigo trisulfate at a pH less than four. Hydrogen peroxide and organic peroxides react very slowly with the indigo reagent. If ozone is measured in less than six hours after adding the reagents, hydrogen peroxide does not interfere.

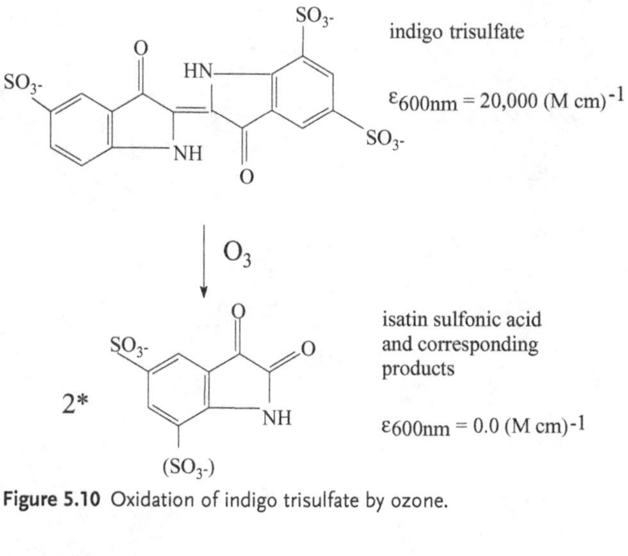

indigo trisulfate

$\varepsilon_{600nm} = 20{,}000 \; (\text{M cm})^{-1}$

isatin sulfonic acid and corresponding products

$\varepsilon_{600nm} = 0.0 \; (\text{M cm})^{-1}$

Figure 5.10 Oxidation of indigo trisulfate by ozone.

advantages: The purchase price for this system is low. The reaction is very fast and selective. Secondary oxidants do not interfere within 4 to 6h.

5.4.1.5 N,N-diethyl-1,4 Phenylenediammonium–DPD (Liquid)

This photometric method requires the addition of iodide and a color indicator N,N-diethyl-1,4 phenylenediammonium (DPD). The method utilizes a two-step reaction to produce a color change in the sample that is proportional to the ozone concentration and can be measured photometrically. First, ozone reacts with iodide which then reacts with the color indicator DPD. The direct oxidation of DPD by ozone is very slow. Therefore, the concentration of ozone is measured indirectly by the oxidation of iodide with ozone (pH = 6).

$$2\,H^+ + O_3 + 2\,I^- \rightarrow I_2 + O_2 + H_2O \tag{5.14}$$

The product iodine forms a radical cation with DPD, which is a red dye (see Figure 5.11). The radical cation $DPD^{+\circ}$ is stabilized by resonance and forms a fairly stable color with one absorption maximum at 510nm and one at 551nm. The concentration of ozone is proportional to the intensity of the dye and can be calculated according to the Lambert–Beer law (5.12).

advantages: see iodometric method
disadvantages: see iodometric method

After adding reagents the sample is only stable for 5 min.

5.4.1.6 Chemiluminescence–CL (Liquid)

In this method, the intensity of light from a chemical reaction is measured. The difference between chemiluminescence and photometric absorption is that instead of measuring the decrease in light intensity due to absorption by the compound of interest, the production of light due to a chemical reaction with the compound

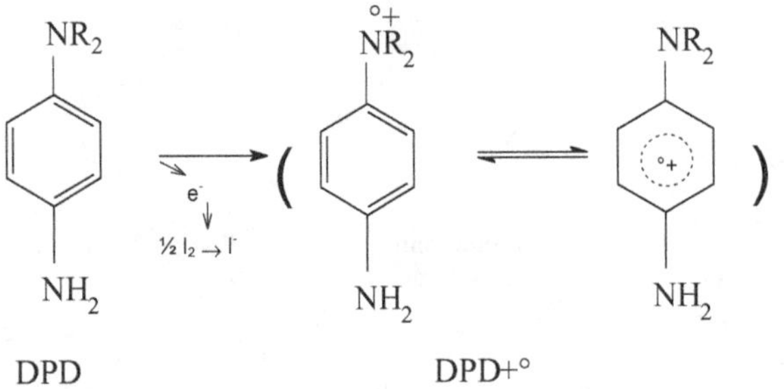

Figure 5.11 Oxidation of N,N -diethyl-1,4 phenylenediammonium -DPD by iodine that has been oxidized by ozone.

Table 5.7 Reagents for CL-FIA [63].

1. Benzoflavin	5. Eosin Y
2. Acridine Yellow	6. Rhodamin B
3. Indigotrisulfate	7. Chromotropic acid
4. Fluorescein	

of interest is measured and it is proportional to the concentration. Often this method is automated with flow injection analysis (FIA). The ozonated water sample is injected into a pure water carrier and mixed with a dye reagent in front of the photodetector. The dye reagent, which is very selective for ozone over other oxidants, undergoes a fast reaction with aqueous ozone and produces chemiluminescence. The intensity is proportional to the concentration of ozone. Table 5.7 gives some examples of possible reagents, listed in ascending order according to the intensity of chemiluminescence produced by contact with ozone.

advantages: continuous method
disadvantages: this method is not easy to handle.

5.4.1.7 Membrane Ozone Electrode (Liquid)

Electrochemical (amperometric) techniques provide the possibility for *in-situ*, continuous and automated measurements of ozone in the liquid. The membrane electrode usually consists of a gold cathode, a silver anode, an electrolyte (e.g., AgBr, K_2SO_4 or KBr) and a teflon membrane. Several companies offer such electrodes in different configurations. The application range and accuracy differs depending on the kind of electrode.

In the presence of ozone in water, ozone diffuses through the membrane into the reaction chamber. The rate of flow is dependent on the partial pressure of ozone. In order to avoid a depletion of ozone molecules at the surface of the membrane, the electrode should be immersed into a continuous flow.

At the gold cathode ozone is reduced into oxygen:

$$\text{cathode:} \quad O_3 + 2\,H^+ + 2\,e^- \rightarrow O_2 + H_2O \tag{5.15}$$

The electrons are produced at the anode by the following process:

$$\text{anode:} \quad 4\,Ag \rightarrow 2\,Ag^+ + 2\,e^- \tag{5.16}$$

The result is a current conduction that is measured and proportional to the ozone concentration.

advantages: continuous method
disadvantages: high purchase price

5.4.2
Practical Aspects of Ozone Measurement

In order to evaluate the results of the experimental work, an ozone balance over the reactor is essential. Measurement of the ozone-gas concentration in the in- and outlet of the reactor system as well as the liquid-ozone concentration are necessary to complete the balance. This is true for both steady-state and nonsteady-state systems. The monitoring of nonsteady-state systems, in which the concentrations change over time, requires more measurements.

Continuous analytical methods (amperometric and UV-Absorption methods) are advantageous. However, sometimes only discontinuous methods (titrimetric and some photometric methods) are available due to expense. In such cases it is important to measure immediately after sampling to avoid the decomposition of ozone and in the case of liquid ozone to avoid degassing. Discontinuous photometric methods requiring the addition of chemicals to the sample can be converted to a continuous method by combination with flow injection analysis (FIA). This analytical technique requires instrumentation and is not easy to handle (Table 5.8).

5.5
Safety Aspects

5.5.1
Vent Ozone Gas Destruction

Destruction of excess ozone in the vent gas is an essential safety precaution in every ozone system. The following techniques are applied: thermal ($T \geq 300\,°C$) or catalytic (manganese or palladium, $T = 40$–$80\,°C$) destruction. Both methods offer reliable ozone destruction. Whereas thermal units require more energy than catalytic units, catalytic units are more sensitive to constituents in the off-gas. The lifetime of the catalyst can be limited by poisoning. Current known poisons are nitrogen oxide and halogen-containing chemicals, as well as water [61]. Condensation of water vapor out of the off-gas can destroy the catalyst material. Since in most ozone applications discussed in this book, water vapor is present, the ozone destruction unit is usually heated between 60–80 °C.

In small-scale laboratory systems with low ozone concentration (up to $10\,\mathrm{g\,m^{-3}}$); ozone can also be destroyed in a packed column filled with granulated activated carbon ($d_p = 1$–$2\,\mathrm{mm}$). This method is not used for off-gases in large-scale systems because of the fire hazard. The activated carbon (AC) reacts exothermically with the ozone. Stoichiometrically, 2.7 g ozone react with 1 g AC to produce CO_2 and O_2 and heat, which destroys not only the ozone but also the structure of the AC. Localized overheating of the AC can cause fires.

In large-scale systems with electrical-discharge ozone generators the recycling of oxygen is a common standard today; in this case drying and compression of the

Table 5.8 Overview of analytical methods for ozone measurement.

Method	Gas	Liquid	Continuous method	Detection limit	Interferences	Advantages ↑ / disadvantages ↓	Reference
Titrimetric Iodometric	+	+		100 µg l^{-1}	Cl_2, Br^-, Mn, H_2O_2, all oxidants with $E_o > 0.54$ eV	Not expensive↑, no selectivity↓, time consuming↓	DIN 38408-G3-1[a] [64] IOA 001/87(F) [65]
Photometric UV/VIS-Absorption	+	+	+	[b]		Easy and simple↑, aromatic components can interfere↓	IOA 002/87 (F) [66] IOA 003/89 (F) [67]
Indigo trisulfate		+	(+) FIA	5 µg l^{-1}	Cl_2, ClO_2, Br^-	Not expensive↑, relatively selective↑, fast reaction↑, stability of the sample after adding reagents <4–6 h↑, secondary products do not interfere↑, need calibration↓, natural color of water does not interfere↓	Bader and Hoigné, 1981[a] [68] DIN 38408-G3-3[a] [69] Hoigné and Bader, 1980[a] [70] IOA 006/89 (F) [71] Standard methods, 1989 [72]
DPD		+	(+) FIA	20 µg l^{-1}	Cl_2, Br^-, Mn, H_2O_2, all oxidants with $E_o > 0.54$ eV	Not expensive↑, no selectivity↓, stability of the sample after adding reagents <5 min↓	DIN 38408-G3-2[a] [73] Gilbert, 1981[a] [74]
CL		+	+	2 µg l^{-1}	Cl_2, $H_2O_2 > 10$ mg/l $MnO_4^- > 2$ mg l^{-1} in the case of Indigo	Not easy↓, instrumentally complex ↓	Chung et al., 1992 [63]
Amperometric Electrode		+	+	6–10 µg l^{-1} [b]		Selective↑, expensive↓	IOA 007/89(F) [75] Smart et al., 1979 [76] Stanley and Johnson, 1979 [77]

a Good description of the procedure.
b Detection limit and precision depend on the systems.

off-gas is required before it can be introduced into the ozone generator again. In lab-scale systems oxygen is normally not recovered.

Even in ELOG applications where theoretically only dissolved ozone is produced, the evolution of very little but highly concentrated ozone gas is possible, so that safety precautions for ozone in the exhaust air must be taken. During the *in-situ* ozone production, the liquid-ozone concentration (c_L) can easily reach the solubility level, depending on the pressure (P) and temperature (T) in the cell. Outgassing can easily occur due to oversaturation of the feed water, especially when the pressure drops suddenly.

Possibilities of explosive gas mixtures exist. In full-scale ground- or waste-water applications for example, volatile organic compounds could be contained in the off-gas. These compounds escape from the reactor due to comparatively high temperatures and condense in the off-gas piping. Care has to be taken to avoid explosive mixtures of these organics and the oxygen/ozone gas. In lab-scale systems, principally the same problem can occur with greasy valves or other equipment bearing organic "contamination".

5.5.2
Ambient Air Ozone Monitoring

Ambient air ozone monitoring coupled with an automatic ozone generator shut off device is recommended for every lab-scale application in order to ensure that the laboratory staff will not be harmed in case of leaks in the piping or reactor. Ozone analyzers measuring ozone at $\lambda = 254$ nm in the appropriate range (0.001–1.000 ppm) are supplied by several companies. When installing them, care has to be taken that the input gas-side of the analyzer is properly installed, for example, the air being free of particles that could form a layer on the inner surface of spectrophotometric cell. Some systems need an ozone-free reference gas for online calibrating, which is normally achieved by sucking the ambient air through a small activated carbon filter.

Problems with shutting off the ozone generator at the legislatively required threshold level, for example, the MAK-value (200 $\mu g\,m^{-3}$), might occur in the case of higher ambient ozone concentrations due to summer smog. Sometimes, the threshold values for a regular laboratory environment are lower than the actual ambient concentrations.

5.6
Common Questions, Problems and Pitfalls

The following section will be helpful to read before starting your own experiments. Since there is so much information available on ozone applications it is hard for the beginner to sort it into essential and unessential categories. Experience makes everything easier, usually it is paid for with energy and mistakes. Here are some of the things various experimenters would have found helpful to know before they started their work (Table 5.9). Everyone's list is individual depending on his or her

5.6 Common Questions, Problems and Pitfalls

Table 5.9 Common questions, problems and pitfalls.

Problem or question	Aspects to be considered	Section
Define system		
Ozone generator (EDOG): Is it possible to produce every c_{Go} desired at constant Q_G?	No, this depends on the EDOG operation characteristics. Choose an appropriate generator!	5.2.1 Figure 5.3
Ozone generator (ELOG): Is it possible to use an electrolytical cell ozone generator in waste-water applications instead of an EDOG so that no mass transfer will be necessary?	No, normally not! The ELOG ozone-production rates are rather small, the specific energy consumption is very high and the ELOG requires deionized water for its *in-situ* production of ozone.	5.2.2
Ozone reactor: Is it possible to use PVC?	Yes, but it is only recommended for waste-water experiments.	5.1, 5.1.2
Ozone reactor off-gas: Can the ozone containing off-gas be handled in a safe way?	Yes, a vent ozone destructor is necessary; several types are available. Beware of explosive mixtures of O_3/O_2 gas and certain organics. Apply a water-filled gas trap after the ozone destructor. This makes it easy to visually control the gas flow or detect leaks. It, furthermore, "smoothes" the gas flow, due to the back pressure of the liquid (height).	5.5.1
Off-gas ozone destruction: Is it also necessary to have an ozone destructor in the system when using an ELOG?	Yes, ozone can desorb from the water being ozonated due to oversaturation and appear in the reactor off-gas.	5.5.1
Ambient air ozone monitor: Is it necessary to check the ambient-air ozone concentration in the lab?	Yes, it is highly recommended to control the lab environment and connect the monitor to the ozone generator so that the ozone production will be shut off in case of leakage in the system	5.5.2
Materials in three-phase systems: Can every solvent or solid be used?	No! Make sure that the solvent is immiscible with water, nonvolatile and nontoxic. Solvents as well as solids must be inert against ozone attack, which some are not (e.g., if the material contains C–C double bonds they will be readily destroyed by ozone)	9.2

Table 5.9 Continued

Problem or question	Aspects to be considered	Section
Select analytical method		
Assess correct c(M): How can unwanted reactions in the sample be stopped?	Ozone has to be destroyed (quenched) by using NaS_2O_3.	5.4.1
Assess correct c_L: Can amperometric probes be used in the ozonation of waste water?	Yes, but often a lot of maintenance is required. A comparison of the measured data with data gathered simultaneously with the indigo method is strongly recommended. Particles, oil, emulsifiers or colored compounds in the waste water will disturb the measurement.	5.4 9.2.4
Determine procedure		
Ozone generator (EDOG): Can an EDOG be operated at very small gas flow rates Q_G?	No, EDOGs normally need a minimum gas flow rate $Q_{G,min}$ (see Figure 5.3), below this, c_{Go} might vary considerably. An experiment with constant operation parameters will be difficult to obtain. It is recommended to avoid this region!	5.2.1.5
Optimization of mass transfer: Should k_La be optimized by varying Q_G? (especially important in systems where mixing is exclusively provided by the gas flow rate, e.g., bubble columns, where Q_G is the only variable parameter)	No, in systems employing an EDOG (at a constant ozone production rate) two parameters will change at the same time and cause adverse effects on the oxidation-rate: an increase in Q_G normally increases k_La, but decreases c_{Go}	5.2.1.5
Use of scavengers: Can *tert*-butyl alcohol (TBA) be applied without constraints in reaction kinetic measurements?	No, TBA influences the k_La-value considerably (*alpha* factor).	6.2.2
Ozone decomposition: Is it necessary to know the ozone-*decomposition* rate (exactly) in waste-water treatment studies?	No, not in every case. Often the reactions of ozone with organic compounds occur in the liquid film (fast reactions), so that c_L is (approx.) zero and ozone *decomposition* cannot occur. Generally, measure the dissolved-ozone concentration c_L!	6.2, 7

5.6 Common Questions, Problems and Pitfalls | 157

Table 5.9 Continued

Problem or question	Aspects to be considered	Section
pH control in the ozone reactor: What can be done to prevent a pH drop (due to the production of organic acids) in the ozonated water?	In batch tests buffers are normally used. They keep the pH very stable, but induce a higher ionic strength in the water compared with no buffer.	3, 6, 7.3.3
	Remember side-effects when using a buffer solution: buffers contain substances that influence the radical chain reactions (phosphates!) or the $k_L a$ (e.g., high concentrations of sulfates); keep their concentration as low as possible! Especially conduct $k_L a$-measurements with the same buffer solution (without compound (M))!	
	In continuous-flow experiments a feasible way is to apply a pH-control unit that continuously doses NaOH. The slight scatter of the pH normally observed will not have negative influence on the oxidation results.	
Stoichiometric factor z: Which method should be applied for determining the stoichiometry of a direct reaction between ozone and compound (M)?	The following method is proposed: dissolve compound M as well as ozone in independent vessels, mix the two solutions making sure that $c(M)_o \geq 4-10 c_{L_o}$ and let them react until ozone is completely used.	–
	Define z as mol O_3 per mol M eliminated; sometimes it is defined inversely!	
Evaluate data and assess results		
Use of reaction kinetic data from the literature: Are kinetic data (e.g., $k_D(M)$) from the literature useful for comparisons with my own data and may they be used for predictions of my own results?	Yes, if all data necessary to describe the entire process have been assessed and reported; comparisons are only possible on this basis. Consider that mass transfer might influence the (apparent) disappearance kinetics of M in the specific system. Keep in mind, that the k_D (and k_R) values do not take into account the effect of mass transfer enhancement on the removal rate of M ($r(M)$).	4 7 6.2 8
	In general, predictions require good models.	

Table 5.9 *Continued*

Problem or question	Aspects to be considered	Section
Explanation of widely varying compound removal rates (r(M)): What are the reasons for the wide variations in r(M) frequently being observed in different studies on the (semibatch) ozonation of one and the same compound?	Mass transfer with simultaneous reactions is a complex matter. Most probably, the results reported were obtained in systems where mass transfer was limiting the oxidation to different extents. Mass transfer might have been enhanced by the reactions and/or the reactions might have occurred in different kinetic regimes.	6.2
Calculate mass transfer: Is it advisable to use $k_L a$-values calculated purely from the literature?	No, this is not advisable at all! Use the literature (and this book) to screen the ranges and measure it in the actual system with the actual (waste) water!	6.2 6.3
Assess mass transfer correctly: Are there any influences on $k_L a$, which are often forgotten before starting the experiments?	Yes, the influence of chemical factors is often forgotten, although it cannot be neglected. A general problem arises from the fact that $k_L a$-measurements are normally performed with "pure" water and not with the actual water under consideration. Be especially aware of the action of	6.2.2
	1. surface-active compounds, e.g., phenols, TBA, surfactants, etc. 2. "unimportant" compounds being not oxidizable by molecular ozone, but influencing gas-dispersion in the reactor, e.g., sulfates in buffer solutions which hinder bubble coalescence and increase $k_L a$.	6.2.2
	The enhancement factor E was defined to assess mass-transfer enhancement due to (ozone) gas absorption into a liquid accompanied by simultaneous reaction	6.2 Figure 6.5

background, but perhaps these points will help the beginner to minimize his or her mistakes.

The short answers in this list only provide an overview of the aspects to be considered. Details will be found in the cited chapters or sections of this book.

References

1 Janknecht, P., Wilderer, P.A., Picard, C. and Larbot, A. (2001) Ozone-water contacting by ceramic membranes. *Separation and Purification Technology*, 25, 341–346.

2 Heng, S., Yeung, K.L., Djafer, M. and Schrotter, J.-C. (2007) A novel membrane reactor for ozone water treatment. *Journal of Membrane Science*, 289, 67–75.

3 Mori, Y., Oota, T., Hashino, M., Takamura, M. and Fujii, Y. (1998) Ozone-microfiltration system. *Desalination*, 117, 211–218.

4 Pines, D.S., Min, K-N, Ergas, S.J. and Reckhow, D.A. (2005) Investigation of an ozone membrane contactor system. *Ozone: Science and Engineering*, 27, 209–217.

5 Jansen, R.H.S., de Rijk, J.W., Zwijnenburg, A., Mulder, M.H.V. and Wessling, M. (2005) Hollow-fiber membrane contactors – a means to study the reaction kinetics of humic substance ozonation. *Journal of Membrane Science*, 257, 48–59.

6 Phattaranawik, J., Leiknes, T. and Pronk, W. (2005) Mass-transfer studies in flat-sheet membrane contactor with ozonation. *Journal of Membrane Science*, 247, 153–167.

7 DVGW (1999) *Technical Rule – Code of Practice W625: Anlagen zur Erzeugung und Dosierung von Ozon (Plants for the Production and Dosage of Ozone)*, DVGW German Technical and Scientific Association for Gas and Water, Bonn.

8 Saechting, H. (1995) *Kunststoff-Taschenbuch, 26. Ausgabe*, Carl Hanser Verlag, München, Wien, ISBN 3-446-17885-4.

9 Lenntech Water Treatment & air purification Holding B.V. (2008) Ozone Technology Ozone Generator, http://www.lenntech.com/ozone-technology.htm (accessed 8 December 2008).

10 Masschelein, W.J. (1994) Towards one century application of ozone in water treatment: scope – limitations and perspectives, in *Proceedings of the International Ozone Symposium "Application of Ozone in Water and Wastewater Treatment" May 26–27, Warsaw, Poland* (ed. A.K. Bin), International Ozone Association, pp. 11–36.

11 MKS Instruments Inc. (2008) LIQUOZON® Ozonated Water Delivery Subsystems and Accessories, http://www.mksinst.com/product/Product.aspx?ProductID=300 (accessed 25 September 2008).

12 Anseros (2008) Ozone Generator COM-AD-08 (L-com-ad-01-d.doc), www.anseros.de (accessed 5 December 2008).

13 Ozonia Ltd. Switzerland (2005) www.ozonia.com (accessed 8 June 2007).

14 Krost, H. (1995) Ozon knackt CSB. *WLB Wasser, Luft und Boden*, 5, 36–38.

15 Pontiga, F., Soria, C., Castellanos, A. and Skalny, J.D. (2002) A study of ozone generation by negative corona discharge through different plasma chemistry models. *Ozone: Science and Engineering*, 24, 447–462.

16 Kogelschatz, U., Eliasson, B. and Hirth, M. (1988) Ozone generation from oxygen and air: discharge physics and reaction mechanisms. *Ozone: Science and Engineering*, 10, 367–378.

17 Magara, Y., Itoh, M. and Morioka, T. (1995) Application of ozone to water treatment and power consumption of ozone-generating systems. *Progress in Nuclear Energy*, 29, 175–182.

18 Motret, O., Hibert, C. and Pouvesle, J.M. (2002) Ozone production by an ultra-short triggered dielectric barrier discharge, geometrical considerations. *Ozone: Science and Engineering*, 24, 202–213.

19 Stanley, B.T. (1999) Feedgas for modern High-Performance Ozone Generators, www.ozonia.com (accessed 8 June 2007).

20 Manning, T. (2000) Production of ozone in an electrical discharge using inert gases as catalysts. *Ozone: Science and Engineering*, 22, 53–64.

21 Haverkamp, R.G., Miller, B.B. and Free, K.W. (2002) Ozone production in a high frequency dielectric barrier discharge generator. *Ozone: Science and Engineering*, 24, 321–328.

22 Samoilovitch, V.G. (1994) The possibility of increasing efficiency of an ozonizer, *Proceedings of the International Ozone Symposium "Application of Ozone in Water*

and Wastewater Treatment" May 26–27, Warsaw, Poland (ed. A.K. Bin), International Ozone Association, pp. 235–242.

23 Horn, R.J., Straughton, J.B., Dyer-Smith, P. and Lewis, D.R. (1994) Development of the criteria for the selection of the feed gas for ozone generation from case studies, in *Proceedings of the International Ozone Symposium "Application of Ozone in Water and Wastewater Treatment"* May 26–27, Warsaw, Poland (ed. A.K. Bin), pp. 253–262.

24 Wronski, M., Samoilovitch, V.G. and Pollo, I. (1994) Synthesis of NOX during ozone production from the air, in *Proceedings of the International Ozone Symposium "Application of Ozone in Water and Wastewater Treatment"* May 26–27, Warsaw, Poland (ed. A.K. Bin), pp. 263–272.

25 Stargate International Inc. (2006) http://www.stargateinternational.com/ozone/products/comparison.php (accessed 10 November 2006).

26 Hakiai, K., Ihara, S., Satoh, S. and Ch, Y. (1999) Characterisitcs of ozone generation by a diffuse glow discharge at atmospheric pressure using a double discharge method. *Electrical Engineering in Japan*, **127** (2), 8–14.

27 Gottschalk, C. (1997) Oxidation organischer Mikroverunreinigungen in natürlichen und synthetischen Wässern mit Ozon und Ozon/Wasserstoffperoxid. Dissertation. Sharon Verlag Aachen (Germany).

28 ASTeX Sorbios GmbH (1996) *Semozon 90.2 HP, Ozone Generator Operation Manual*, Ver. 1.5 e., ASTeX SORBIOS GmbH, Berlin, Germany.

29 Fischer, W.G. (1997) Electrolytical Ozone-Production for Super-Pure Water Disinfection, Pharma International 2/1997: (Sonderdruck).

30 G.E.R.U.S. mbH (2006) Diachem®-SPE-Technology – A novel method of ozone generation, Wasser Berlin 2006, see also: http://www.gerus-online.de/englisch/dokumente/gerozon_en.pdf (8 December 2008).

31 Da Silva, L.M., Franco, D.V., Forti, J.C., Jardim, W.F. and Boodts, J.F.C. (2006) Characterisation of a laboratory electrochemical ozonation system and its application in advanced oxidation process. *Journal of Applied Electrochemistry*, **36**, 523–530.

32 Levenspiel, O. (1999) *Chemical Reaction Engineering*, 3rd edn, John Wiley & Sons, Inc., New York, Singapore.

33 Beltrán, F.J. (2004) *Ozone Reaction Kinetics for Water and Wastewater Systems*, Lewis Publishers, CRC Press Company, Boca Raton, London. New York, Washington D.C., ISBN: 1-56670-629-7.

34 Martin, N., Martin, G. and Boisdon, V. (1994) Modelisation of ozone transfer to water using static mixers. *Proceedings of the International Ozone Symposium "Application of Ozone in Water and Wastewater Treatment"* May 26–27, 1994, Warsaw, Poland, pp. 293–313.

35 Rüütel, P.I.L., Lee, S.-Y., Barratt, P. and White, V. (1998) *Efficient Use of Ozone with the CHEMOX™-SR Reactor*, Air Products and Chemicals, Inc., Knowledge Paper No. 2.

36 Noda, S., Messaoudi, B. and Kuzumoto, M. (2002) Generation of highly ozonized water using a microporous hollow fiber module. *Japanese Journal of Applied Physics*, **41**, 1315–1323.

37 Marinas, B.J., Liang, S. and Aieta, E.M. (1993) Modeling Hydrodynamics and ozone residual distribution in a pilot-scale ozone bubble-diffusor contactor. *Journal American Waterworks Association*, **85** (3), 90–99.

38 Stockinger, H. (1995) Removal of Biorefractory Pollutants in Wastewater by Combined Ozonation-Biotreatment. Dissertation. ETH No 11 063, Zürich.

39 Huang, W.H., Chang, C.Y., Chiu, C.Y., Lee, S.J., Yu, Y.H., Liou, H.T., Ku, Y. and Chen, J.N. (1998) A refined model for ozone mass transfer in a bubble column. *Journal Environmental Science and Health*, **A33**, 441–460.

40 Lin, S.H. and Peng, S.E.E (1997) Performance characteristics of a packed-bed ozone contactor. *Journal Environmental Science and Health*, **A32**, 929–941.

41 Ta, C.T. and Hague, J. (2004) A two-phase computational fluid dynamics model for ozone tank design and troubleshooting in water treatment. *Ozone: Science and Engineering*, **26**, 403–411.

42 Cockx, A., Do-Quang, Z., Liné, A. and Roustan, M. (1999) Use of computational fluid dynamics for simulating hydrodynamics and mass transfer in industrial ozonation towers. *Chemical Engineering Science*, **54**, 5085–5090.

43 Smeets, P.W.M.H., van der Helm, A.W.C., Dullemont, Y.J., Rietveld, L.C., van Dijk, J.C. and Medema, G.J. (2006) Inactivation of Escherichia coli by ozone under bench-scale plug-flow and full-scale hydraulic conditions. *Water Research*, **40**, 3239–3248.

44 Gottschalk, C., Schweckendiek, J., Beuscher, U., Hardwick, S., Kobayashi, M. and Wikol, M. (1998) Production of high concentrations of bubble-free dissolved ozone in water. *The Fourth International Symposium on ULTRA CLEAN PROCESSING OF SILICON SURFACES, UCPSS '98 September 21–23, Oostende, Belgium*, pp. 59–63.

45 Muroyama, K., Yamasaki, M., Shimizu, M., Shibutani, E. and Tsuji, T. (2005) Modeling and scale-up simulation of U-tube ozone oxidation reactor for treating drinking water. *Chemical Engineering Science*, **60**, 6360–6370.

46 El-Din, M.G. and Smith, D.W. (2001) Maximizing the enhanced ozone oxidation of kraft pulp mill effluents in an impinging-jet bubble column. *Ozone: Science and Engineering*, **23**, 479–493.

47 Gaddis, E.S. and Vogelpohl, A. (1992) The impinging-stream reactor: a high performance loop reactor for mass transfer controlled chemical reactions. *Chemical Engineering Science*, **47**, 2877–2882.

48 Sotelo, J.L., Beltrán, F.J. and Gonzales, M. (1990) Ozonation of aqueous solutions of resorchinol and phloroglucinol 1: stoichiometry and absorption kinetic regime. *Industrial Engineering and Chemical Research*, **29**, 2358–2367.

49 Beltrán, F.J., Encinar, J.M. and Garcia-Araya, J.F. (1995) Modelling industrial waste waters ozonation in bubble contactors: scale-up from bench to pilot plant. *Proceedings, 12th Ozone World Congress, May 15–18, 1995, International Ozone Association, Lille, France*, Vol. 1, pp. 369–380.

50 Beltrán, F.J. and Alvarez, P. (1996) rate constant determination of ozone-organic fast Reactions in water using an agitated cell. *Journal Environmental Science and Health*, **A31**, 1159–1178.

51 Levenspiel, O. and Godfrey, J.H. (1974) A gradientless contactor for experimental study of interphase mass transfer with/ without reaction. *Chemical Engineering Science*, **29**, 1123–1130.

52 Sunder, M. and Hempel, D.C. (1997) Oxidation of tri- and perchloroethene in aqueous solution with ozone and hydrogen peroxide in a tube-reactor. *Water Research*, **31**, 33–40.

53 Craik, S.A., Finch, G., Lepare, J. and Chandrakanth, M.S. (2002) The effect of ozone gas-liquid contacting conditions in a static mixer on microorganism reduction. *Ozone: Science and Engineering*, **23**, 91–103.

54 Calderbank, P.H. (1967) *Mass Transfer in Mixing, Theory and Practice* (ed. J.B. Gray), Academic Press, New York, London.

55 Mueller, J.A., Boyle, W.C. and Pöpel, H.J. (2002) *Aeration: Principles and Practice*, CRC Press, p. 353.

56 Saupe, A. (1997) *Sequentielle chemisch-biologische Behandlung von Modellabwässern mit 2,4-Dinitrotoluol, 4-Nitroanilin und 2,6-Dimethylphenol unter Einsatz von Ozon* Dissertation Fortschritt-Berichte VDI, Reihe 15 Nr. 189, VDI-Verlag, Düsseldorf, Germany.

57 Hemmi, M., Krull, R. and Hempel, D.C. (1999) Sequencing batch reactor technology for the purification of concentrated dyhouse liquors. *The Canadian Journal of Chemical Engineering*, **77**, 948–954.

58 Baerns, M., Hofmann, H. and Renken, A. (1992) *Chemische reaktionstechnik, Lehrbuch der technischen Chemie Band 1*, Georg Thieme, Verlag, Stuttgart, New York.

59 Saupe, A. and Wiesmann, U. (1998) Ozonization of 2,4-dinitrotoluene and 4-nitroaniline as well as improved dissolved organic carbon removal by sequential ozonization-biodegradation. *Water Environment Research*, **70**, 145–154.

60 Tsugura, H., Watanabe, T., Shimazaki, H. and Sameshima, S. (1998) Development of a monitor to simultaneously measure dissolved ozone and organic matter in

ozonated water. *Water Science and Technology*, **37**, 285–292.
61 Wallner, F. (2007) High concentration photometric ozone measurement primer. *Ozone News*, **35** (1), 20–26.
62 Levine, S., Seiwert, J., Lohr, J., Brammer, U., Fittkau, J. (2005) Ozone concentration sensor, Patent Application Publication US 2005/0200848 A1.
63 Chung, H.-K., Bellamy, H.S. and Dasgupta, P.K. (1992) Determination of aqueous ozone for potable water treatment application by chemiluminescence flow-injection analysis. A feasibility study. *Talanta*, **39**, 593–598.
64 DIN (1993) 38408-G3-1. German Standard Methods for the Examination of Water, Waste Water and Sludge; Gaseous Components (Group G); Determination of Ozone (G3), Beuth Verlag GmbH, GmbH.
65 IOA (1987) 001/87(F). International Ozone Association Standardisation Committee; c/o Cibe; 764 Chaussée de Waterloo B-1180 Brussels, International Ozone Association, Paris, France, European-African Group.
66 IOA (1987) 002/87(F). International Ozone Association Standardisation Committee; c/o Cibe; 764 Chaussée de Waterloo, B-1180 Brussels, International Ozone Association, Paris, France, European-African Group.
67 IOA (1987) 003/87(F). International Ozone Association Standardisation Committee; c/o Cibe; 764 Chaussée de Waterloo, B-1180 Brussels, International Ozone Association, Paris, France, European-African Group.
68 Bader, H. and Hoigné, J. (1981) Determination of ozone in water by the indigo method. *Water Research*, **15**, 449–461.
69 DIN (1993) 38408-G3-3. German Standard Methods for the Examination of Water, Waste Water and Sludge; Gaseous Components (Group G); Determination of Ozone (G3), Beuth Verlag GmbH, GmbH.
70 Hoigné, J. and Bader, H. (1980) Bestimmung von ozon und chlordioxid in wasser mit der indigo methode. *Vom Wasser*, **55**, 261–279.
71 IOA (1987) 006/89(F). International Ozone Association Standardisation Committee; c/o Cibe; 764 Chaussée de Waterloo, B-1180 Brussels, International Ozone Association, Paris, France, European-African Group.
72 Lenore S. C. and Greenberg A. E. (1989) *Standard Methods for Examination of Water and Wastewater*, 17th edn, American Public Health Association, American water Works Association, Water Pollution Control Federation, Washington, pp. 4/162-4/165.
73 DIN (1993) 38408-G3-2. German Standard Methods for the Examination of Water, Waste Water and Sludge; Gaseous Components (Group G); Determination of Ozone (G3), Beuth Verlag GmbH, GmbH.
74 Gilbert, E. (1981) Photometrische Bestimmung niedriger Ozonkonzentrationen im Wasser mit Hilfe von Diäthyl-p-phenylendiamin (DPD), gas-wasser-fach (gwf) wasser/abwasser 122, pp. 410–416.
75 IOA (1989) 007/89(F). International Ozone Association Standardisation Committee; c/o Cibe; 764 Chaussée de Waterloo, B-1180 Brussels, International Ozone Association, Paris, France, European-African Group.
76 Smart, R.B., Dormond-Herrera, R. and Mancy, K.H. (1979) In situ voltammetric membrane ozone electrode. *Analytical Chemistry*, **51**, 2315–2319.
77 Stanley, J.H. and Johnson, J.D. (1979) Amperiometric membrane electrode for measurement of ozone in water. *Analytical Chemistry*, **51**, 2144–2147.

6
Mass Transfer

Ozone is a gas at standard temperature and pressure. It is normally generated onsite with electrical power from the oxygen in air or, especially when high concentrations are needed, from pure gaseous oxygen. Consequently, the ozone-containing gas has to be brought into contact with the water or waste water to be ozonated. An efficient mass transfer from gas to liquid is required. The theory and practical aspects of mass transfer are the subjects of this chapter.

First, this chapter will provide some basics on mass transfer, including theoretical background on the (two-) film theory of gas absorption and the definition of overall mass-transfer coefficients $K_L a$ (Section 6.1). Section 6.2 offers an overview of the main parameters of influence. This includes especially the influence of simultaneous chemical reactions on mass transfer, since it is an important topic for ozone mass transfer. Some methods for predicting mass-transfer coefficients are also presented in Section 6.2. These basics will be followed by a description of the common methods for the determination of ozone mass-transfer coefficients (Section 6.3) including practical advice for the performance of the appropriate experiments. Emphasis is laid on the design of the experiments so that true mass-transfer coefficients are obtained.

6.1
Theory of Mass Transfer

When material is transferred from one phase to another across a separating interface, resistance to mass transfer causes a concentration gradient to develop in each phase (Figure 6.1). This concentration difference is the driving force for mass transfer. The molar flux in each phase N is proportional to this concentration gradient and can be described with a proportionality constant k, a mass-transfer coefficient, and a linear concentration gradient. For example, the molar flux into the liquid phase can be written:

$$N = k_L (c_{Li} - c_L) \tag{6.1}$$

Ozonation of Water and Waste Water. 2nd Ed. Ch. Gottschalk, J.A. Libra, and A. Saupe
Copyright © 2010 WILEY-VCH Verlag GmbH & Co. KGaA, Weinheim
ISBN: 978-3-527-31962-6

6 Mass Transfer

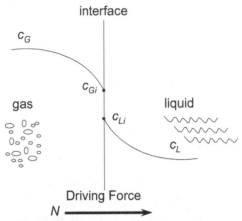

Figure 6.1 Concentration gradients at the interface between gas and liquid.

N = molar flux into one phase
k_L = liquid film mass-transfer coefficient
c_{Li} = liquid concentration in equilibrium with the gas concentration at the interface
c_L = concentration of compound in the bulk liquid

The various theories that exist to describe the mass transfer in one phase are briefly discussed in the following section. However, when material is transferred across an interface between two phases, the total resistance to mass transfer must be considered. Lewis and Whitman [1] proposed that this can be described by the sum of the resistances in each phase. They called this concept the *two-film theory*. As Treybal [2] pointed out, their two-film theory does not depend on which model is used to describe the mass transfer in each phase, therefore, the "two-resistance" theory would be a more appropriate name. It would also cause less confusion, since one of the important theories to describe mass transfer in just one phase has a similar name; the *film theory*. The following section first discusses mass transfer in one phase before returning to mass transfer between two phases, that is, the "two-resistance" theory in Section 6.2.

6.1.1
Mass Transfer in One Phase

Mass transfer in one phase depends on molecular diffusion as well as fluid motion. All current theories, that is, film, penetration, and surface renewal, postulate that a laminar film develops at the interface. They all assume that the major resistance to mass transfer occurs in this laminar film at the interface, while the resistance in the bulk fluid is negligible (Figure 6.2). Fick's first law of diffusion for steady-state diffusion forms the basis for these theories proposed to describe

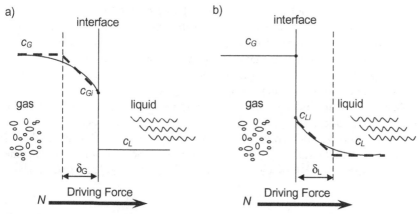

Figure 6.2 Concentration gradient at the interface in the (a) gas laminar film and (b) liquid laminar film.

the mass transfer from the phase boundary through this laminar film to the bulk liquid.

$$N = -D \frac{dc}{dx} \quad (6.2)$$

D = molecular diffusion coefficient
c = molar concentration of compound to be transferred
x = location relative to interface

It is assumed that the mass transfer within the film is at steady state, that is, there is no accumulation of mass within the film. Another general assumption for the following discussion is that the liquid phase is a closed system, for example, that it is a batch.

The theories vary in the assumptions and boundary conditions used to integrate Fick's law, but all predict the mass-transfer coefficient k within the laminar film is proportional to some power of the molecular diffusion coefficient D^n, with n varying from 0.5 to 1:

$$k \propto \frac{D^n}{\delta} \quad (6.3)$$

k = film mass-transfer coefficient
D = molecular diffusion coefficient
n = 0.5–1.0; depending on system turbulence
δ = width of film

In the film theory, the concentration gradient is assumed to be at steady state and linear over the width δ of the film (Figure 6.2) [3]. The width of the film δ is dependent on the turbulence in the system. However, the time of exposure of a fluid to mass transfer may be so short that the steady-state gradient of the film theory does not have time to develop. The penetration theory was proposed to account for a limited, but constant time that fluid elements are exposed to mass transfer at the surface [4]. The surface renewal theory brings in a modification to allow the time of exposure to vary [5].

Postulating that n is dependent on the turbulence in the system, Dobbins [6] proposed that under sufficiently turbulent conditions, n approaches 0.5 (surface renewal or penetration theory), while under laminar or less-turbulent conditions n approaches 1.0 (film theory). Thus, the selection of the value for n to predict the mass-transfer coefficient should depend on the degree of turbulence in the system.

6.1.2
Mass Transfer between Two Phases

The mass-transfer flux N out of one phase is the product of the film coefficient and the concentration gradient in the film, and is equal to the flux into the second phase (Figure 6.3a):

$$N = k_G(c_G - c_{Gi}) = k_L(c_{Li} - c_L) \tag{6.4}$$

The concentrations of the diffusing material in the two phases immediately adjacent to the interface c_{Li}, c_{Gi} are generally unequal, but are usually assumed to be related to each other by the laws of thermodynamic equilibrium. This is discussed in detail in the next section (Section 6.1.3).

However, the experimental determination of the film coefficients k_L and k_G is very difficult. When the equilibrium distribution between the two phases is linear, overall coefficients K, which are more easily determined by experiment, can be used. Overall coefficients can be defined from the standpoint of either the liquid phase or gas phase. Each coefficient is based on a calculated overall driving

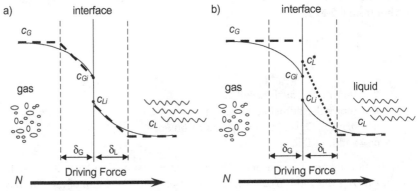

Figure 6.3 Two-film or two-resistance theory with (a) linear concentration gradients in each film and (b) approximation with an overall driving force.

force Δc, defined as the difference between the bulk concentration of one phase (c_L or c_G) and the equilibrium concentration (c_L^* or c_G^*) corresponding to the bulk concentration of the other phase (Figure 6.3b). When the controlling resistance is in the liquid phase, the overall mass-transfer coefficient K_L is generally used:

$$N = k_G(c_G - c_{Gi}) = k_L(c_{Li} - c_L) = K_L(c_L^* - c_L) \tag{6.5}$$

c_L^* = liquid concentration in equilibrium with the bulk gas concentration.

This simplifies the calculation in that the concentration gradients in the film and the resulting concentrations at the interface (c_{Li} or c_{Gi}) need not be known. To obtain the molar flow rate, the interfacial surface area A must be known:

$$n = N \cdot A = K_L A(c_L^* - c_L) \tag{6.6}$$

n = molar flow rate
A = interfacial surface area

Often, we are interested in the specific mass-transfer rate into the liquid m, expressed as the mass flow rate per unit volume, so the volumetric interfacial surface area, a, defined as transfer surface area per volume of liquid, must be introduced, as well as the conversion to mass units:

$$m = N \cdot \frac{A}{V_L} \cdot MW = K_L a(c_L^* - c_L) \tag{6.7}$$

m = specific mass-transfer rate
$a = \dfrac{A}{V_L}$ = volumetric interfacial surface area
V_L = volume of liquid
MW = molecular weight of transferred compound (e.g., MW (O_3) = 48 g mol^{-1})
c = concentration in mass units

The transfer interface produced by most of the mass-transfer apparatus considered in this book is in the form of bubbles. Measuring the surface area of swarms of irregular bubbles is difficult, although advances have been made in measurement techniques. For example, the distribution of ozone-gas bubble sizes can be rapidly and accurately measured with a 2D laser particle dynamics analyzer [7]. In general, however, the difficulty in determining the interfacial area is overcome by not measuring it separately, but rather lumping it together with the mass-transfer coefficient and measuring $K_L a$ as one parameter.

6.1.3
Equilibrium Concentration for Ozone

For dilute nonreacting solutions, Henry's Law is used to describe the linear equilibrium distribution of a compound between the gas and liquid phases. For any

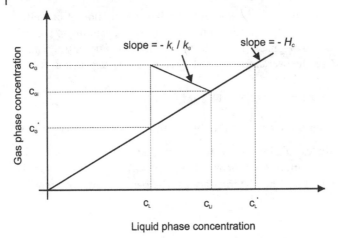

Figure 6.4 Overall and interfacial concentration differences (after Sherwood et al. [8]).

gaseous compound this means that the solubility of a gas in a liquid is directly proportional to its partial pressure in the gas above the liquid (Figure 6.4). The equilibrium distribution between the interfacial concentrations and the bulk liquid and gas phases can then be written:

$$H_C = \frac{c_G - c_{Gi}}{c_L^* - c_{Li}} = \frac{c_{Gi} - c_G^*}{c_{Li} - c_L} \tag{6.8}$$

H_C = dimensionless Henry's Law constant

if the function passes through the origin, it simplifies to

$$H_C = \frac{c_{Gi}}{c_{Li}} = \frac{c_G^*}{c_L} = \frac{c_G}{c_L^*} \tag{6.9}$$

and the equilibrium concentration c_L^* or c_G^* can be calculated from H_C and the bulk concentration.

The concept of equilibrium distribution is another area where names can cause much confusion. The equilibrium distribution of a compound between the gas and liquid phase has been expressed in various forms, that is, Bunsen coefficient ß, solubility ratio s, Henry's Law constant expressed as dimensionless H_C, or with dimensions H. These are summarized in Table 6.1 along with equations showing the relationships between them. Another more general term to describe the equilibrium concentrations between two phases is the partition coefficient, denoted by K. It is often used to describe the partitioning of a compound between two liquid phases.

Not only have various names been used for the same concept, Morris [9] found a variety of values for the equilibrium concentration of ozone in water reported in

Table 6.1 Equilibrium distribution of ozone, definitions and calculations.

Parameter	Definition	Equation (T in [K])
Bunsen coefficient β (dimensionless)	Normal volume of ozone (calculated for NTP)[a] dissolved per volume of water at T, when the partial pressure of ozone in the gas phase is one standard atmosphere	$\beta = \dfrac{V_G}{V_L}$ $\beta = \dfrac{(273.15/T)}{H_C}$ $= s \cdot (273.15/T)$
Solubility ratio s (dimensionless)	Ratio (wt/wt) of the equilibrium liquid concentration to the ozone concentration in the gas at one standard atmosphere total gas pressure (function of T)	$s = \dfrac{1}{H_C} = \dfrac{c_L^*}{c_G}$
Henry's Law constant H_C (dimensionless)	Ratio of the ozone concentration in the gas to the equilibrium concentration in the liquid, (inverse of the solubility ratio s; function of T)	$H_C = \dfrac{c_G}{c_L^*} = \dfrac{1}{s}$
Henry's Law constant H (dimensional) atm L mol^{-1} or Pa L mol^{-1}	Ratio of the ozone partial pressure in the gas and the molar dissolved-ozone concentration in equilibrium[b] (function of T)	$H = \dfrac{p(O_3)}{c_L^*}$

a NTP ≡ normal temperature and pressure: $p = 10^5$ Pa = 100 kPa = 1 bar; $T = 273.15$ K = 0 °C.
b Also can be given as atm (mol fraction)$^{-1}$ or Pa (mol fraction)$^{-1}$; 1 atm = 101 325 Pa.

the literature. Based on the data from more than nine authors (data is not shown in Figure 6.5), he suggested that a linear correlation between the solubility ratio of ozone s (which is the inverse of the dimensionless Henry's Law constant H_C), and the temperature can be used to obtain a first estimate for the solubility ratio of ozone in water:

$$\log_{10} s = -0.25 - 0.013\, T\,[°C] = 3.302 - 0.013\, T\,[K] \tag{6.10}$$

He cautions that the correlation may underestimate the true solubility, since due to ozone decomposition some authors might have failed to achieve true equilibrium or steady state in their work. In addition, the ionic strength μ as well as the type of ions are not considered in the equation, both of which have an effect on solubility. Figure 6.5 shows the correlation in comparison with a variety of experimental values measured in solutions of various ionic strengths. The correlation generates comparatively high values, especially below $T = 20$ °C.

Using the value from the correlation $s = 0.3098$ for $T = 20$ °C, the equilibrium concentration of ozone in water can be calculated from the relationship in Table 6.1. For example, we can calculate c_L^* for a gas concentration commonly achieved with modern electrical discharge ozone generators – approximately

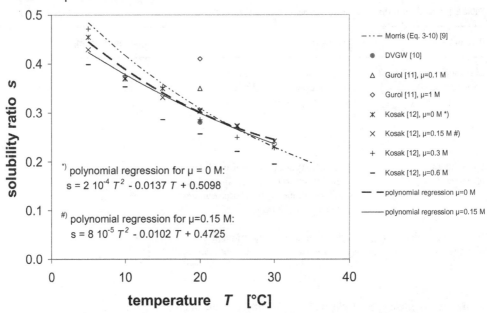

Figure 6.5 Ozone solubility ratio s as a function of the water temperature (T = 5–30 °C).

20%wt O_3 in O_2, which corresponds to c_G = 298.0 g m^{-3} at NTP (see Table 4.1): $c_L^* = 0.3098 \cdot 298.0$ g m^{-3} = 92.3 g m^{-3}, while the polynomial regression of the values of Kosak-Channing and Helz [12] for a distilled water with μ = 0 l mol^{-1} yields $c_L^* = 0.3158 \cdot 298.0$ g m^{-3} = 94.1 g m^{-3} and for an ionic strength of μ = 0.15 l mol^{-1} (Na_2SO_4), similar to the one of drinking water, gives a value of: $c_L^* = 0.3005 \cdot 298.0$ g m^{-3} = 89.5 g m^{-3}.

That is approximately ten times higher than the solubility of oxygen in water for a similar oxygen gas concentration (23.135% wt in air).

Similarly, the following correlation for the Henry's Law constant (with H in Pa (mol fraction)$^{-1}$) has no specific reference to the ionic strength of the ozone-water system. The correlation for H as a function of temperature (T measured as °C) was developed by Watanabe et al. [13].

$$H = \frac{p(O_3)}{x_L(O_3)} = 1.59 \times 10^8 + 1.63 \times 10^7 \, T \tag{6.11}$$

with:

$x_L(O_3)$ as liquid mole fraction of ozone

and yields a value of H = 4.85 × 10^5 kPa (mol fraction)$^{-1}$ at T = 20 °C being at least of the order of magnitude of the values that were reported by Sotelo et al. [14] who, in a more comprehensive study, considered the effect of ionic strength on the

Table 6.2 General equations for the dimensional Henry's Law constant H (kPa [mol fraction]$^{-1}$) (after Sotelo et al.[14]).

Type of salt in solution	Henry's Law constant H [kPa (mol fr.)$^{-1}$]^a	T (°C)	pH (−)	Ionic strength μ (mol l^{-1})^b
Na$_3$PO$_4$	$H = 1.03 \cdot 10^9 \, e^{\left(\frac{-2,118}{T}\right)} e^{(0.961\mu)} \cdot c_{OH^-}^{0.012}$	0–20	2–8.5	10^{-3}–5.0×10^{-1}
Na$_3$PO$_4$ and Na$_2$CO$_3$	$H = 4.67 \cdot 10^7 \, e^{\left(\frac{-1,364.5}{T}\right)} e^{(2.98\mu)}$	0–20	7.0	10^{-2}–10^{-1}
Na$_2$SO$_4$	$H = 1.76 \cdot 10^6 \, e^{(0.033\mu)} \cdot c_{OH^-}^{0.062}$	20	2–7	4.9×10^{-2}–4.9×10^{-1}
NaCl	$H = 4.87 \cdot 10^5 e^{(0.48 \cdot \mu)}$	20	6.0	4.0×10^{-2}–4.9×10^{-1}
NaCl and Na$_3$PO$_4$	$H = 5.82 \cdot 10^5 e^{(0.42 \cdot \mu)}$	20	7.0	5.0×10^{-2}–5.0×10^{-1}

a Temperature expressed in (K).
b For comparison: distilled water: $\mu \approx 0$; σ (conductivity) $\leq 5 \times 10^{-6}$ S m^{-1}; drinking water: $\mu \approx 0.007$; $\sigma \approx 0.05$ S m^{-1}; seawater: $\mu \approx 0.7$; $\sigma \approx 5$ S m^{-1}; values for drinking water in Berlin, Germany: $\sigma \approx 0.07$ S m$^{-1} \rightarrow \mu \approx 0.01$.

equilibrium distribution in detail. They measured Henry's Law constant H in the presence of several salts, that is, buffer solutions frequently used in ozonation experiments. Based on an ozone mass balance in a stirred-tank reactor and employing the two-film theory of gas absorption followed by an irreversible chemical reaction, they developed equations for the Henry's Law constant as a function of temperature, pH and ionic strength (Table 6.2). In this study, much care was taken to correctly analyze the ozone decomposition due to changes in the pH, assuming a second-order rate law as well as to achieve the steady-state experimental concentration at every temperature in the range considered (0 °C ≤ T ≤ 20 °C). Both values were used to calculate the true equilibrium concentration.

The effect of ionic strength was found to depend strongly on the type of salt used. Chloride as well as sulfate ions were shown to have practically no influence on the ozone solubility at constant pH and temperature. But it is important to keep in mind that salts may considerably influence the gas/liquid mass transfer of ozone due to the hindrance of bubble coalescence. Phosphate alone, and even more in a mixture with sodium bicarbonate, had a pronounced effect on solubility (Figure 6.6).

In a sodium phosphate solution of $\mu = 0.15$ mol l^{-1}, at $T = 20$ °C, pH = 7, and $p(O_3) = 2.0$ kPa, the experimental values of the Henry's Law constant averaged $H = (5.86 \pm 0.29) \times 10^5$ kPa (mol fraction)$^{-1}$ while $H = 7.14 \times 10^5$ kPa (mol fraction)$^{-1}$ is the calculated value. This shows that there are deviations between the calculated and the experimental values in the range of ±20%. Nevertheless, the equations

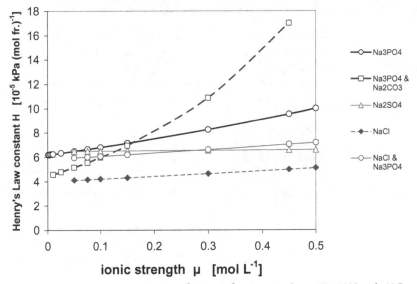

Figure 6.6 Henry's Law constant H as a function of ionic strength μ at $T = 20\,°C$ and pH 7 (exception: Na_2SO_4 at pH 6; correlations from Sotelo et al. [14], see also Table 6.2).

developed by Sotelo et al. [14] are recommended for general use in experiments with these buffer solutions or solutions of the indicated salts.

In a study of the effect of pH on the ozone volumetric mass-transfer coefficient $k_L a$ and the Henry's Law constant H in deionized water with a conductivity of 1 to $5 \times 10^{-4}\,S\,m^{-1}$ ($\mu \approx 0\,mol\,l^{-1}$), Kuosa et al. [15] found that H is only marginally dependent on pH in the range of 4 to 9 ($H = 7$ to $9 \; 10^6\,kPa\,l\,mol^{-1} = 71$ to $92\,atm\,l\,mol^{-1}$; $T = 21\,°C$) and that the Henry's Law constants were in the same range as those in the literature, for example, $H = 74\,atm\,l\,mol^{-1}$ at $T = 21\,°C$, pH = 4 compared to $H = 78\,atm\,l\,mol^{-1}$ at $T = 20\,°C$, pH = 3.4, $\mu = 0$ in the work of Kosak-Channing and Helz [12]. $k_L a$ and H were determined simultaneously from the experimental results by parameter estimation using a nonlinear optimization method.

6.1.4
Two-Film Theory

The connection between the film mass-transfer coefficients and the overall mass-transfer coefficients is provided by the two-film theory from Lewis and Whitman [1]: the total resistance to mass transfer is the sum of the resistances in each phase.

$$R_T = R_L + R_G = \frac{1}{K_L a} = \frac{1}{k_L a} + \frac{1}{H_C \cdot k_G a} \tag{6.12}$$

Table 6.3 Mass-transfer control.

When ... (ratio of individual mass-transfer coefficients and Henry's Law constant)	Then ... (overall mass-transfer coefficient defined by)	Mass transfer controlled by ...
$k_L a \ll k_G a\, H_C$	$K_L a = k_L a$	Liquid-side resistance only
$k_L a \cong k_G a\, H_C$	$K_L a \cong k_L a\, R_L / R_T$	Liquid- and gas-phase resistance

Rearranging Equation 6.12 yields an equation relating the overall mass-transfer coefficient to the individual film coefficients:

$$K_L a = \frac{k_L a}{1 + \dfrac{k_L a}{k_G a \cdot H_C}} = k_L a \frac{R_L}{R_T} \tag{6.13}$$

R_T = total resistance
R_L = liquid phase resistance
R_G = gas phase resistance.

In cases where the major resistance is in the liquid phase, the ratio $R_L/R_T \cong 1$ and the simplification can be made that the overall coefficient is equal to the liquid film coefficient. Which resistance dominates has to be determined from the ratio $k_L a / (k_G a\, H_C)$ (Table 6.3). For compounds with a low H_C, such as semivolatile organic compounds, both resistances can be important [16]. In oxygen transfer the liquid-side resistance dominates and $K_L a = k_L a$ and no information about $k_G a$ is needed. This is also true for most of the cases in ozone mass transfer, unless there is strong mass-transfer enhancement by very *fast* or *instantaneous* reactions of ozone with a dissolved compound in the film or at the gas/liquid interface [17]. Mass-transfer enhancement will be discussed in the following section.

6.2
Parameters That Influence Mass Transfer

Many parameters affect the mass transfer between two phases. They can be separated into parameters that affect the driving force for the transfer – the concentration gradient between the two phases – and those that affect the overall mass-transfer coefficient. Of course, some parameters can influence both. The parameters that affect the concentration gradient between the two phases have already been touched upon in Section 6.1.3 in the discussion on the equilibrium concentration. Here, we will look at those influencing the mass-transfer coefficient.

They can be divided into process parameters (e.g., flow rates, energy input), physical parameters (e.g., density, viscosity, surface tension) and the reactor geometry. For example, the important parameters for $K_L a$ in stirred-tank reactors are:

$$K_L a = f\left(\frac{P}{V_L}, v_G; g, v_L, \rho_L, v_G, \rho_G, D_L, \sigma_L, \text{Si}, H_C; \text{reactor geometry}\right) \quad (6.14)$$

process parameters
P = power
V_L = reactor volume
v_S = superficial gas velocity
g = gravitational constant

physical parameters
v = kinematic viscosity
ρ = density
σ = surface tension
Si = coalescence behavior of the bubbles
D = diffusion coefficient
H_c = Henry's Law constant

Which process parameters are used, depends on the reactor system. For the above example of a stirred-tank reactor, the energy inputs via the stirrer P as well as the superficial gas velocity v_G are considered in Equation 6.14. In a bubble column the energy input into the system is expressed by the superficial velocities of the gas v_S and liquid v_L, which also consider the energy input from recirculation flow rates, if any.

The reactor geometry plays an important role in the fluid mixing patterns and, therefore, on the mass-transfer coefficient. If experimental results are to be used for designing a full-scale process, it is important that the geometry of the reactor be similar at both scales. Here again, which geometry parameters are considered, depends on the reactor type. Important for all types is the ratio of reactor height to diameter H/D, while the stirrer geometry size and placement is decisive for stirred reactors. Details can be found in the literature; for example, Houcine et al. [18] deal with which geometry parameters should be considered in stirred-tank reactor designs, while Wiesmann et al. [19] describe the method of scaling up based on dimensional analysis as well as Cussler [20] who has also tabulated various correlations for mass-transfer devices.

In order to be able to predict the mass-transfer rate for the reaction system, correlations that consider the above process and physical parameters are helpful and are looked at in Section 6.2.2. Since the physical properties of the water to be treated can vary strongly over a wide range, the effects of physical properties on mass transfer are discussed in more detail in Section 6.2.3. Often, measurements must be made to develop a correlation or to adapt existing correlations to the reaction system. Therefore, methods to determine mass-transfer coefficients are presented in Section 6.3.

But before looking at such correlations we have to introduce another important influence on mass transfer that is missing in the above list: chemical reactions. A chemical reaction can change the ozone-concentration gradient that develops in the laminar film, which in turn can increase the mass-transfer rate. The increase can be quite large, commonly up to three times higher than that due to just physical-mass transfer. However, instead of adding "reactions" to our list of parameters in Equation 6.14, we use a convention that is common in most literature on ozone mass transfer and in the rest of this book: the mass-transfer coefficient $k_L a$ is defined to describe the mass-transfer rate without reaction, and the enhancement factor E is introduced to describe the increase due to the chemical reaction. This topic will be discussed in Section 6.2.1.

Another convention frequently used in this book and elsewhere is the simplification that the major mass transfer resistance lies in the liquid phase, therefore, the overall mass-transfer coefficient is equal to the film coefficient: $K_L a = k_L a$. This is also based on the assumption that the mass-transfer coefficient describes physical absorption of ozone or oxygen, since the presence of a chemical reaction can also change the gradient in the gas phase. It is further assumed that the concentration gradient can be described by the difference between the liquid concentration in equilibrium with the bulk gas phase c_L^* and the bulk liquid concentration c_L.

6.2.1
Mass Transfer with Simultaneous Chemical Reactions

A chemical reaction can change the ozone-concentration gradient that develops in the laminar film. The size of the effect depends on the relative rates of reaction and mass transfer. Fast reactions can cause a steeper gradient to develop in the liquid laminar film. This is illustrated in Figure 6.7. Gradients are shown for four

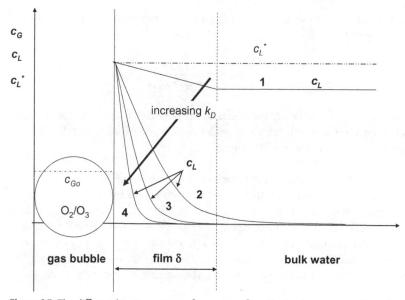

Figure 6.7 The different kinetic regimes of mass transfer with simultaneous reaction of ozone.

different regions or regimes of reaction rates, varying from slow (1), moderate (2), fast (3) to instantaneous (4). It is also possible that a fast reaction can cause a gradient to develop in the gas phase at the interface. These changes would result in different film and overall mass-transfer coefficients from those without reaction. Instead of changing the mass-transfer coefficients, though, the enhancement factor E is introduced to describe the increase in mass transfer due to the chemical reaction.

Therefore, Equation 6.5 for the mass-transfer flux of ozone transferred out of the gas phase is modified to contain a term for enhancement.

$$N = Ek_L(c_L^* - c_L) \tag{6.15}$$

where the enhancement factor E is defined as the ratio between the actual flux with chemical reaction and the flux due to mass transfer from physical absorption alone at the gas/liquid interface ($x = 0$):

$$E = \frac{-D_L(dc_L/dx)_{x=0}}{k_L(c_L^* - c_L)} \tag{6.16}$$

We can rewrite Equation 6.15 in terms of the specific mass-transfer rate m (see 6.7). And recalling the assumption that the mass transfer through the film is at steady state, we can set the amount transferred equal to the amount reacting:

$$m = N \cdot \frac{A}{V_L} \cdot MW(O_3) = E \cdot k_L a(c_L^* - c_L) = r(O_3) \tag{6.17}$$

Rearranging Equation 6.17, results in an easy to remember expression for the enhancement factor E:

$$E = \left(\frac{\text{rate with reaction}}{\text{rate for mass transfer alone}}\right) = \frac{r(O_3)}{k_L a(c_L^* - c_L)} \tag{6.18}$$

The convention of separating the effects of the physical parameters and chemical reaction on the mass transfer allows wider application of basic correlations and measurements made for model systems.

6.2.1.1 Interdependence of Mass Transfer and Chemical Reaction

We see from Equation 6.18 that the degree of enhancement depends upon the relative rates of reaction and mass transfer. These two rates can be, in fact, interdependent; the speed of one can control the speed of the other. For instance, the oxidation reaction can only proceed at a certain rate if sufficient ozone is transferred, while on the other hand a certain mass-transfer rate can only be achieved if a reaction is present to reduce the ozone concentration in the liquid phase.

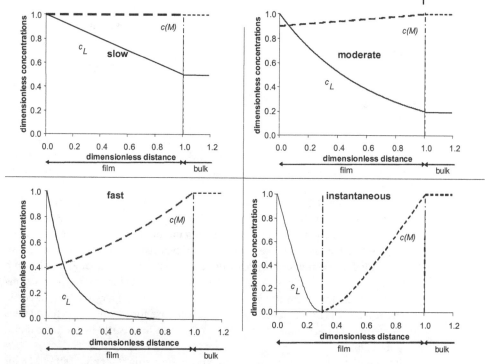

Figure 6.8 Concentration profiles of ozone and target compound M in the liquid film for the four kinetic regimes.

We can visualize this interdependence by considering the concentration profiles of ozone and a target compound M that can develop in the liquid film for various rates of mass transfer and reaction (Figure 6.8). If M is a slow reacting compound, a linear concentration gradient of the liquid-ozone concentration c_L will develop in the film unaffected by the reaction. This is illustrated in the upper left graph in Figure 6.7 ("slow"). A comparatively high concentration of ozone will be present in the bulk liquid, where the reaction takes place. The concentration of M in the film is virtually constant and the same as the concentration in the bulk liquid. No mass-transfer enhancement occurs ($E = 1$).

If M reacts more readily with ozone, a moderate, fast or instantaneous reaction can occur causing the gradients shown in the other parts of Figure 6.8. The concentration gradient of c_L in the liquid film becomes steeper as the kinetic regime goes from slow to moderate to fast or instantaneous reactions. The reaction moves further into the film as is indicated by the profiles of $c(M)$. For instantaneous reactions, there is a reaction plane within the film where both concentrations, c_L and $c(M)$, are at zero. We see then that the concentration gradients of the reactants and location of the reaction is dependent on both the mass-transfer rate and the reaction rate. At relatively high reaction rates, mass transfer is enhanced by the steeper gradient, while the reaction rate becomes dependent on the rate of ozone

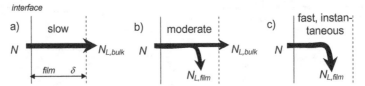

Figure 6.9 Mass-transfer fluxes between the gas and liquid phases for the various kinetic regimes.

transferred into solution. Therefore, for fast and instantaneous reactions, mass transfer often becomes rate controlling, so that the reactions develop entirely in the film.

We would expect that the mathematical description of the mass-transfer flux must change for the various regimes, since the distribution of the total flux N between the film and bulk changes depending on the reaction rates. This is shown in Figure 6.9. For slow reactions, no (measurable) reaction occurs in the film, so that the total ozone flux transferred from the gas phase N is equal to the flux into the bulk liquid $N_{L,bulk}$ (Figure 6.9a). In contrast, for fast or instantaneous reactions the total ozone flux from the gas phase reacts in the film, so $N = N_{L,film}$ and no ozone reaches the liquid bulk (Figure 6.9c). The case for the moderate regime lies in between. The reactions occur partly in the film and partly in the bulk liquid (Figure 6.9b).

Since the model of mass-transfer enhancement using the enhancement factor alone cannot be used to determine how much of the compound is reacting in the film, Benbelkacem and Debellefontaine [21] developed a mathematical model with this capability. It can be used to describe mass transfer with simultaneous reaction for all kinetic regimes, although it was developed in particular for the moderate regime to determine that part of ozone which reacts within the liquid film and that which reacts in the liquid bulk. The following description is confined to a short overview of the model. The reader is referred to the original articles [21], [22], and the comprehensive book on ozone kinetics by Beltrán [23] for more details.

The model considers the three molar fluxes that can occur when mass transfer and reactions are present (shown in Figure 6.9b). The flux from the gas phase absorbed into the liquid phase N has already been described above using Equation 6.15. In order to describe the flux out of the laminar film into the bulk liquid $N_{L,bulk}$, they introduced a new parameter, the depletion factor D:

$$N_{L,bulk} = Dk_L(c_L^* - c_L) \qquad (6.19)$$

The depletion factor is analogous to the enhancement factor E and equal to the ratio between the flux out of the laminar film into the bulk liquid and the flux due to mass transfer from physical absorption alone at the film/bulk liquid surface ($x = \delta$):

$$D = \frac{-D_L(dc_L/dx)_{x=\delta}}{k_L(c_L^* - c_L)} \qquad (6.20)$$

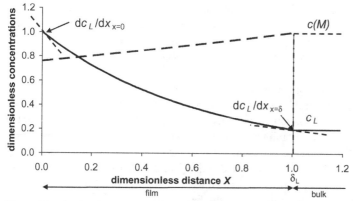

Figure 6.10 Graphical representation of the concentration gradient for the determination of E (at $x = 0$) and D (at $x = \delta$) (adapted from: Benbelkacem and Debellefontaine [21]).

Thus, the flux reacting within the film, $N_{L,film}$ can be written as the difference between the other two fluxes:

$$N_{L,film} = N - N_{L,bulk} = (E - D)k_L(c_L^* - c_L) \qquad (6.21)$$

The two factors E and D are graphically represented in Figure 6.10 for the moderate regime. The concentration gradient of ozone at the respective points in the laminar film determine the enhancement factor ($x = 0$) and the depletion factor ($x = \delta$). An important feature of this model is that it can be used for all kinetic regimes. For example, for fast or instantaneous reactions, the gradient is zero at $x = \delta$ so D is also zero; all ozone reacts within the film and no ozone enters the bulk liquid.

The discussion above on the concentration profiles for the various kinetic regimes (Figure 6.8) used changes in the hypothetical reactivity of the compound M with ozone to illustrate what occurs in each regime. A change from one kinetic regime to another can also be caused by a change in the concentration of the reactants. This situation often occurs in semibatch reactors during waste-water treatment, where the concentrations of the reactants change over the reaction time. The initially high pollutant concentration decreases as ozonation continues. Due to the decreasing concentrations of reacting compounds and thus decreasing rates of ozone consumption, the liquid bulk concentration of ozone will rise. As a consequence, the kinetic regime may vary over time, going from instantaneous and/or fast to moderate and finally to slow. How much variation occurs depends on the initial concentrations and how long the oxidation is continued. Rapp and Wiesmann observed this change by calculation of the actual enhancement factor $E\ k_L a$ as a function of the ozone dose during the oxidation of two dyes, Reactive Black 5 and Indigo, which are frequently used in textile processing [24].

For waters with relatively low initial concentrations of target compounds, for example, in drinking-water ozonation, reactions usually take place in the slow-kinetic regime, even for highly reactive compounds. In contrast, comparatively slow-reacting compounds at high concentrations can also cause enhancement. As Beltrán pointed out an interesting example for this situation is ozone decomposition at pH > 12 (i.e., $c(OH^-) > 10^{-2}$ (mol l^{-1}; $k_D = 70\,$l mol^{-1}s^{-1}) ([23], Section 7.1, p. 152). In this pH range the decomposition reaction can be a moderate or even fast reaction.

The authors of the depletion model demonstrated these changes in the regime over time by applying their model to describe the experimental results from two investigations on ozonation in the moderate regime [21, 22]. Two unsaturated dicarboxylic acids were studied at similar operating conditions, however, they differed in their reaction constants by a factor of ten, fumaric acid with $k_D = 1.5 \times 10^5\,$l mol^{-1}s^{-1} and maleic acid with $k_D = 1.2 \times 10^4\,$l mol^{-1}s^{-1}. For the more reactive compound, fumaric acid, the results of the model showed that 70% of the compound reacted in the film within the first 25 min, starting with an enhancement factor of 3 and reaching the value of one after 25 min. The less-reactive compound maleic acid reacted mainly in the bulk. Nevertheless, approximately 30% did react in the film initially with an enhancement factor of 1.27, decreasing over time.

So we see that the interdependence of the mass transfer and reaction rates can make the description of mass transfer in the presence of a chemical reaction a challenge. Fortunately, the interdependence is not only a curse. It can be exploited to expand the gain of knowledge from an experiment. For example, under the right conditions, information on the mass-transfer coefficients $k_L a$, E and D factors and/or the rate constants k_D for the direct reaction can be gathered simultaneously. This must be considered for each kinetic regime separately. In the following discussion the practical aspects to be considered in each of the four kinetic regimes when seeking such information are briefly explained.

Before moving on though, we would like to point out that this has been a very brief introduction to mass transfer with simultaneous reaction. An in-depth mathematical description of this topic is beyond the scope of this book, and the reader is referred to further literature on mass transfer in general (Cussler [20]) in addition to the work cited above (Benbelkacem and Debellefontaine [21] and coauthors [22]). Furthermore, the fundamentals of fast gas–liquid reactions were first described by Hatta [25]. Detailed descriptions were published by Charpentier [17], Levenspiel [26] and Levenspiel and Godfrey [27]. Beltrán and coworkers were the first who systematically applied these fundamentals to the ozonation of waste water [28–30] and a comprehensive work was published in 2004 by Beltrán [23].

6.2.1.2 Effect of Kinetic Regime on Determination of Mass-Transfer Coefficients

Regime 1 – (Very) Slow Reactions Developing Entirely in the Liquid Bulk The mass-transfer rate is (very) high and/or ozone reacts (very) slowly with the pollutants. A small difference between the solubility level of ozone c_L^* and the bulk liquid-ozone concentration c_L is sufficient for ozone transfer into the liquid along the linear concentration gradient. Besides the general influences of pressure and temperature, the overall rate of ozone consumption and thus, the reaction rate of the target compound M, is exclusively influenced by chemical parameters, for example, the concentrations of ozone and M, pH, and the molecular structure of the target compound. The reaction is controlled by chemical kinetics. This situation is often found in drinking-water ozonation, where (very) small concentrations of M react with ozone at considerably slow rates. Even ozone decomposition, when occurring at pH < 12, develops in this region and is often the predominating direct reaction among other possible direct reactions in drinking-water ozonation [23]. As a consequence, mass transfer is independent of chemical reaction and both $k_L a$ and k_D can be determined independently.

Regime 2 – Moderate Reactions in the Bulk and in the Film For higher reaction rates and/or lower mass-transfer rates, the ozone concentration decreases considerably inside the film. Both chemical kinetics *and* mass transfer are rate controlling. The reaction takes place inside *and* outside the film at a comparatively low rate. The ozone consumption rate within the film is lower than the ozone transfer rate due to convection and diffusion, resulting in the depletion of part of the ozone within the film as well as presence of dissolved ozone in the bulk liquid.

This regime can be used to determine the enhancement and depletion factors as well as the reaction rate constant by applying the depletion model from Benbelkacem and Debellefontaine [21]. They are found by fitting the model to the experimental data. For example, the direct reaction rate constant for fumaric acid found by this method [21] agreed well with the value in Hoigné and Bader [31]. However, no method exists to determine $k_L a$ in this regime. It must be determined under other experimental conditions.

Regime 3 – Fast Reactions in the Film Ozone is entirely consumed inside the liquid film, so that no ozone can escape to the bulk liquid, that is, $c_L = 0$. Here, the enhancement factor is defined as:

$$E = \frac{r(O_3)}{k_L a c_L^*} \tag{6.22}$$

The concentration gradient shown in Figure 6.8 for $c(M)$ decreases sharply inside the film, leveling off at a constant value in the film, yielding $dc(M)/dx = 0$ at the gas/liquid interface [17]. The mass transfer is rate controlling. Due to the reactions in the film, oxidation products are formed, which can also react in the film or diffuse out of it into the bulk liquid. The reaction rate constant k_D for the direct reaction of ozone with the pollutant M can be experimentally determined if $k_L a$ is known and a pseudo-first-order reaction can be verified, that is, the dissolved-ozone concentration is zero. This procedure in *heterogeneous* systems has been extensively exploited by Beltrán and coworkers for various fast-reacting organic compounds, such as phenolics or dyes [30, 32, 33], for which the determination of k_D is very difficult in *homogenous* systems.

Regime 4 – Instantaneous Reactions at a Reaction Plane Inside the Film For very high reaction rates and/or (very) low mass-transfer rates, ozone reacts immediately at or very close to the surface of the bubbles. The reaction is no longer dependent on ozone transfer through the liquid film k_L or the reaction constant k_D, but rather on the specific interfacial surface area a and the gas-phase concentration. Here, the resistance in the gas phase may be important. If the reaction plane is within the liquid film, both film-transfer coefficients as well as a can play a role. The enhancement factor can increase to a high value and is typically $E \gg 3$.

If a reaction develops in this kinetic regime due to appropriate choice of $c(M)$ and c_G, this situation can be used to determine the k_La-value of the system [33], or if the specific interfacial surface a in the reactor is known (e.g., from independent measurements) k_L can be determined. This method has been used in combination with fast direct reactions of organic compounds with ozone to determine both k_La and k_D ([32, 33]; see also Section 6.3).

Mass transfer in most drinking-water treatment processes generally occurs in *regime 1*, with (very) slow reaction rates. The concentration of pollutants and consequently the oxidation rates are very low. The process is completely controlled by chemical kinetics. In waste-water treatment, the concentration of pollutants is often higher by a factor of 10 or more. In this case, ozonation often takes place in *regime 3 or 4*, with considerable mass-transfer limitation. This has to be considered in the kinetics of waste water ozonation.

6.2.2
Predicting the Mass-Transfer Coefficient

Correlations based on dimensional analysis with the variables in Equation 6.14 would allow mass-transfer rates to be easily predicted, for example, in scaling-up lab results to full scale or for changes in the liquid properties. However, no correlations have been developed with this complexity.

Scale-up factors have been developed for changes in density, viscosity, surface tension and correlations using these factors have been successful for certain reactor geometries, that is, stirred-tank reactors, and well-defined systems, that is air/water [34]:

$$k_La\left(\frac{v_L}{g^2}\right)^{1/3} = f\left(\left[\frac{P}{V_L\rho_L(g^4v_L)^{1/3}}\right]; \left[\frac{v_S}{(gv_L)^{1/3}}\right]; Sc; \frac{\rho_G}{\rho_L}; \frac{v_G}{v_L}; \sigma^*; Si^*\right) \quad (6.23)$$

where:

Sc = Schmidt number, $\dfrac{v}{D}$

σ^* = dimensionless surface tension, $\dfrac{\sigma}{\rho_L(v^4g)^{1/3}}$

Si^* = coalescence number, not yet defined.

Similar factors have been developed for bubble columns, which includes the concept of gas hold-up ε_G, the fraction of the reactor liquid volume occupied by the gas dispersed in the liquid phase. The number of such factors can be reduced when comparing the mass transfer of just one compound in the same liquid/gas system, for example, for oxygen or ozone transfer in clean water/air systems the above relationship reduces to the first three terms.

Some useful correlations that can be used for a first approximation of the k_La s or ε_{GS} in laboratory-scale ozone reactors can be found in Dudley [35] for bubble columns, and in Libra [16] for STRs. Both compare various correlations found in the literature, empirical as well as those based on theoretical or dimensional analysis, to their own experiments. Dudley concluded that correlations based on theoretical considerations performed better than those developed by curve fitting.

Although constituents in water may not effect a noticeable change in density or viscosity, they may drastically change the mass-transfer coefficient due to changes in surface tension or bubble coalescence behavior for which no reliable correlations exist. Empirical correction factors have been introduced to deal with this problem [36], two of which will be discussed here in detail. In general, these empirical correction factors are used to "correct" the mass-transfer coefficient measured in clean water for temperature changes Θ, or matrix changes α.

The disadvantages of using empirical correction factors, which lump many parameters together, becomes clear when one considers that α and Θ have been found to change depending on not only the concentration and type of contaminants, but also on the hydrodynamics of the system. Clearly, a better understanding of the relationship between physical properties and k_La and the quantification of these physical properties in (waste) water is necessary, so that correlations based on dimensional analysis can be made. However, from the practical point of view, the empirical correction factors have proven their worth, when measured and used appropriately.

The empirical correction factors are developed from comparing two mass-transfer coefficients, thus, it is essential that both are measured correctly. Brown and Baillod [37] point out that the α value from the ratio of two incorrectly measured mass-transfer coefficients, apparent mass-transfer coefficients, is different from the α of true mass-transfer coefficients.

6.2.2.1 Theta Factor–Correction Factor for Temperature

Temperature affects all the physical properties relevant in mass transfer: viscosity, density, surface tension, and diffusivity. The empirical factor most often used to account for temperature changes in all these parameters is the *theta* factor, Θ,

$$k_La_{20} = k_La_T \cdot \Theta^{(20-T)} \tag{6.24}$$

$k_La_{20} = k_La$ at 20 °C
$k_La_T = k_La$ at temperature T (°C)
Θ = temperature correction factor.

In reviewing the literature on temperature corrections, Stenstrom and Gilbert [36] found values for Θ range from 1.008 to 1.047, and suggested $\Theta = 1.024$ should be used, representing an accuracy of ±5%.

6.2.2.2 Alpha Factor–Correction Factor for Water Composition

Historically, the *alpha* factor, α, was developed from oxygen mass-transfer studies in the aerated basins of municipal waste-water treatment plants. It thus denotes the ratio of the mass-transfer coefficient for oxygen measured in the waste water (WW) to that measured in tap water (TP).

$$\alpha_{O2} = \frac{k_L a_{WW}}{k_L a_{TP}} \tag{6.25}$$

More generally speaking, it denotes the ratio of the $k_L a$ measured in a "dirty" water for certain conditions, that is, energy input, reactor geometry, etc. compared to that found in "clean" water, meaning almost free of any organic or inorganic contamination, under the same conditions. In waste water the constituents in the liquid phase are highly variable depending on its source. Little to no changes in density or viscosity may be measured, but the mass-transfer coefficient for oxygen may be drastically changed, for example, by more than a factor of two to three [36, 38]. This change may be due to changes in surface tension or bubble-coalescence behavior, and may as well – interdependently – influence or be influenced by the hydrodynamic conditions of the system.

6.2.3
Influence of Water Constituents on Mass Transfer

The possible changes in the mass-transfer rate due to changes in the composition of the (waste) water must be considered when performing experiments with the intent to scale-up the results. Examples are given below to illustrate which compounds can affect the mass-transfer coefficient and to what degree. The effects can be generally divided into changes in the bubble coalescence or in surface tension. It is important to keep in mind, though, that if the compounds react with ozone, not only are the hydrodynamics of the system dependent on the type of reactor used, but also their concentration in the reactor. The changes in the mass-transfer rate in a batch reactor, where the concentrations change over time, will be very different from those found in a CFSTR, in which the reactor concentration is equal to the effluent concentration and constant over time at steady state.

6.2.3.1 Change in Bubble Coalescence
Substituted phenols as well as phenol itself are typical constituents of (bio-)refractory waste waters and can increase $\alpha(O_2) > 3$ [38]. Gurol and Nekouinaini [38] studied the influence of these compounds in oxygen transfer measurements and attributed this effect to the hindrance of bubble coalescence in bubble swarms, which increases the interfacial area a. When evaluating the effect of these phenols on the ozone mass-transfer rate, it is important to note that these substances react fast with ozone (direct reaction, $k_D = 1.3 \times 10^3 \, l\,mol^{-1}\,s^{-1}$, pH = 6–8, $T = 20\,°C$ [31]).

The same effect was also observed for *tertiary* butyl alcohol (TBA), a substance that is frequently used as a radical scavenger in kinetic experiments (see Section 7.4). Depending on its concentration (c(TBA) = 0–0.6 mM) alpha values up to $\alpha(O_2) = 2.5$ were measured [38]. Since TBA is hard to oxidize by molecular ozone in contrast to phenols ($k_D(TBA) \approx 3 \times 10^{-3} \text{l mol}^{-1} \text{s}^{-1}$, pH = 7 [39]), the effect of TBA on $k_L a$ can also be studied in mass-transfer experiments with ozone. In a study on the effect of reaction medium on ozone mass transfer in pulp-bleaching applications, it was found that TBA enhanced ozone transfer to the liquid phase. This resulted in a higher specific ozone absorption η_{O3} as well as increased efficiency of delignification relative to the quantity of ozone consumed [40].

The hydrodynamic conditions in the reactor system play a large role in coalescence. Bubbles coalesce if the contact time between the bubbles is larger than the coalescence time [41]. Since different reactors have different available contact times, the degree of coalescence inhibition produced by a certain concentration of an organic compound depends on the type of reactor and aerator used. The greater the possibility of coalescence, the greater is the effect. Due to a generally higher tendency for bubble coalescence at higher gas-flow rates, *alpha* factors in the presence of these compounds have been found to increase with increasing gas-flow rates [38].

The importance of a correct evaluation of $k_L a(O_3)$ or $k_L a(O_2)$ was confirmed in a study on the simulation of (semi-)batch ozonation of phenol [42]. It was shown that a close match between the measured and the calculated data was only obtained when $k_L a(O_2)$ was measured as a function of the residual phenol concentration. The oxygen mass-transfer coefficient was observed to change from $k_L a(O_2) = 0.049\,\text{s}^{-1}$ at $c(M) = 50\,\text{mg l}^{-1}$ phenol to $k_L a(O_2) = 0.021\,\text{s}^{-1}$ at $c(M) = 5.0\,\text{mg l}^{-1}$ phenol.

6.2.3.2 Changes in Surface Tension

Further organic compounds that can severely affect ozone or oxygen mass-transfer coefficients are surface-active agents or surfactants. Small amounts of surfactants can potentially cause a large change due to their ability to lower the surface tension at relatively low bulk concentrations by adsorbing strongly at the gas/liquid interface. Many studies of the effect of surfactants on mass transfer have found mass transfer to decrease with decreasing surface tension. However, reports of increased mass transfer have also been made. This can be explained by looking at the two ways surfactants can affect mass transfer, by changing the film mass-transfer coefficient k_L or the interfacial area a.

The decrease in the *alpha* factor to values below $\alpha = 1$ can be due to a decrease in either k_L or a or both. Two theories are commonly used to explain the reduction in k_L: the barrier effect and the hydrodynamic effect. In the barrier theory, the presence of the surfactants at the phase interface creates an additional resistance to mass transfer due to diffusion through the surfactant layer. In the hydrodynamic theory, the layer of surfactant molecules at the gas/liquid interface depresses the hydrodynamic activity [38].

An increase in the *alpha* factor to values above $\alpha = 1$ is most probably due to an increase in the interfacial area a. The reduced surface tension σ can cause smaller primary bubbles to form at the aerator or the layer of surfactant at the interface can inhibit bubble coalescence. Just what effect the surfactant will have depends on the hydrodynamic conditions in the reactor [16, 43, 44] and on the surfactant itself (e.g., anionic, nonionic) [45]. In studies with an anionic surfactant in a CFSTR, Libra [16] found that the oxygen mass-transfer coefficient $k_La(O_2)$ was reduced for moderately turbulent regions due to the dampening of interfacial turbulence by the adsorbed layer of surfactant on the bubble/water interface. The lower the surface tension (i.e., the higher the surfactant concentration), the larger the decrease in $k_La(O_2)$. As power increased, $k_La(O_2)$ recovered to the values found in tap water; the increased turbulence caused increased surface renewal at the bubble/water interface, thereby annulling the effect of the surfactant. In the highly turbulent region, $k_La(O_2)$ increased significantly. The inhibition of coalescence by the surfactant increased the interfacial area.

However, not all surfactants inhibit coalescence. Some, especially nonionic surfactants (commonly used as antifoaming agents), are well known to increase coalescence, decreasing the interfacial area a [34, 45]. Comparing the effects of two anionic and a nonionic surfactants, Wagner [45] found that although α decreased as σ decreased for each surfactant, it was not possible to develop a general correlation between α and σ.

The influence of surfactants is predominantly of importance in waste-water ozonation studies where often comparatively high concentrations of such compounds occur. However, similar effects can occur in drinking- or ground-water ozonation applications. This was shown for the decomposition of the organic phosphate pesticide diazinon (phosphorotoic acid o,o-diethyl-o-[6-methylethyl)-4-pyrimidinyl]ether) in aqueous solution by ozonation. This compound was found to considerably affect the surface tension of the aqueous solution, even at low concentrations ($c(M) \leq 10\,\mathrm{mg\,l^{-1}}$) and, thus, also influenced the oxidation mechanism [46].

6.3
Determination of Mass-Transfer Coefficients

The methods to determine mass-transfer coefficients can be grouped according to whether the concentration of the transferred compound changes over time:

- nonsteady-state methods;
- steady-state methods.

Which experimental method should be used depends on the system to be investigated: the type of reactor and how it will be operated, if ozone or oxygen will be transferred, and if clean or process water is to be used for the measurement.

In general, it is preferable to perform mass-transfer measurements using the same reaction system and mode of operation that will be used in the ozonation experiments. In that way, the factors discussed in the previous sections (e.g., α, E, D) will be similar in the mass-transfer experiments to those in the oxidation experiments. However, it may not be practical or possible to use the same system. In such cases, it is important to take into account that in batch reactors changes in the compounds and in the speed of reaction occur as the oxidation reactions proceed that may not take place in continuous-flow systems, especially at steady state. The various factors (e.g., α, E, D) can vary over time in batch systems. If these are not evaluated separately, an apparent $k_L a$ will be measured that is only valid for a limited range.

Nonsteady-state methods are generally simpler and faster to perform if $k_L a$ is to be determined in clean water without reaction. Steady-state methods are often applied for investigations of the mass-transfer coefficient under real process conditions with chemical reactions or biological activity, for example, in wastewater treatment systems. In that way operating conditions similar to the normal process conditions can be used, the reaction rate can be held constant and continuous-flow processes need not be interrupted.

The most common and appropriate methods used to determine the mass-transfer coefficient and the problems inherent in each are presented in the following sections. The methods are discussed from a practical viewpoint for the direct determination of the ozone mass-transfer coefficient. If ozone cannot be used as the transferred species, for example, because of fast reactions that cause mass-transfer enhancement, etc., then the oxygen mass-transfer coefficient $k_L a(O_2)$ can be used to indirectly determine the ozone mass-transfer coefficient $k_L a(O_3)$. This is first described below and then the special aspects of using oxygen instead of ozone in the mass-transfer experiments are referred to in the following sections whenever necessary or of general importance.

6.3.1
Choice of Direct or Indirect Determination of $k_L a(O_3)$

If possible, the determination of the mass-transfer coefficients should be conducted with the same water or waste water and gas in which the oxidation reactions will later be performed. Unfortunately, it may not be possible to use ozone in the real process water. For instance, when organic substances are present that are (easily) oxidized by molecular ozone, mass-transfer enhancement may occur during such measurements. As pointed out in Section 6.2.1 $k_L a$ cannot be determined independently in the moderate kinetic regime. However, various possibilities exist to obtain a $k_L a(O_3)$ for the system:

- $k_L a(O_3)$ can be measured with ozone in clean water, and the correction factor α can be measured using O_2/real water
- $k_L a(O_2)$ can be measured in real water and corrected for $k_L a(O_3)$ using the ratio of the diffusion coefficients described below. α can be measured by comparing $k_L a(O_2)$ in clean water.

An additional area to be evaluated when using ozone is the effect of indirect reactions on mass transfer. Highly reactive hydroxyl radicals can be produced, changing the composition and therefore the alpha factor of the system over time. Changes in pH-values may cause complications due to variations in the ozone decomposition rate. Unfortunately, even in clean water, a high pH > 12 may affect ozone mass-transfer measurements and its effect should be evaluated [23]. Determination of oxygen-transfer coefficients may be preferable in such cases too.

The ozone mass-transfer coefficient can be determined indirectly using the oxygen mass-transfer coefficient $k_L a(O_2)$ and a ratio of the diffusion coefficients:

$$k_L a_{O3} = \left(\frac{D_{O3}}{D_{O2}}\right)^n \cdot k_L a_{O2} = 0.867 \cdot k_L a_{O2} \qquad (6.26)$$

Using experimentally determined values of the diffusion coefficients for ozone $D(O_3) = 1.76 \times 10^{-9}\,m^2\,s^{-1}$ [47] and oxygen $D(O_2) = 2.025 \times 10^{-9}\,m^2\,s^{-1}$ [48] results in the above factor of 0.867 for $n = 1$. The power can vary from 0.5 to 1.0 depending on the hydrodynamic conditions in the reactor as discussed in Section 6.1.1. For ozonation in bubble columns, n is generally assumed to be 1.0. The diffusion coefficients are valid for the system "gas/(clean) water" at $T = 20\,°C$.

Other values for $D(O_3)$ and $D(O_2)$ can be found in the literature, and depending on which ones are used the factor can vary considerably. For example, using diffusion coefficients for ozone derived from theoretical considerations, for example, Wilke and Chang (1955) [49] or Scheibel (1958) cited in Reid et al. [50] (Table 4.4), with the same $D(O_2)$ as mentioned above will result in factors for $D(O_3)/D(O_2)$ in the range of 0.864 and 0.899 and would thus result in different values for $k_L a(O_3)$. This matter was also treated in the work of Rapp and Wiesmann [24].

6.3.2
General Experimental Considerations and Evaluation Methods

Experimental determination of the mass-transfer coefficient is based on the appropriate mass balance on the specific reactor used. The closer the reactor system is to ideal conditions, the simpler the mass-balance model for evaluating the experimental results can be. For example, if mixing in a stirred-tank reactor deviates too far from ideality, $k_L a$ as well as the concentrations are no longer uniform throughout the reactor. The methods and mass balances as described below cannot be used. Instead, a more complicated model that can describe the mixing zones in the reactor would be necessary [21, 23, 51].

Reactor systems, in which the following general assumptions are valid, are preferable:

- Both gas and liquid phase are ideally mixed.
- Negligible gas transfer occurs at the liquid surface.
- The liquid and gas-flow rates to the reactor are constant.
- There is no net change in gas-flow rate in and out of the reactor, so that $Q_{G\ in} = Q_{G\ out} = Q_G$.

Then, the following equations, written for absorption (of any gas) in a continuous-flow stirred-tank reactor (CFSTR) can be used.

The general mass balance for each phase at nonsteady state, considering convection, mass transfer and reaction (e.g., ozone decomposition), can be written:

liquid phase:

$$V_L \cdot \frac{dc_L}{dt} = Q_L(c_{Lo} - c_L) + k_L a \cdot V_L(c_L^* - c_L) - r_L \cdot V_L \tag{6.27}$$

gas phase:

$$V_G \cdot \frac{dc_G}{dt} = Q_G(c_{Go} - c_G) - k_L a \cdot V_L(c_L^* - c_L) - r_G \cdot V_G \tag{6.28}$$

In the case of steady state dc/dt equals zero, the material balance for each phase can be combined for a total balance on the reactor, which can be used as a check on the system, making sure that what goes in, either reacts or comes out:

$$Q_G(c_{Go} - c_G) - r_G \cdot V_G = Q_L(c_{Lo} - c_L) - r_L \cdot V_L \tag{6.29}$$

The validity of the above assumptions should be evaluated for each reaction system before they are used. For example, when large amounts of ozone are absorbed, there may be a net change in the gas-flow rate over the reactor. Usually though it may be neglected for low ozone consumption rates, especially when air is used to generate ozone. Typically, ozone gas concentrations of only 2 to 4% by weight are reached, so changes due to ozone absorption are negligible. Similarly, this usually can be neglected in oxygen-transfer experiments. For example, when nitrogen is used in $k_L a(O_2)$ measurements to produce oxygen-free water, the volume of nitrogen desorbed approximately equals the volume of oxygen absorbed. However, this change can be easily included in the calculations by using the method described by Redmon et al. [52] or the ASCE Standard [53].

The assumption of ideally mixed phases can be checked by determining the residence-time distribution in the reactor (e.g., [26, 54, 55]). Often the liquid phase is found to be close to ideally mixed. For the gas phase, though, this may not be valid in reactors with high ratios of H/D. The gas concentration may then change over the height of a reactor as the bubbles rise. This must be considered in the balance – both the change in the gas-phase concentration c_G as well as its effect on the equilibrium concentration c_L^*.

6.3.2.1 Equilibrium Concentration c_L^*

The value for the equilibrium concentration c_L^* – the liquid-ozone concentration in equilibrium with the ozone gas concentration c_G – is calculated from the gas-phase ozone concentration by applying Henry's Law (6.9). In general, even for ideally mixed gas phases, it is important to consider the operating conditions when choosing c_L^*, since it is essential for calculating the correct $k_L a$. It is evident that the value can change throughout the reactor as the gas phase becomes depleted if the gas phase is not ideally mixed; but even for ideally mixed systems, the value calculated for the effluent gas will be lower than that calculated for the influent gas. The effect of reactor temperature and pressure, especially the hydrostatic pressure, must also be considered when determining c_L^*. For instance, a deep water column will increase c_L^*.

In small laboratory-scale ozonation reactors, such as STRs, a mean value can be calculated:

$$\overline{c_L^*} = \frac{c_{Go} - c_G}{2 \cdot H_C} \tag{6.30}$$

In larger and especially higher reactors, such as bubble columns, a constant value of c_L^* can normally not be assumed. A differential mass balance over the height of the column is necessary, where c_L^* is calculated as a function of the changes in the gas concentration over the column height. An approximation often used in this case for oxygen transfer (Zlokarnik [34]) is the so-called "logarithmic concentration difference", which can be used for the concentration gradient $(c_L^* - c_L)$ and defined as:

$$\overline{(c_L^* - c_L)} = \frac{c_{Lo}^* - c_{Le}^*}{\ln \dfrac{c_{Lo}^* - c_L}{c_{Le}^* - c_L}} \tag{6.31}$$

c_{Lo}^* = equilibrium concentration corresponding to the influent gas
c_{Le}^* = equilibrium concentration corresponding to the effluent gas and
c_L = dissolved concentration in the reactor.

It is also possible to determine the equilibrium concentration or the Henry's Law constant directly from mass-transfer experiments. The choice of procedure depends on which compound is being transferred and which method is being used to measure $k_L a$. For example, when using the nonsteady-state method (6.36 below), c_L^* may be found by fitting its value to the data, either by changing the value used for c_L^* by hand until a straight line is achieved in the logarithmic graph or with the program used for nonlinear regression. In both cases, the data should be truncated below 20% of c_L^*, since the initial concentrations are often erroneous due to probe lag, etc. (see Section 6.3.6).

It is important to note that for ozone, the steady-state ozone concentration in the liquid $c_{L\infty}$, even in clean water, is usually not equal to c_L^* due to ozone decomposition. c_L^* can be determined experimentally for ozone in the presence of a reaction from steady-state experiments if the $k_L a$ and decomposition rate constant k_d are known, using the steady state version of Equation 6.27 for semibatch operation:

$$0 = k_L a (c_L^* - c_{L\infty}) - k_d c_{L\infty} \tag{6.32}$$

Rearranging Equation 6.32, we can solve for c_L^*:

$$c_L^* = \frac{(k_L a + k_d)}{k_L a} c_{L\infty} \tag{6.33}$$

Or the equation can be rearranged to determine the Henry's Law constant after dividing both sides by the gas-phase concentration:

$$H_c = \frac{\overline{c_G}}{c_L^*} = \frac{\overline{c_G}}{c_{L\infty}} \frac{k_L a}{(k_L a + k_d)} \tag{6.34}$$

using the logarithmic average of the gas concentration to account for changes in the gas-phase concentration between influent and effluent:

$$\overline{c_G} = \frac{c_{Go} - c_{Ge}}{\ln \dfrac{c_{Go}}{c_{Ge}}} \tag{6.35}$$

6.3.3
Nonsteady-State Methods Without Mass-Transfer Enhancement

The nonsteady-state methods described in this section are all based on no or negligible reactions taking place in the system. If reactions are present, the treatment of the mass balances becomes more complicated since

- the reaction rate is most often a function of the reactant concentrations, which then also changes over time;
- or mass-transfer enhancement can occur.

Other methods such as those described in Section 6.3.5 to evaluate the data with mass-transfer enhancement are then necessary.

6.3.3.1 Batch Model
The most common approach in the laboratory is to use a batch setup (with respect to the liquid) where ozone-free (clean) water is gassed with the ozone/air or ozone/oxygen mixture. The change in the liquid-ozone concentration over time is measured with an ozone probe or an online photometer. The mass balance reduces to:

$$\frac{dc_L}{dt} = k_L a \cdot (c_L^* - c_L) - r_L \tag{6.36}$$

It is important to note that even in clean water ozone decomposition (which can be considered a special ozone "reaction") cannot be prevented at pH ≥ 4. In order to achieve a situation where no reactions occur ($r_L = 0$), the ozone mass-transfer experiments are mostly conducted at pH = 2. The use of higher pH values in ozone mass-transfer experiments require that the ozone decomposition rate (r_L) has to be known or experimentally assessed [14] (see also Section 7.4). Fortunately, changes in $k_L a$ due to mass-transfer enhancement from ozone decomposition can be neglected at pH < 12 [23].

6.3.3.2 Experimental Procedure

In batch nonsteady-state experiments the reactor is first filled with the water, in which the gas is to be absorbed. The water is allowed to come to the desired operating temperature. Then, the ozone is removed from the water by vacuum degassing or by gassing with N_2 to at least 0.10 mg l^{-1} O_3. Through the use of a three-way valve, the gas is switched to the ozone/air or ozone/oxygen mixture and the change in the dissolved-ozone concentration over time is recorded, preferably with a computer. The ozone generator should be started and running before the gas is switched, so that c_{Go} is constant during the experiments.

The mass-transfer coefficient can be found using the mass balance given by Equation 6.27, simplified for $Q_L = 0$, and, if ozone decomposition is negligible, $r_L = 0$. The integrated mass balance is then:

$$\ln\frac{c_L^* - c_L}{c_L^* - c_{Lo}} = k_L a \cdot t \quad \text{or} \quad \frac{c_L^* - c_L}{c_L^* - c_{Lo}} = e^{k_L a \cdot t} \tag{6.37}$$

The data can be evaluated using any commonly available nonlinear regression program or with a linear regression, in which $k_L a$ is the slope from the plot of the natural log of the ratio of the concentration differences versus time. Linearity of the logarithmic values over one decade is required for the validity of the measurement. Of course, the assumptions inherent in the model must apply to the experimental system, especially in respect to completely mixed gas as well as liquid phases and reactions are negligible. Two common problems are discussed below. Other common pitfalls and problems are discussed later in Section 6.3.6.

The initial data with the fastest change in c_L may not be correct because of possible probe-lag effects. The degree of deviation depends on the response time of the probe and the magnitude of $k_L a$ [56]. The response time is often reported as the time required to reach 90 per cent of full response ($t90$); for example, commonly used laboratory oxygen electrodes reach $t90$ in approximately 10 s. Similar response times have been reported for ozone probes. Probe lag can be taken into consideration in the data-evaluation step with both linear and nonlinear regression by choosing the appropriate start values for c_{Lo} and t_o. For common oxygen probes, truncation of the initial data below 20% of c_L^* is recommended [57].

Another method to account for probe lag is to include the response characteristics of the probe in the mathematical model [51, 59]. In a study on the hydrodynamics and ozone mass transfer in a tall bubble column (h = 4.8 m) which employed an axial dispersion model, it was confirmed that accounting for both the ozone electrodes response characteristics as well as the hydrostatic effect on solubility improved the accuracy in the determination of the values of $k_L a$ in the range of 5×10^{-2} to 2.5×10^{-2} s^{-1} [59].

6.3.3.3 Continuous-Flow Model

In systems with continuous flow, the nonsteady-state approach is a little more complicated. A perturbation in the dissolved-ozone concentration (c_L) is made and the change in c_L over time is measured as the system returns to the steady-state concentration $c_{L\infty}$. In principle, to remove ozone a fast chemical reaction with some substance leaving no oxidation products that can further consume ozone or influence the mass transfer has to be used. For oxygen mass transfer $c_L(O_2)$ can be removed by a chemical reaction with Na_2SO_3 catalyzed with cobalt [16]. Although ozone reacts fast with sulfite (k_D = 0.9 × $10^9 M^{-1} s^{-1}$; calculated for pH = 8; $pK_a(HSO_3^-/SO_3^{2-})$ = 7.2, [39]) no reports on the application of this method for the assessment of $k_La(O_3)$ have been found in the literature. The data are evaluated using the nonsteady state liquid-phase mass balance, Equation 6.27. The integrated form of the equation without reaction is:

$$\ln\left[1 - \frac{c_L - c_{Lo}}{c_{L\infty} - c_{Lo}}\right] = -K_2 \cdot t \tag{6.38}$$

$$K_2 = \frac{Q_L}{V_L} + k_L a \tag{6.39}$$

c_{Lo} = ozone concentration at $t = 0$
$c_{L\infty}$ = ozone concentration at $t = t_\infty$

6.3.3.4 Experimental Procedure

The continuous-flow nonsteady-state measurements can be made after the reactor has reached steady state, which usually takes at least 3 to 5 times the hydraulic retention time under constant conditions. Then, an appropriate amount of the compound to be oxidized (e.g., Na_2SO_3) is injected into the reactor. An immediate decrease in the liquid-ozone concentration to $c_L \approx 0\,mg\,l^{-1}$ indicates that the concentration is correct. Enough sulfite has to be added to keep $c_L = 0$ for at least one minute so that it is uniformly dispersed throughout the whole reactor. Thus, a little more than one mole of sodium sulfite per mole ozone dissolved is necessary. The subsequent increase in c_L is recorded by a computer or a strip chart. The data are evaluated according to Equation 6.39, the slope from the linear regression is $-(Q_L/V_L + k_L a)$.

6.3.4
Steady-State Methods Without Mass-Transfer Enhancement

The steady-state method is often used in continuous-flow operation with reaction, which is often the case in full-scale applications. In laboratory-scale investigations, the steady-state method can be used with a semibatch setup (gas phase continuous) with reaction or a continuous-flow setup (both gas and liquid phases continuous) with or without reaction.

The advantages of the steady-state method are

- little to no changes in the hydrodynamics of an operating system are necessary;

- since concentrations do not change, no dynamic effects have to be considered and concentration measurement is simplified;
- reaction rates for slow reactions in the liquid phase do not have to be known if using the gas-phase balance.

Since we know the mass of ozone transferred has to have reacted or left the system, it is relatively easy to determine the reaction rate for slow reactions, which are controlled by chemical kinetics with this method. For kinetic regimes with mass-transfer enhancement, the two rates, mass transfer and reaction rate are interdependent. Whether $k_L a$ or k_D can be determined in such a system and how depends on the regime. Possible methods are similar to those described below in Section 6.3.5 (see also Levenspiel and Godfrey [27]).

The following discussion assumes any reaction present is in the slow-kinetic regime. It also combines the semibatch and continuous-flow models, since there is so little difference between their application.

6.3.4.1 Semibatch and Continuous-Flow Models

For the calculation of $k_L a$ two methods based on the liquid- and gas-phase mass balances (6.27 and 6.28) are possible. For the case that no reactions take place in the gas phase, and at steady state one obtains:

for the gas phase:

$$k_L a = \frac{Q_G}{V_L} \frac{(c_{Go} - c_G)}{(c_L^* - c_L)} \tag{6.40}$$

for the liquid phase (semibatch):

$$k_L a = \frac{r_L}{(c_L^* - c_L)} \tag{6.41}$$

for the liquid phase (continuous-flow):

$$k_L a = \frac{Q_L}{V_L} \frac{(c_L - c_{Lo}) - r_L V_L}{(c_L^* - c_L)} \tag{6.42}$$

The error associated with the steady-state method becomes large as the liquid-phase concentration approaches the saturation concentration. Care must be taken to avoid this region.

6.3.4.2 Experimental Procedure

The semibatch setup uses a continuous reaction to remove the gas absorbed in the liquid. In oxygen transfer measurements, sulfite (SO_3^{2-}) or hydrazine (N_2H_4) have been used to remove the oxygen transferred (Charpentier, pp. 42–49, [17]) as well as biological reactions [53]. For example, the addition rate of the reactant (SO_3^{2-}

or N_2H_4) is adjusted until the system comes to a steady-state dissolved oxygen concentration of about $2\,\mathrm{mg\,l^{-1}}$. Then, the sulfite addition rate equals the transfer rate. Mass-transfer enhancement must be avoided, which is difficult with hydrazine, so that it is rarely used.

The continuous-flow setup can either use a reaction, similar to semibatch, to remove the gas transferred or operate with two reactors in series. For the two-reactor system, the ozone or oxygen is removed from the liquid in the first reactor by stripping or vacuum degassing. The liquid then flows into the absorber. After having passed through the absorber, the liquid can be recycled or discharged.

The average time required for the reactors to reach steady state is approximately three to five times the hydraulic retention time.

More information on the application of this method for measuring oxygen transfer with biological reactions in full-scale municipal waste-water treatment plants can be found in ASCE [53] and Redmon et al. [52].

The mass-transfer coefficient for ozone can be calculated from both the liquid- and gas-phase mass balances as described by Equations 6.40, 6.41 or 6.42. Difficulties arise with the liquid-phase mass balance if a reaction is present. The reaction rate under the operating conditions investigated must be used, considering especially the c_L prevalent in the system. Since this can be very difficult to assess, and use of inaccurate reaction rates leads to inaccurate k_La, application of both the gas-phase and liquid-phase mass balances is an elegant way to avoid this problem and described below.

6.3.4.3 Simultaneous Determination of k_La and r_L

We can see from the above equations that by combining the information from the steady-state gas and liquid mass balances, we can determine both k_La and the reaction rate r_L. k_La is first determined from the gas balance Equation 6.40 and then used in the liquid balance (6.41) to determine the overall-reaction rate.

For example (semibatch):

$$r_L = k_L a(c_L^* - c_L) = \frac{Q_G(c_{Go} - c_G)}{V_L} \tag{6.43}$$

This combination of the gas and liquid balances is often used to determine the mass-transfer coefficient and the oxygen-uptake rate simultaneously in biological waste-water treatment.

Whether a rate constant can be determined depends on which reactions are taking place. If the experiment is made in clean water and the only reaction is ozone decomposition, the first-order decomposition rate constant for ozone k_d can be calculated by combining Equation 6.43 with the first-order rate law:

$$r_L = k_d \cdot c_L \tag{6.44}$$

So that:

$$k_d = \frac{Q_G}{V_L} \frac{(c_{G_0} - c_G)}{c_L} \tag{6.45}$$

6.3.5
Methods with Mass-Transfer Enhancement

In a *heterogeneous* gas–liquid reactor system, that is, a two-phase system where gas absorption precedes a liquid-phase reaction, the mass-transfer rate has to at least equal the reaction rate. This principle can be used to determine mass-transfer coefficients and/or reaction rate constants for certain kinetic regimes (see Section 6.2.1). To determine the mass-transfer coefficient, the kinetic regime must be instantaneous, and the place of the reaction must be in the film [17, 33]. To determine the ozone direct reaction rate constant k_D, the kinetic regime must be fast and $k_L a$ must be known.

An instantaneous reaction is the fastest reaction possible and no gas is transferred into the liquid bulk. This can be utilized to determine $k_L a$, for example with the reaction of ozone with certain fast-reacting organic compounds. The reaction either develops in a reaction plane located

- directly at the gas/liquid interface, if $c(M)_o \geq c_L^*$ (as molar concentrations) holds; or
- in the liquid film, if $c(M)_o \ll c_L^*$ holds.

The situation is characterized by the fact that both reactants are entirely consumed, so that $c_L = c(M) = 0$ holds in the plane. Only in the latter case can $k_L a$ be determined. In the former case there is no transport of ozone into the liquid film, so that the mass-transfer rate is only determined by $k_G a$ [17]. The reaction rate depends on the mass-transfer rate of ozone and pollutant to the reaction plane in the liquid film, but not on the reaction rate constant. Whether the reaction develops instantaneously in the liquid film depends on the experimental conditions, especially on the values of the influent ozone gas concentration c_{G_0} or the applied ozone partial pressure $p(O_3)$ and the initial concentration of M $c(M)_o$. For example, the reaction tends toward instantaneous for low $p(O_3)$ and high $c(M)_o$.

Ozonation experiments to determine $k_L a$ from such an instantaneous reaction should preferably be conducted in a so-called agitated cell in which both phases are perfectly mixed and the transfer area is determined by the geometry of the constructed interface between the gas and the liquid in the system (see Figure 5.5 [27]). The method has also been used in stirred-tank reactors, but these reactors have two drawbacks [60, 62, 63]:

- the specific interfacial area is unknown; and
- the ozone partial pressure at the reactor outlet changes with time.

If an agitated cell is used, both problems are overcome because the transfer area is known and $p(O_3)$ is practically constant due to continuous dosing and complete mixing of the gas phase.

For example, Beltrán and Alvarez [33] successfully applied a semibatch agitated cell for the determination of k_L, $k_L a$, and the rate constants of synthetic dyes, which react very fast with molecular ozone (direct reaction, $k_D = 5 \times 10^5$ to $1 \times 10^8 \, l\,mol^{-1}\,s^{-1}$). In conventional stirred-tank reactors operated in the semibatch mode the mass-transfer coefficient for ozone $k_L a(O_3)$ was determined from an instantaneous reaction of ozone and 4-nitrophenol [28] as well as ozone and resorchinol (1,3-*di*hydroxybenzene) or phloroglucinol (1,3,5-*tri*hydroxy-benzene) [60]. In the latter study, comparisons were made with $k_L a$-values from independent measurements using the nonsteady-state method for oxygen absorption in high-purity water or the steady-state method for oxygen absorption in cuprous chloride. The three different methods showed a high similarity and low standard deviations of the determined $k_L a$–values ($k_L a(O_3) = 0.0018 \pm 8.8 \times 10^{-5}\,s^{-1}$) were calculated.

6.3.5.1 Experimental Procedure

The determination of $k_L a$ from an instantaneous reaction is rather complex and the experimental procedure complicated, requiring an extensive knowledge of the theoretical background. Since it is not within the scope of this book to go into the necessary details, the basic experimental procedure is summarized in Figure 6.11, and only a few remarks shall be made here. For complete information the reader is referred to the original literature in which the nonsteady-state method has been applied in ozonation experiments (e.g., [28, 33, 60]) and/or to the literature explaining the basics behind this method (e.g., [17, 23, 27]).

First, $k_L a$ is determined from an instantaneous reaction. Then, k_D from a fast reaction using the known $k_L a$. In order to avoid the influence of the gas-side resistance the experiments have to be conducted with an initial molar concentration of pollutant M far below the solubility level of ozone (related to the input gas concentration respectively the ozone partial pressure). In the study of Beltrán and Alvarez [33] an instantaneous reaction of ozone and phenol developed with $c(M)_o < 0.5\,mM$ and $p(O_3) > 500\,Pa \approx 6.1\,mmol\,l^{-1}$ gas ($T = 20\,°C$). All parameters were held constant while testing pairs of $c(M)_o$ and $p(O_3)$, for which the concentration change over time was measured. The instantaneous kinetic regime must be verified for each run [33].

6.3.6
Problems Inherent to the Determination of Mass-Transfer Coefficients

In the previous sections, the most common methods to determine the mass-transfer coefficient have been described. Each method is accompanied by assumptions and experimental errors that must be considered for the sensitivity and error analyses. In the following section, an overview of the problems inherent in each method is given so that the experimenter can evaluate potential problems in their results.

6.3.6.1 Nonsteady-State Method

A variety of problems encountered with the measurement of oxygen mass-transfer coefficients by using the nonsteady-state method are well known and well understood. Libra [16] presents a comprehensive discussion of these problems for

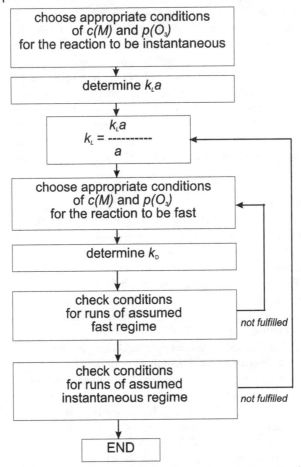

Figure 6.11 Experimental procedure for the determination of $k_L a$ from instantaneous reactions and k_D from fast reactions of organic compounds with ozone (after: Beltrán and Alvarez, [33]).

oxygen mass-transfer measurements. The most important problems associated with nonsteady-state methods are summarized in Table 6.4.

The magnitude of error that can be caused by neglecting these problems depends on the system used. Most of these problems are not serious in the case of small $k_L a$s, for example, for the range normally found in waste-water treatment plants (0.001-0.005 s^{-1}). For example, Brown and Baillod [37] found that the error caused by neglecting gas-phase depletion was less than 10% for $k_L a < 0.0025$ s^{-1}. However, the problem grows larger as the mass-transfer coefficient increases.

In order to overcome these problems especially for processes with large $k_L a(O_2)$ values, the following recommendations are made [51, 57]:

- use pure oxygen combined with vacuum degassing of the liquid; or
- use an appropriate model for the gas phase oxygen concentration;

Table 6.4 Problems inherent to the determination of mass-transfer coefficients with the nonsteady-state method.

Problematic step of experimental procedure	Conflict with model assumptions	Explanation	Remark
Degassing the water by purging with nitrogen gas (N_2)	c_G and c_L^* = constant over time	Dilution of the gas phase by N_2 just after the beginning of reaeration	Underestimation of $k_L a$'s of the order of 40% [63] or even 50% was observed [64], depending on the magnitude of $k_L a$
Reaeration, especially with low Q_G	Ideally mixed gas phase; c_L^* = spatially constant	Local concentration gradients (of c_G and c_L^*), probable surface aeration	Can cause major differences if the gas space above the liquid is large and Q_G small
Addition of sodium sulfite and cobalt catalyst	Low to moderate ionic strength and coalescing system	High ionic strength and noncoalescing system → change in interfacial area	This method generally yields incomparable values to low ionic strength waters especially for large $k_L a$s [16]
Oxygen probe	Immediate response to changes in c_L	Oxygen probe lag τ	Various models have been proposed to describe the oxygen probe lag and can be used for ozone probes [51, 58]

- compensate for the probe time constant with either a model or by truncating the initial data in the reaeration test below 20% of c_L^*.

The first recommendation, to use a one-component gas, is not possible in ozonation systems, since ozone generators can only achieve approximately 20% wt in an ozone/oxygen gas mixture, equivalent to 14.3% vol. Therefore, when measuring $k_L a$s in highly efficient transfer devices, for example, $k_L a > 0.01\,\text{s}^{-1}$, which can be found in industrial applications for ozonation, it is important to use an appropriate model for the gas-phase concentration. In addition, the model should include probe dynamics.

6.3.6.2 Steady-State Method

The steady-state method avoids many of the sources of experimental error associated with the nonsteady-state method; however, it has its own distinct sources of error. Because data analysis of the nonsteady-state method uses the form of the response curve (with nonlinear regression) or the slope (with linear regression), the absolute value of the concentration is not important. A systematic error in measuring O_2 or O_3 concentration does not affect the $k_L a$ value. In the steady-state method, error in the values of Q_G, Q_L, V_L, c_G, results in an error in $k_L a$ of the same magnitude, that is, 1% → 1%. However, error in the liquid concentration and equilibrium concentration is magnified, especially in the high $k_L a$ range since they are in the denominator (see 6.40–6.42). The difference between c_L^* and c_L becomes small as the mass-transfer rate increases, unless the liquid flow rate is increased. As the difference approaches the magnitude of the error in the concentrations, the error in $k_L a$ explodes. Error considerations and physical constraints on the maximum flow rate limit the range for the mass-transfer rate that can be determined with this method. Just what this range is must be determined from a sensitivity analysis. The results of such calculations can be found in [16].

References

1 Lewis, W.K. and Whitman, W.G. (1924) Principles of gas absorption. *Industrial and Engineering Chemistry*, **16**, 1215–1219.

2 Treybal, R.E. (1968) *Mass Transfer Operations*, 2nd edn, McGraw-Hill, New York.

3 Nernst, W. (1904) Theorie der Reaktionsgeschwindigkeit in heterogenen systemen. *Zeitschrift für Physikalische Chemie*, **47**, 52–55.

4 Higbie, R. (1935) The rate of absorption of a pure gas into a still liquid during short periodics of exposure. *Transactions of the American Institute of Chemical Engineers*, **31**, 365–388.

5 Danckwerts, P.V. (1970) *Gas Liquid Reactions*, McGraw-Hill, New York.

6 Dobbins, W.E. (1964) Mechanism of gas absorption by turbulent liquids. Proceedings of the International Conference Water Pollution Research, London, Pergamon Press, London., pp. 61–76.

7 Zhou, H. and Smith, D.W. (2000) Ozone mass transfer in water and wastewater treatment: experimental observations using a 2D laser particle dynamics analyzer. *Water Research*, **34**, 909–921.

8 Sherwood, T.K., Pigford, R.L. and Wilke, C.R. (1975) *Mass Transfer*, McGraw-Hill, New York.

9 Morris, J.C. (1988) The aqueous solubility of ozone–a review. *Ozone News*, **1**, 14–16.

10 DVGW (1987) *Ozon in der Wasseraufbereitung–Begriffe, Reaktionen, Anwendungsmöglichkeiten, Wasserversorgung, Wasseraufbereitung–Technische Mitteilung Merkblatt W 225*, DVGW-Regelwerk, Eschbom.

11 Gurol, M.D. and Singer, P.C. (1982) Kinetics of ozone decomposition: a dynamic approach. *Environmental Science and Technology*, **16**, 377–383.

12 Kosak-Channing, L.F. and Helz, G.R. (1983) Solubility of ozone in aqueous solutions of 0–0.6 M ionic strength at 5–30°C. *Environmental Science and Technology*, **17**, 145–149.

13 Watanabe, K., Kinugasa, I. and Higaki, K. (1991) Ozone absorption in bubble column. *Memoir of Niihama National College of Technology*, **27**, 48–52.

14 Sotelo, J.L., Beltrán, F.J., Benitez, F.J. and Beltrán-Heredia, J. (1989) Henry's law constant for the ozone-water system. *Water Research*, **23**, 1239–1246.

15 Kuosa, M., Laari, A. and Kallas, J. (2004) Determination of Henry's coefficient and mass transfer for ozone in a bubble column at different ph values of water. *Ozone: Science and Engineering*, **26**, 277–286.

16 Libra, J.A. (1993) *Stripping of Organic Compounds in an Aerated Stirred Tank Reactor Forstschritt-Berichte*, VDI Reihe 15: Umwelttechnik Nr. 102, VDI-Verlag, Düsseldorf, Germany.

17 Charpentier, J.C. (1981) *Mass-transfer Rates in Gas-liquid Absorbers and Reactors, Advances in Chemical Engineering*, Vol. 11, Academic Press, New York, pp. 3–133.

18 Houcine, I., Plasari, E. and David, R. (2000) Effects of the stirred tank's design on power consumption and mixing time in liquid phase. *Chemical Engineering Technology*, **23**, 605–613.

19 Wiesmann, U., Choi, I.S. and Dombrowski, E.-M. (2007) *Fundamentals of Biological Wastewater Treatment*, Wiley-VCH Verlag GmbH, Weinheim, ISBN: 978-3-527-31219-1.

20 Cussler, E.L. (1997) *Diffusion: Mass Transfer in Fluid Systems*, 2nd edn, Cambridge University Press, New York.

21 Benbelkacem, H. and Debellefontaine, H. (2003) Modeling of a gas-liquid reactor in batch conditions. Study of the intermediate regime when part of the reaction occurs within the film and part within the bulk. *Chemical Engineering and Processing*, **42**, 723–732.

22 Benbelkacem, H., Cano, H., Mathe, S. and Debellefontaine, H. (2003) Maleic acid ozonation: reactor modeling and rate constants determination. *Ozone Science and Engineering*, **25**, 13–34.

23 Beltrán, F.J. (2004) *Ozone Reaction Kinetics for Water and Wastewater Systems*, CRC Press LLC, Boca Raton, FL, ISBN 1-56670-629-7.

24 Rapp, T. and Wiesmann, U. (2007) Ozonation of C.I. reactive black 5 and indigo. *Ozone: Science and Engineering*, **29**, 493–502.

25 Hatta, S. (1932) Absorption velocity of gases by liquids. II. Theoretical considerations of gas absorption due to chemical reactions. *Technology Reports of the Tōhoku Imperial University*, **10**, 119–135.

26 Levenspiel, O. (1972) *Chemical Reaction Engineering*, John Wiley & Sons, Inc., New York.

27 Levenspiel, O. and Godfrey, J.H. (1974) A gradientless contactor for experimental study of interphase mass transfer with/without reaction. *Chemical Engineering Science*, **29**, 1123–1130.

28 Beltrán, F.J., Gomez, V. and Duran, A. (1992) Degradation of 4-nitrophenol by ozonation in water. *Water Research*, **26**, 9–17.

29 Beltrán, F.J., Encinar, J.M. and Garcia-Araya, J.F. (1992) Absorption kinetics of ozone in aqueous o-cresol solutions. *Canadian Journal of Chemical Engineering*, **70**, 141–147.

30 Beltrán, F.J., Encinar, J.M., Garcia-Araya, J.F. and Alonso, M.A. (1992) Kinetic study of the ozonation of some industrial wastewaters. *Ozone Science and Engineering*, **14**, 303–327.

31 Hoigné, J. and Bader, J. (1983) Rate constants of the reaction of ozone with organic and inorganic compounds in water–II. Dissociating organic compounds. *Water Research*, **17**, 185–194.

32 Beltrán, F.J., Encinar, J.M. and Garcia-Araya, J.F. (1993) Oxidation by ozone and chlorine dioxide of two distillery wastewater contaminants: gallic acid and epicatechin. *Water Research*, **27**, 1023–1032.

33 Beltrán, F.J. and Alvarez, P. (1996) Rate constant determination of ozone-organic fast reactions in water using an agitated cell. *Journal Environmental Science and Health A*, **31**, 1159–1178.

34 Zlokarnik, M. (1978) Eignung und Leistungsfähigkeit von Volumenbelüftern für biologische Abwasserreinigungsanlagen. *Korrespondenz Abwasser*, **27**, 194–209.

35 Dudley, J. (1995) Mass transfer in bubble columns: a comparison of correlations. *Water Research*, **29**, 1129–1138.

36 Stenstrom, M.K. and Gilbert, R.G. (1981) Review paper: effects of alpha, beta and theta factor upon the design specification and operation of aeration systems. *Water Research*, **15**, 643–654.

37 Brown, L.C. and Baillod, C.R. (1982) Modelling and interpreting oxygen transfer data. *Journal Environmental Engineering Divison, ASCE*, **108** (4), 607–628.

38 Gurol, M.D. and Nekouinaini, S. (1985) Effect of organic substances on mass transfer in bubble aeration. *Journal Water Pollution Control Federation*, **57**, 235–240.

39 Hoigné, J., Bader, J., Haag, W.R. and Staehelin, J. (1985) Rate constants of reactions of ozone with organic and inorganic compounds in water–III inorganic compounds and radicals. *Water Research*, **19**, 173–183.

40 Cogo, E., Albet, J., Malmary, G., Coste, C. and Molinier, J. (1999) Effect of reaction medium on ozone mass transfer and applications to pulp bleaching. *Chemical Engineering Journal*, **73**, 23–28.

41 Drogaris, G. and Weiland, P. (1983) Coalescence behavior of gas bubbles in aqueous solutions of n-alcohols and fatty acids. *Chemical Engineering Science*, **38**, 1501–1506.

42 Gurol, M.D. and Singer, P.C. (1983) Dynamics of the ozonation of phenol–II mathematical model. *Water Research*, **16**, 1173–1181.

43 Mancy, K.H. and Okun, D.A. (1960) Effects of surface-active agents on bubble aeration. *Journal Water Pollution Control Federation*, **32**, 351–364.

44 Eckenfelder, W.W. and Ford, D.L. (1968) New concepts in oxygen transfer and aeration, in *Advances in Water Quality Improvements* (eds E.F. Gloyna and W.W. Eckenfelder), Univ. of Texas Press, pp. 215–236.

45 Wagner, M. (1991) Einfluss oberflächenaktiver Substanzen auf Stoffaustauschmechanismen und Sauerstoffeintrag. Dissertation, Schriftenreihe Institut für Wasserversorgung, Abwasserbeseitigung und Raumplanung der TH Darmstadt 53.

46 Ku, Y., Chang, J.-L., Shen, Y.-S. and Lin, S.Y. (1998) Decomposition of diazinon in aqueous solution by ozonation. *Water Research*, **32**, 1957–1963.

47 Johnson, P.N. and Davis, R.A. (1996) Diffusivity of ozone in water. *Journal of Chemical and Engineering Data*, **41**, 1485–1487.

48 St-Denis, C.E. and Fell, C.J. (1971) Diffusivity of oxygen in water. *Canadian Journal of Chemical Engineering*, **49**, 885.

49 Wilke, C.R. and Chang, P. (1955) Correlation of diffusion coefficients in dilute solutions. *American Institute of Chemical Engineering Journal*, **1**, 264–270.

50 Reid, R.C., Prausnitz, J.M. and Sherwood, T.K. (1977) *The Properties of Gases and Liquids*, 3rd edn, McGraw-Hill, New York.

51 Linek, V., Vacek, V. and Benes, P. (1987) A critical review and experimental verification of the correct use of the dynamic method for the determination of oxygen transfer in aerated vessels to water electrolyte solutions and viscous liquids. *The Chemical Engineering Journal*, **34**, 11–34.

52 Redmon, D., Boyle, W.C. and Ewing, L. (1983) Oxygen transfer efficiency measurements in mixed liquor using off-gas techniques. *Journal Water Pollution Control Federation*, **55**, 1338–1347.

53 ASCE (1997) *Standard Guidelines for in Process Oxygen Transfer Testing ASCE 18-96*, American Society of Civil Engineers, New York.

54 Huang, W.H., Chang, C.Y., Chiu, C.Y., Lee, S.J., Yu, Y.H., Liou, H.T., Ku, Y. and Chen, J.N. (1998) A refined model for ozone mass transfer in a bubble column. *Journal Environmental Science and Health A*, **33**, 441–460.

55 Lin, S.H. and Peng, C.F. (1997) Performance characteristics of a packed-bed ozone contactor. *Journal Environmental Science and Health A*, **32**, 929–941.

56 Philichi, T.L. and Stenstrom, M.K. (1989) The effects of dissolved oxygen probe lag on oxygen transfer parameter estimation. *Journal Water Pollution Control Federation*, **61**, 83–86.

57 ASCE (2006) *Measurement of Oxygen Transfer in Clean Water ASCE 2-06*, American Society of Civil Engineers, New York.

58 Dang, N.D.P., Karrer, D.A. and Dunn, I.J. (1977) *Biotechnology and Bioengineering*, **19**, 853.

59 Biń, A.K., Duczmal, B. and Machniewski, P. (2001) Hydrodynamics and ozone mass transfer in a tall bubble column. *Chemical Engineering Science*, **56**, 6233–6240.

60 Beltrán, F.J. and Gonzales, M. (1991) Ozonation of aqueous solutions of resorchinol and phloroglucinol–3. Instantaneous kinetic study. *Industrial Engineering and Chemical Research*, **30**, 2518–2522.

61 Sotelo, J.L., Beltrán, F.J. and Gonzales, M. (1990) Ozonation of aqueous solutions of resorchinol and phloroglucinol 1 stoichiometry and absorption kinetic regime. *Industrial Engineering Chemical Research*, **29**, 2358–2367.

62 Sotelo, J.L., Beltrán, F.J., Gonzales, M. and Garcia-Araya, J.F. (1991) Ozonation of aqueous solutions of resorchinol and phloroglucinol, 2 kinetic study. *Industrial Engineering Chemical Research*, **30**, 222–227.

63 Chapman, C.M., Gilibaro, L.G. and Nienow, A.W. (1982) A dynamic response technique for the estimation of gas-liquid mass-transfer coefficients in a stirred vessel. *Chemical Engineering Science*, **37**, 1891–1896.

64 Osorio, C. (1985) Untersuchung des Einflusses der Flüssigkeitseigenschaften auf den Stoffübergang Gas/Flüssigkeit mit der Hydrazin-Oxidation. Dissertation. Universität Dortmund 1–109.

7
Reaction Kinetics

In Chapter 2 we discussed the reaction mechanisms, the individual reaction steps that take place in ozonation and AOPs. In this chapter we want to look at the reaction kinetics, which describes how fast the reaction takes place and what influences the reaction. Knowledge of kinetic parameters, such as reaction order n and reaction rate constant k, helps us to assess the feasibility of using ozonation to treat waters and to design an appropriate reactor system. It can help us to understand how a reaction can be influenced, so that a treatment process can be optimized. Kinetic parameters are also necessary for use in scientific models, with which we further improve our understanding of the chemical processes we are studying.

First, we will review the basic concepts of kinetics, discussing in detail reaction order (Section 7.1) and reaction rate constants (Section 7.2) with emphasis on the practical aspects of determining them for oxidation processes. This lays the foundation for the discussion of which operating parameters influence the reaction rate and how (Section 7.3). These influences are illustrated with results from current publications, with special emphasis on analyzing the common and apparently contradictory trends.

7.1
Reaction Order

In general, a reaction can be described as follow:

$$\alpha A + \beta B + \gamma C \rightarrow \delta D + \varepsilon E + \zeta F \tag{7.1}$$

where $\alpha, \beta, \gamma, \delta, \varepsilon, \zeta$ are the stoichiometric coefficients for the reactants A, B, C and the products D, E and F.

The overall reaction for the oxidation of a target compound with ozone is split into two different reactions – into the direct and into the indirect reaction:

$$\text{Direct:} \quad \alpha M + \beta O_3 \rightarrow \gamma M_{oxid1} \tag{7.2}$$

$$\text{Indirect:} \quad \varepsilon M + \zeta OH^\circ \rightarrow \eta M_{oxid2} \tag{7.3}$$

Ozonation of Water and Waste Water. 2nd Ed. Ch. Gottschalk, J.A. Libra, and A. Saupe
Copyright © 2010 WILEY-VCH Verlag GmbH & Co. KGaA, Weinheim
ISBN: 978-3-527-31962-6

In order to be able to determine how fast the reaction occurs, we need an equation that describes the dependence of the reaction rate on the concentrations of the reactants. The following differential rate equation (also called rate law or rate expression) can be used to describe the reaction rate of a nonvolatile compound A:

$$\frac{dc(A)}{dt} = -k c(A)^{n_A} c(B)^{n_B} c(C)^{n_C} \tag{7.4}$$

$c(A), c(B), c(C)$ concentration of the compound A, B, C
k reaction rate constant
n reaction order with respect to compound

The concentration of each compound is raised to a power n, which is called the order of the reaction with respect to the considered reactant. The total order of the reaction is the sum of the orders for each reactant.

$$n_\Sigma = n_A + n_B + n_C \tag{7.5}$$

n_A, n_B and n_C are determined empirically from experimental results and are not necessarily related to the stoichiometric coefficients α, β or γ (see 7.1). Reactions, in which the order of reaction corresponds with the stoichiometry of the reaction, are called elementary reactions. For example:

$$1\,A + 2\,B \rightarrow 1\,D \tag{7.6}$$

$$\frac{dc(A)}{dt} = -k c(A)^1 c(B)^2 \tag{7.7}$$

In nonelementary reactions, the reaction order and stoichiometric coefficients differ. Often, only a single reaction is observed, but in reality a sequence of elementary reactions occurs. The amount of intermediates formed is negligible and, therefore, not detectable. One famous example is the reaction between hydrogen and bromine. The overall reaction can be described as:

$$H_2 + Br_2 \rightarrow 2\,HBr \tag{7.8}$$

which has a rate expression:

$$\frac{dc(HBr)}{dt} = \frac{2 k_2 k_3 (k_1/k_5)^{0.5} c(H_2) c(Br_2)^{0.5}}{k_3 + k_4 c(HBr)/c(Br_2)} \tag{7.9}$$

We can see from the complicated equation that the reaction is nonelementary. The following chain of reactions occurs:

$$Br_2 \rightarrow 2\,Br^\circ \tag{7.10}$$

$$H_2 + Br^\circ \rightarrow HBr + H^\circ \quad k_2 \quad (7.11)$$

$$H^\circ + Br_2 \rightarrow HBr + Br^\circ \quad k_3 \quad (7.12)$$

$$H^\circ + HBr \rightarrow H_2 + Br^\circ \quad k_4 \quad (7.13)$$

$$2\,Br^\circ \rightarrow Br_2 \quad k_5 \quad (7.14)$$

It is easy to see from this example why the rate law must be determined experimentally, rather than from the chemical reaction. The rate depends on the slowest, rate-determining step of the mechanism, not on the overall reaction.

In general, knowing the order of a reaction does not allow us to determine the stoichiometry of the reaction, nor vice versa. Since the order refers to an empirically found rate expression, it need not be an integer. If it is a fraction, the reaction is most likely nonelementary, but we have no clue to the stoichiometry of the reaction. If the reaction order is an integer, it may or may not be an elementary reaction. Fortunately, it is often not necessary to know the stoichiometry of a reaction exactly, or even the reaction order for all compounds to design a reactor system.

Since the ozonation of a compound M involves both the direct and indirect reaction pathways, the general rate Equation 7.4 has to be modified to include both reactions. The equation can be written specifically for the ozone reaction, substituting the compounds M, O_3 and OH° for A, B and C, respectively:

$$r(M) = -\frac{dc(M)}{dt} = k_D c(M)^{n_{1M}} c(O_3)^{n_{O3}} + k_R c(M)^{n_{2M}} c(OH^\circ)^{n_{OH}} \quad (7.15)$$

where $c(M)$, $c(O_3)$ and $c(OH^\circ)$ represent the concentration of the pollutant, ozone and the hydroxyl radical, and k_D and k_R are the reaction rate constants for the direct and indirect reactions.

Ideally, the concentrations of O_3 and OH° can be regarded as constant over time. Only the concentration of compound M changes, and the rate Equation 7.15 can be further simplified. The reaction can be regarded as pseudo n_M-th order:

$$r(M) = -\frac{dc(M)}{dt} = k'c(M)^{n_M} \quad (7.16)$$

k': reaction rate coefficient

The reaction rate constant also changes. It becomes a pseudo-rate coefficient k', where the concentrations and reaction orders of the reactants O_3 and OH° are lumped together with the reaction rate constant of compound M. Therefore, it is dependent on the concentrations of O_3 and OH°. For the case that $n = 1$, it becomes a pseudo-first-order rate coefficient and has the dimension (time)$^{-1}$.

In general the dimensions of the reaction rate constants for the nth-order reaction are:

(concentration)$^{1-n}$(time)$^{-1}$.

The oxidation of compounds with ozone is often first order with respect to the oxidant (O_3 or $OH°$) as well as to the pollutant M, leading to a second-order rate equation [1, 2]

$$r(M) = -\frac{dc(M)}{dt} = k_D c(M) c(O_3) + k_R c(M) c(OH°) \tag{7.17}$$

However, the reaction order for each reactant must be determined experimentally to develop a correct rate equation. This requires measuring the disappearance of one reactant independent of the other reactant concentrations. This is the case, for example, when the concentration of ozone in the bulk liquid is so large compared to the pollutant M that its concentration remains almost constant. A further requirement for determining kinetic parameters in general, is that the reaction rate should be independent of the mass-transfer rate.

These requirements are easy to fulfil for (very) slow reactions by using a continuously sparged semibatch reactor to replenish the ozone concentration. Such a reaction regime is often found in drinking-water ozonation. In contrast, mass-transfer limitations often occur in waste-water applications. More effort is necessary to find operating conditions where mass transfer does not affect the reaction rate, or if this is not possible, more complicated methods than the ones presented below must be used to determine the reaction order (see [3] for more details).

Care must be taken that these requirements are met over the whole experiment. Especially when ozonating waste water in the semibatch mode, the reaction regime can change over time as the concentration of pollutant changes. Often, no dissolved ozone can be measured in the bulk liquid ($c_L \cong 0$) at the beginning of the experiment. The pollutant is normally present at a high concentration, causing (very) fast reactions with molecular ozone, and consequently, the direct reaction may occur inside the liquid film. The reaction rate is then limited by the mass-transfer rate (Section 6.2), and the reaction order and constant cannot be measured as such. Instead, measurement in this regime produces pseudo-first-order kinetic parameters that are specific only to those experimental conditions. As the semibatch ozonation continues, and the pollutant concentration decreases, the reaction regime moves from mass-transfer to chemical-kinetic controlled. Only then are the reaction order and constant independent of the mass-transfer rate.

A very prominent, but somewhat confusing, example for the determination of reaction order is the process of ozone decay in "clean water". The large variation in reaction order found for ozone decay by various authors shows that the determination of reaction order can be rather complicated (see Table 7.1). From a chemical point of view, this radical chain process – as shown in detail in Chapter 2 – is foremost a function of pH, or more accurately the hydroxide ion concentration. The main reason for the different reaction orders found, is that ozone decay is part of a chain-reaction mechanism involving hydroxyl radicals ($OH°$), similar to the example of bromide shown above, but much more complex.

A systematic dependence of reaction order on temperature and pH is not visible, n varies between one and two. Different experimental conditions and/or missing

Table 7.1 A comparison of reported reaction orders for the decay rate of ozone in phosphate-buffered solutions of demineralized water.

Reference	T in °C	pH	n with respect to O_3
Stumm, 1954[a]	0.2–19.8	7.6–10.4	1
Kilpatrick et al., 1956[a]	25	0–6.8	1.5
		8–10	2
Rankas, 1962[b]	5–25	5.4–8.5	1.5
Hewes and Davis, 1971[a]	10–20	2–4	2
		6	1.5–2
		8	1
Kuo et al., 1977 [57]	15–35	2.2–11	1.5
Sullivan, 1979[b]	3.5–60	0.5–10	1
Gurol and Singer, 1982 [4]	20	2.2–9.5	2
Staehelin and Hoigné, 1982 [8]	20	8–10	1
Sotelo et al., 1987 [61]	10–40	2.5–9	1.5–2
Minchew et al., 1987 [5]		6.65	2
Grasso and Weber, 1989 [54]		5–9	1
Gottschalk, 1997 [7]	20	7	1–2
Kuosa et al., 2005 [58]	21	7–10	1.12
Mizuno et al., 2007 [60]	15–30	4–7	2

a Taken from Minchew et al., 1987 [5].
b Taken from Gurol and Singer, 1982 [4].

details about these conditions as well as different analytical methods make a comparison of these results impossible. Staehelin and Hoigné [6] proposed a possible explanation for the second-order reaction ($n = 2$). Since in "clean" water ozone not only reacts with the hydroxide ions but also with the intermittently produced hydroxyl radicals (see Chapter 2), it behaves like a promoter and the decay rate increases with the square of the of the liquid-ozone concentration. This is supported by the results obtained by Gottschalk [7], who found a second-order decay rate in deionized water, while in Berlin tap water, which contains about $4\,mg\,l^{-1}$ DOC and $4\,mmol\,l^{-1}$ total inorganic carbon, the decay rate was first order. Staehelin and Hoigné [8] also found first order in complex systems.

7.1.1
Experimental Procedure to Determine the Reaction Order n

The reaction order with respect to a target compound can be determined in many different ways. The three commonly used methods described here require that reaction conditions, that is, temperature, pH, etc., be held constant and intermediates must not influence the reaction. They are also based on the method of excess – ozone is in excess and its concentration is constant during the experiment.

7.1.1.1 Half-Life Method

An important concept is the half-life of the reaction τ, sometimes written as $t_{1/2}$, the time needed to decrease the concentration to one half of the initial value $c(A)_o$. It is related to the order of reaction and the rate constant through the following equations:

$$n = 1: -\frac{dc(A)}{dt} = k'c(A)^n \Rightarrow \tau = \frac{\ln 2}{k'} \tag{7.18}$$

$$n \neq 1: \frac{dc(A)}{dt} = -k'c(A)^n \Rightarrow \tau = \frac{2^{n-1} - 1}{k'(n-1)c(A)_o^{n-1}} \tag{7.19}$$

For first-order reactions the "half-life" only depends on the rate constant, while for $n \neq 1$, it is also a function of the initial concentration $c(A)_o$.

In order to determine the reaction rate a minimum of two different initial concentrations of A are used. The decrease in A over time is measured. The half-life $\tau = t_{1/2}$ is defined as the time where $c(A) = c(A)_o/2$. The order can be determined with the following equation: (see Figure 7.1a and b)

$$n = 1 - \frac{\log \tau_1 - \log \tau_2}{\log c(A_1)_o - \log c(A_2)_o} \tag{7.20}$$

7.1.1.2 Initial Reaction Rate Method

Again, the decrease in the concentration of A over time is measured for two different initial concentrations. The initial reaction rate r for $t \to 0$ is calculated with the help of a tangent at the steepest region of the curve, intersecting c_o at $t = 0$ and the following equation (see Figure 7.1c):

$$r(A)_i = \frac{\Delta c(A)_i}{\Delta t}, t \to 0 \tag{7.21}$$

The order of the reaction for compound A can be determined as follows:

$$n = \frac{\log r(A_1) - \log r(A_2)}{\log c(A_1)_o - \log c(A_2)_o} \tag{7.22}$$

This method works for relatively slow reactions. For fast reactions, the initial rate measured will have a large range of uncertainty.

7.1.1.3 Trial and Error

A rate equation that describes the experimental points with the best fit is chosen. The differential and integrated rate equations for the various reaction orders are found in Table 7.2. The best fit is easy to find by comparing the linear regression coefficients for the appropriate *xy*pairs. The *x*-axis is always the time *t*.

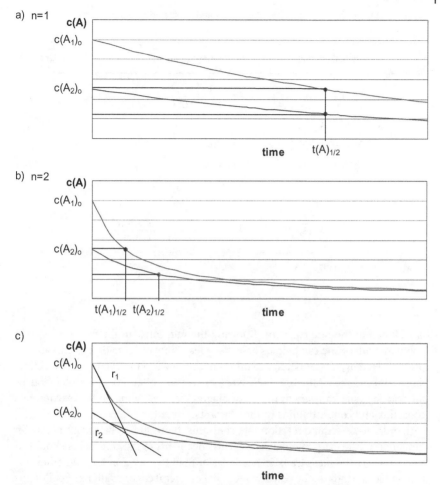

Figure 7.1 Method of half-life (a and b) and initial reaction rate method (c).

Alternatively, the order can also be obtained graphically by trial and error. For example: to find the reaction order with respect to compound A with

a) differential equations – plot $\ln(-dc(A)/dt)$ versus $\ln c(A)$; slope = n
b) integrated equations – plot $1/c(A)^{n-1}$ versus t. Vary n until the plot is linear. The slope is equal to the rate constant k.

7.2
Reaction Rate Constants

The rate constant k is a constant of proportionality relating the rate of the reaction to the reactant concentrations. The temperature dependency of the reaction rate

Table 7.2 Order of reaction with their differential and integrated equation for concentration over time.

n	Differential equation	Integrated equation	Unit of k^a
n	$dc(A)/dt = -k\, c(A)^n$	$kt = \dfrac{1}{n-1}\left[\dfrac{1}{c(A)^{n-1}} - \dfrac{1}{c(A)_o^{n-1}}\right], n \neq 1$	$M^{(1-n)} s^{-1}$
0	$dc(A)/dt = -k$	$kt = c(A)_o - c(A)$	$M s^{-1}$
0.5	$dc(A)/dt = -k\, c(A)^{0.5}$	$kt = 2\left(c(A)_o^{0.5} - c(A)^{0.5}\right)$	$M^{0.5} s^{-1}$
1	$dc(A)/dt = -k\, c(A)$	$kt = \ln c(A)_o/c(A)$	s^{-1}
2	$dc(A)/dt = -k\, c(A)^2$	$kt = \dfrac{1}{c(A)} - \dfrac{1}{c(A)_o}$	$M^{-1} s^{-1}$
2	$dc(A)/dt = -k\, c(A)\, c(B)$	$kt = \dfrac{1}{c(A)_o - c(B)_o}\ln\dfrac{c(A)c(B)_o}{c(B)c(A)_o}$	$M^{-1} s^{-1}$

a   M = molar = mol l^{-1}.

is included in the rate constant. Knowing the rate constant for the reaction of a pollutant with ozone can help us assess the feasibility of ozonation as a treatment process. Hereby, it is important to differentiate between the two types of reaction, direct and indirect. Although both types of reaction often occur simultaneously, the reaction rates can vary by orders of magnitude, depending on the reaction rate constants and concentrations of the reactants present.

In order to determine reaction rate constants, there should be no mass-transfer limitation in the experimental setup. This means operating conditions should be chosen where the mass-transfer rate is faster than the reaction rate. Whether this is possible depends on the reaction regime. For typical concentrations of pollutants in drinking water (c(M) < 10^{-4} M) – even in the case of AOPs using incident UV-radiation and a hydrogen peroxide concentration lower than 10^{-3} M – the kinetic regime is likely to be slow, so that mass-transfer limitations can be avoided. In a fast-kinetic regime mass-transfer limitations are unavoidable and a more rigorous and complex procedure, which takes the mass-transfer limitation into consideration, is necessary [3, 9] (Section 6.4).

In general, the method of determining the reaction rate constants is based on knowing the reaction order for each reactant. When the order has been determined, the equation appropriate for the reaction order is chosen (Table 7.2) and the reaction rate constants are calculated from the linearization of the observed concentration decrease over time (usually the slope, see Table 7.2). However, since the reactions are complicated and often occur simultaneously, there are many experimental parameters to be considered when planning the experiments to determine the reaction order and rate constants. Some of the most important are summarized below.

7.2.1
Determination of Rate Constants

For the determination of the direct rate constant k_D, the indirect reactions have to be suppressed. This is generally done by inhibiting any reactions between the hydroxyl radicals and the target compounds through the addition of substances that do not react with molecular ozone (or only very slowly), but quickly scavenge the hydroxyl radicals produced.

The various methods are: addition of *tert*-butanol (TBA), decrease in the pH (i.e., an addition of H^+ ions), use of n-propanol, methylmercury (pH > 4) or bicarbonate (HCO_3^-, pH > 7). The concentration of these scavengers has to be kept as low as possible, in order to exclude any direct reactions with these substances. For experimental details see Hoigné and Bader [1], Staehelin and Hoigné [6] and Andreozzi *et al.* [10].

To inhibit the indirect reaction, Beltrán *et al.* [11] found that it is sometimes not sufficient to use a low pH-value, even as low as pH = 2. Comparing the reaction rate of atrazine with ozone at pH = 2 with and without *tert*-butanol, they observed a decrease in the reaction rate in the presence of TBA. This means that there were radical reactions even at this extremely low pH.

Care has to be taken that the changes to the water matrix, for example, to inhibit the indirect reaction, do not change the direct reaction rate. The addition of TBA, for example, can change the mass-transfer rate (see Section 6.2). A change in pH can also affect the reaction rate of compounds that can dissociate. Hoigné and Bader measured the direct reaction rate of molecular ozone with organic solutes that do not dissociate [1], which can dissociate [2] and with inorganic compounds [12]. In general, the compounds are more reactive in their dissociated form, which was attributed to the electrophilic character of the ozone molecule.

Further results of interest can be found in Yao and Haag [13], who determined the kinetics of direct ozonation of several organic trace contaminants in deionized water with 50 mM phosphate buffer and 10 mM TBA.

Another aspect to be considered is that ozone also decays during the experiments. To obtain a constant ozone concentration in a system where ozone decay occurs, the reactor can be operated semibatch. Gaseous ozone is continuously sparged to the reactor and after the ozone concentration reaches steady state, the investigated compound is injected into the reactor. Another possibility is to measure the ozone decay rate independently and take this into account in the calculations.

A method that can be applied universally for direct and indirect reactions is the method of competition kinetics that was first described by Hoigné and Bader [14] for *homogeneous* systems. Fast reactions with relatively high k_D-values above approximately $10^4 M^{-1} s^{-1}$ can best be measured by this method. An ozone-containing fluid is rapidly mixed with the fluid containing the pollutants M_1 and M_2 and the disappearance of both is tracked over time [15, 13]. The reaction rate constant of one of them must be known, serving as a reference compound, the other one can then be determined. Under the assumption that both components react pseudo-first order in a batch system, the unknown reaction rate constant k_{M2} can be calculated by the following equation:

$$k_{M_2} = \frac{\ln\dfrac{c(M_2)_t}{c(M_2)_o}}{\ln\dfrac{c(M_1)_t}{c(M_1)_o}} k_{M_1} \qquad (7.23)$$

A prerequisite for the use of this method in *heterogeneous*, gas-sparged semibatch systems is that the reaction does not develop in the *instantaneous* regime [16, 17]. Phenol has often been used as a reference compound (M_1) yielding a pH and temperature-dependent value of $k_D = 1300\,M^{-1}s^{-1}$ at pH = 2, T = 22 °C [15] and $k_D = 10\,000\,M^{-1}s^{-1}$ at pH = 4, T = 25 °C (Li et al., 1979).

Rate constants for the reaction of hydroxyl radicals with different compounds were determined by Haag and Yao [18] and Chramosta et al.[19]. In the study of Haag and Yao [18] all hydroxyl radical rate constants were determined using competition kinetics. The measured rate constants demonstrate that $OH°$ is a relatively nonselective radical toward C–H bonds, but is least reactive with aliphatic polyhalogenated compounds. Olefins and aromatics react with nearly diffusion-controlled rates.

A comparison of direct (k_D) and indirect (k_R) reaction rate constants of important micropollutants in drinking water is given in Table 7.3. In general, the reaction rate constants of the direct reaction are between 1 and $10^3\,l\,mol^{-1}s^{-1}$, and the indirect reaction rate constants are between 10^8 and $10^{10}\,l\,mol^{-1}s^{-1}$ [1, 2].

Table 7.3 Examples of reaction rate constants for direct and indirect reaction of well-known drinking-water contaminants (micropollutants) [13, 18, 20].

Pollutant	Reaction rate constant k in $M^{-1}s^{-1}$	
	k_D	k_R
Dibromomethane	–	$0.4–1.1 \times 10^9$
1,1,2 Trichloroethane	–	$0.13–0.35 \times 10^9$
Lindane	<0.04	$4.2–26 \times 10^9$
Phthalates	0.14–0.2	4×10^9
Simazine	4.8	2.8×10^9
Atrazine	6–24	2.6×10^9
2,4-D	2.4	5×10^9
Sulfamethoxazole (antibiotic)	2.5×10^6	5.5×10^9
Carbamazepine (antiepileptic)	3×10^5	8.8×10^9
Diclofenac (antiphlogistic)	1×10^6	7.5×10^9
17α-Ethinylestradiol (ovulation inhibitor)	7×10^9	9.8×10^9

Knowledge of rate-constant values can be put to practical use, for example, to evaluate whether treatment with ozone or AOP is feasible, and which reaction, direct or indirect, is more likely to be successful. By comparing k_D and k_R for a target compound, it is possible to get an idea of which process can proceed faster. The last four compounds mentioned in the table are pharmaceuticals that are responsible for ecotoxicological issues in the aquatic environment such as building microbial resistance and feminization of higher organisms [21]. They have been detected quite often in waste water [22] and have shown very large direct rate constants with both ozone and the hydroxyl radical [20]. Therefore, ozonation seems to be a promising treatment process that can be used to reduce them drastically.

Furthermore, a comparison of the rate constants of all the compounds present in the water can be helpful to estimate the relative treatment time needed for each compound. For instance, you would expect the indirect reaction of OH° with sulfamethoxazole to take almost twice as long as that with 17α-ethinylestradiol for the same molar concentrations.

However, to predict how fast the reactions will proceed, thus allowing an estimate of reactor size, or to estimate which reaction, direct or indirect, will dominate in a reaction system, all parameters in Equation 7.17 must be considered. The concentration of the reactants is especially important. The differences in reactant concentrations (i.e., M, O_3 and OH°) can be many orders of magnitude. At high concentrations of the target compound M, as is often the case for waste waters (mg/l), ozone may react at the gas/liquid interface so that the dissolved ozone concentration is low to nondetectable. There is little to no production of OH° so the direct reaction dominates. In contrast, at low target compound concentrations typical for drinking water (ng/l or μg/l), the direct oxidation kinetic is often negligible. The dissolved ozone decays to OH° before coming in contact with a target molecule. The indirect reaction dominates.

Therefore, knowledge of independent rate constants for each pathway is useful to predict competition effects. Unfortunately, some data continue to be generated that fail to distinguish between the direct ozone reaction and the hydroxyl radical chain reaction. Such rate constants are very system specific and of limited value for process feasibility analysis and design.

7.3
Parameters that Influence the Reaction Rate

In the following section we look more closely at parameters that influence the chemical kinetics of ozonation and how they do it. But first it is important to recall that the addition of compounds to waters can have many unintentional effects on the chemistry of the solution. For example, if salts are added to the water to investigate their effect on the oxidation reactions, they can change the pH, ozone solubility and the coalescence behavior of the bubbles, indirectly affecting both the chemical reaction kinetics and the mass-transfer rate. The addition of higher concentrations of organics can change the surface tension, again influencing the mass-transfer rate. Moreover, this influence can vary over the course of an experiment, due to the oxidation of the surface-active compound. The production of the acidic oxidation products can also cause the pH to vary with time. Neglecting to consider these effects could lead to a false interpretation of experimental results. Therefore, in order to determine the effect of a certain parameter on the chemical reaction kinetics, all other influences must be ruled out or held constant.

7.3.1
Concentration of Oxidants

7.3.1.1 Direct Reactions
Normally, a second-order reaction is assumed for all direct reactions of organic compounds (M) with ozone, with the rate dependent on the concentration of ozone, as well as on that of the compound, to the first power.

In general, an increase in the ozone concentration in the liquid bulk causes an increase in the oxidation rate of the substrate [23–29]. A linear correlation between the oxidation rate and liquid-ozone concentration was found by Gottschalk [7] and Adams and Randtke [24] for the oxidation of atrazine in drinking-water ozonation studies.

In special cases, for example when the reaction is mass-transfer limited, where no liquid ozone can be measured and where the ozone mass-transfer rate is equal to the reaction rate ($E = 1$), the ozone dose rate can be used to describe the amount of ozone available for reaction. Gottschalk [7] was able to correlate the oxidation rate of an organic substrate (atrazine) with the ozone dose rate and applied dose rate.

7.3.1.2 Indirect or Hydroxyl Radical Reactions
The same observation was made for hydroxyl radical reactions occurring in advanced oxidation processes involving ozone. Increasing the ozone dose rate will increase the reaction rate (UV/O_3 [11, 30, 31], H_2O_2/O_3 [7, 23, 25, 32]). To be able to make a good comparison in AOPs, the concentration of the second oxidant has to be constant. Furthermore, it is important that there is enough ozone in the liquid and that the reaction is not limited by the ozone transfer from the gas into the liquid phase.

For example, by varying the dose rate of hydrogen peroxide while holding the ozone dose rate constant, the dependency of the oxidation rate on the dose ratio $F(H_2O_2)/F(O_3)$ can be investigated. The optimal dose ratio, that is, the dose ratio resulting in the fastest reaction rate, has often been found to lie between 0.5 and 1.4 moles H_2O_2/mol O_3 [7, 26, 31, 32]. The different values of the optimal dose ratio can be explained by considering the expected stoichiometry for OH° formation from ozone and hydrogen peroxide:

$$2\,O_3 + H_2O_2 \rightarrow 2\,OH° + 3\,O_2 \tag{7.24}$$

According to this reaction, a molar ratio of 0.5 or a weight ratio of 0.35 moles H_2O_2 per mole O_3 is necessary. This ratio was found in "clean systems" such as deionized water with a low concentration of buffer. In ground water or water with high scavenger concentrations, the optimal dose ratio was higher. Here, the chain reaction is influenced by other components.

There are several factors that may influence this stoichiometry:

- H_2O_2 can act as a free radical scavenger itself;
- O_3 can react directly with $OH°$, consuming O_3 and $OH°$;
- O_3 and $OH°$ may be consumed by other constituents (scavengers).

In order to measure the optimal dose ratio it is suggested that the concentration of one of the oxidant be held constant while the other is varied. The influence of parameters is only measurable if there is no mass-transfer limitation.

7.3.2
Temperature Dependency

For any reaction the rate constant is a temperature-dependent term. The rate constant k has been found to be well represented by Arrhenius' law:

$$k = A' \exp(-E_A / \Re T) \qquad (7.25)$$

A' frequency factor
E_A activation energy in $J\,mol^{-1}$
$\Re$ ideal gas law constant ($8.314\,J\,mol^{-1}\,K^{-1}$)
T temperature in K

If the Arrhenius function is valid, the plot of $\ln k$ versus T^{-1} shows a straight line; and the slope is $-E_A/\Re$. When determining the activation energy for an ozone reaction, it is important to keep in mind that by increasing the temperature of the water, the solubility of ozone decreases. The same liquid-ozone concentration should be used at the various temperatures, which can be a problem in systems with fast reactions. Simplifying the temperature dependency, one could say that the increase of the temperature by 10 °C will double the reaction rate, the so-called van't Hoff rule.

In the case of a mainly direct pathway the activation energy is in the range of 35–50 kJ mol^{-1} and 5–10 kJ mol^{-1} with a radical pathway [33].

7.3.3
Influence of pH

The relevance of the pH-value was already seen in the chain reaction of ozone, especially in the initiation step. It also plays an important role in all the acid–base equilibrium by influencing the equilibrium concentrations of the dissociated/

nondissociated forms. This is especially important for the scavenger reaction with inorganic carbon, which will be discussed further in Section 7.3.4.

The decomposition of ozone is catalyzed by the hydroxide ion. Ozone dissociates in the presence of OH^- to $HO_2^°/O_2^{-°}$. Further decomposition via the ozonide anion radical $O_3^{-°}/HO_3^°$ results in the formation of $OH^°$. They may react with organic substrates, radical scavengers (HCO_3^-, CO_3^{2-}) or ozone itself.

The results reported in the literature show that in natural and synthetic water (deionized water with or without buffer) an increase in the pH-value increases the reaction rate [28, 34–37]. Gottschalk [7] found a direct proportionality between the reaction rate and the OH^- concentration. In deionized water with added scavenger [7, 40] an optimum in the reaction rate was observed at about pH = 8. The positive effect caused by the increased OH^- concentration was counteracted by the strong scavenger potential of carbonate as the pH increased above 8. At higher concentrations of scavenger (>2–3 mmol l^{-1}) this effect was no longer observed, which was explained by the constant potential of scavenger in this concentration range [7].

For the combined oxidation processes, the effect of pH is even more complex. Experimental results have shown a steady increase in the reaction rate of micropollutants with increasing pH, as well as optima at various pH values.

On the one hand, the equilibrium between H_2O_2/HO_2^-, which plays an important role in all the previously discussed AOPs, shifts toward HO_2^-. The result is that for the O_3/UV and O_3/H_2O_2 processes, a higher amount of initiator is present, leading to an increase in the amount of $OH^°$ present. For the UV combined processes it is important that HO_2^- absorbs more UV-light at 254 nm than H_2O_2, again the amount of initiators increases (see Chapter 2). On the other hand HO_2^- is known to act as a scavenger itself. If inorganic carbon is present in the water, the effect of the stronger scavenger potential of CO_3^{2-} compared to HCO_3^- is added (see Section 7.3.4).

7.3.4
Influence of Inorganic Carbon

Inorganic carbon can also influence the total reaction rate by acting as a scavenger for hydroxyl radicals, whereas ozone itself does not react with carbonate or bicarbonate [40]. The reaction of $OH^°$ with inorganic carbon proceeds according to the following mechanisms:

$$HCO_3^- + OH^° \rightarrow H_2O + CO_3^{-°} \tag{7.26}$$

$$CO_3^{2-} + OH^° \rightarrow OH^- + CO_3^{-°} \tag{7.27}$$

Not much is known about reactions of the carbonate radical with organic compounds; they seem to be almost unreactive. The carbonate radical, though, has been found to react with hydrogen peroxide [38]:

$$CO_3^{-°} + H_2O_2 \rightarrow HCO_3^- + HO_2^° \quad k = 8 \times 10^5 \text{ M}^{-1} \text{ s}^{-1} \tag{7.28}$$

Table 7.4 Reaction rate constants for the reaction of hydroxyl radical with inorganic carbon.

Reference	HCO_3^- k_R (l mol^{-1} s^{-1})	CO_3^{2-} k_R (l mol^{-1} s^{-1})
Hoigné and Bader, 1976 [14]	1.5×10^7	20×10^7
Masten and Hoigné, 1992 [43]		42×10^7
Buxton et al., 1986	0.85×10^7	
Buxton et al., 1988 [52]		39×10^7

The reaction rate constants for the two forms of inorganic carbon are summarized in Table 7.4. Mizuno et al. [39] review further possible reactions with inorganic carbon in their publication. A comparison of the reaction rate constants shows that carbonate is a much stronger scavenger than bicarbonate. This indicates that the pH-value, which influences the form and concentration of inorganic carbon present (pK$_a$ (HCO_3^-/CO_3^{2-}) = 10.3), is of major importance in determining the effect inorganic carbon has on the reaction rate.

Although these reaction rate constants are relatively low in comparison with the reaction rate constants of organic compounds with OH° (see Table 7.3), the reaction of inorganic carbon with OH° is not negligible, since it is usually present in drinking water at comparatively high concentrations [7, 14].

By increasing the concentration of inorganic carbon, that is, increasing the concentration of scavengers, the reaction rate of organic target compounds with OH° will be decreased. On the other hand, this also reduces the decay of ozone. Thus, the direct oxidation of the organic substrates becomes more important, with the consequence of a lower total reaction rate. This decrease in reaction rate with increasing inorganic carbon concentration has often been shown in the literature, also for the AOPs [7, 11, 37, 40–44].

The effect on the reaction rate is relatively high at low concentrations of inorganic carbon. However, above 2 mmol l^{-1} for ozonation and about 3 mmol l^{-1} for the ozone/hydrogen peroxide process, the decrease in the reaction rate is negligible [7, 45] found this plateau at 1.5 mmol l^{-1} of inorganic carbon for ozonation.

7.3.5
Influence of Inorganic Salts

The presence of inorganic salts can theoretically affect ozonation in two ways – by acting as a scavenger or promoter in the radical-chain mechanism or by influencing the mass-transfer rate. The effect of salt on mass transfer and ozone solubility was discussed in Chapter 6. The increase in ionic strength through the addition of salt decreases the coalescence of bubbles, thus increasing the interfacial surface area and the mass-transfer rate. If the reaction is mass-transfer limited, this can increase the reaction rate. However, only the influence on the chemical kinetics is of interest in this section.

The participation of inorganic salts in the radical-chain mechanism could cause a shift in the relative importance of the indirect and direct reaction with the target compound, as described above for inorganic carbon. This change in the selectivity of the oxidation reaction could affect the total reaction rate. However, Boncz et al. [47] concluded from their investigations on the effect of high salt concentrations on the oxidation of substituted benzoic acids that the effect of sulfate, phosphate, nitrate, chloride on oxidation selectivity is generally very small. This is in line with Hoigne's observation that their scavenging effect can generally be neglected [57]. In their summary of the literature on this topic, Boncz et al. point out that the various effects reported have usually been small.

In order to avoid influences from pH and mass transfer, comparative experiments with the pH controlled at 2.4 and 10.4 were made with each salt at 0.5 M and without. Although small effects were found for phosphate and sulfate at low pH and for nitrate at high pH, Boncz et al. suggest that the design of ozonation systems at high salt concentrations can often be based on reaction-kinetic data collected at low ionic strength.

7.3.6
Influence of Organic Carbon on the Radical Chain-Reaction Mechanism

Organic carbon can react as a scavenger and/or a promoter. This depends on the kind of organic carbon and its concentration [46, 48]. In the treatment of micropollutants in drinking water, where the concentrations of the target compounds are in the range of micromoles, even a few milligrams of DOC can exert strong influence on the indirect reaction mechanism in the treatment of ground or surface waters, for example, low concentrations of humic acids.

Primary radicals are produced according to the following general reaction:

$$2\,R + 2\,OH° \rightarrow 2\,HR° + O_2 \tag{7.29}$$

In many cases the primary radicals react quickly with dissolved oxygen and form peroxy-radicals, which initiate further oxidation processes.

$$HR° + O_2 \rightarrow HRO_2° \tag{7.30}$$

For this reaction a surplus of oxygen is necessary [44]. Three different pathways exist for further reactions [49]:

- the backreaction

$$HRO_2° \rightarrow HR° + O_2 \tag{7.31}$$

- homolysis to a hydroxyl-radical and carbonyl rest

$$HRO_2° \rightarrow RO + OH° \tag{7.32}$$

Table 7.5 Reaction rate constants of OH° with DOC in natural water.

Reference	DOC k_R in $l\,mg^{-1}\,s^{-1}$
Liao and Gurol, 1995 [59]	1.6×10^4
Novell et al., 1992	1.7×10^4
Haag and Yao, 1992 [18]	2.3×10^4
Hoigné, 1998 [55]	2.4×10^4
De Laat et al., 1995 [53]	2.5×10^4
Kelly, 1992 [56]	10^5

- heterolysis with the formation of an organic cation and a superoxide radical

$$HRO_2^\circ \rightarrow HR^+ + O_2^{\circ -} \tag{7.33}$$

In particular, humic acid in natural waters can react as scavenger or promoter depending on its concentration [40, 50]. Xiong and Graham found that the fastest ozonation of atrazine in deionized water buffered at pH = 7.5 (KH_2PO_4/Na_2HPO_4) takes place with a concentration of humic acid of $1\,mg\,l^{-1}$ (approx. $0.5\,mg\,l^{-1}$ DOC [51]). With higher concentrations of humic acid, the oxidation rate was lower. Various authors have studied the oxidation rate of DOC found in natural systems by OH°. Table 7.5 gives some examples of the reaction rate constants determined.

A comparison with other organic compounds gives the impression that the reactivity is relatively low (see Table 7.3). During oxidation, however, the parameters used to characterize the organic carbon, that is, concentration of the individual compound, DOC and COD, decrease with different velocities. For example, the first step in the oxidation of an organic compound usually adds oxygen, thus modifying the compound, so that analytical identification of the compound shows elimination. The oxidation reduces the COD of the compound somewhat, but does not mineralize the organic compound, yielding no reduction in DOC. The values in Tables 7.3 and 7.5, therefore, describe different degrees of oxidation: elimination (Table 7.3) and mineralization (Table 7.5). This is important to keep in mind when using reaction rate constants to design a reactor system. If the treatment goal is mineralization and not just transformation of the organic carbon, then the disappearance of the parent compound or decrease in COD is not sufficient.

References

1 Hoigné, J. and Bader, H. (1983) Rate constants of reactions of ozone with organic and inorganic compounds in water – I. Non dissociated organic compounds. *Water Research*, **17**, 173–183.

2 Hoigné, J. and Bader, H. (1983) Rate constants of reactions of ozone with organic and inorganic compounds in water–II. Dissociated organic compounds. *Water Research*, **17**, 184–195.

3 Beltrán, F.J. (2004) *Ozone Reaction Kinetics for Water and Wastewater Systems*, Lewis Publisher, Boca Raton, London, New York, Washington D.C.

4 Gurol, M.D. and Singer, P.S. (1982) Kinetics of ozone decomposition: a dynamic approach. *Environmental Science & Technology*, **16**, 377–383.

5 Minchew, E.P., Gould, J.P. and Saunders, F.M. (1987) Multistage decomposition kinetics of ozone in dilute aqueous solutions. *Ozone: Science & Engineering*, **9**, 165–177.

6 Staehelin, J. and Hoigné, J. (1985) Decomposition of ozone in water in the presence of organic solutes acting as promoters and inhibitors of radical chain reactions. *Environmental Science & Technology*, **19**, 1206–1213.

7 Gottschalk, C. (1997) Oxidation organischer Mikroverunreinigungen in natürlichen und synthetischen Wässern mit Ozon und Ozon/Wasserstoffperoxid. Dissertation. TU-Berlin, Shaker Verlag, Aachen.

8 Staehelin, J. and Hoigné, J. (1982) Decomposition of ozone in water: rate of initiation by hydroxide ions and hydrogen peroxide. *Environmental Science & Technology*, **16**, 676–681.

9 Beltrán, F.J. (1997) Theoretical aspects of the kinetics of competitive first order reactions of ozone in the O_3/H_2O_2 and O_3/UV oxidation processes. *Ozone: Science & Engineering*, **19**, 13–37.

10 Andreozzi, R., Caprio, V., D'Amore, M.G., Insola, A. and Tufano, V. (1991) Analysis of complex reaction networks in gas-liquid systems, the ozonation of 2-hydroxypyridine in aqueous solutions. *Industrial Engineering and Chemical Research*, **30**, 2098–2104.

11 Beltrán, F.J., García-Araya, J.F. and Acedo, B. (1994) Advanced oxidation of atrazine in water–I. Ozonation. *Water Research*, **28**, 2153–2164.

12 Hoigné, J., Bader, H., Haag, W.R. and Staehelin, J. (1985) Rate constants of reactions of ozone with organic and inorganic compounds in water–III. Inorganic compounds and radicals. *Water Research*, **19**, 993–1004.

13 Yao, C.C.D. and Haag, W.R. (1991) Rate constants for direct reactions of ozone with several drinking water contaminants. *Water Research*, **25**, 761–773.

14 Hoigné, J. and Bader, H. (1976) Ozonation of water: role of hydroxyl radical reactions in ozonation processes in aqueous solutions. *Water Research*, **10**, 377–386.

15 Hoigné, J. and Bader, H. (1979) Ozonation of water: oxidation competition values of different types of waters used in Switzerland. *Ozone: Science & Engineering*, **1**, 357–372.

16 Gurol, M.D. and Nekouinaini, S. (1984) Kinetic behaviour of ozone in aqueous solutions of substituted phenols. *Industrial Engineering Chemical Fundamentals*, **23**, 54–60.

17 Beltrán, F.J., Encinar, J.M. and García-Araya, J.F. (1993) Oxidation by ozone and chlorine dioxide of two distillery wastewater contaminants: gallic acids and epicatechin. *Water Research*, **27**, 1023–1032.

18 Haag, W.R. and Yao, C.C.D. (1992) Rate constants for reaction of hydroxyl radicals with several drinking water contaminants. *Environmental Science & Technology*, **26**, 1005–1013.

19 Chramosta, N., De Laat, J., Doré, M., Suty, H. and Pouilot, M. (1993) Étude de la Dégradation de Triazine par O_3/H_2O_2 et O_3 Cinétique et sous produit d'Oxydation. *Water Supply*, **11**, 177–185.

20 Huber, M., Canonica, S., Park, G.-Y. and von Gunten, U. (2003) Oxidation of pharmaceuticals during conventional oxidation and advanced oxidation processes (AOP). *Environmental Science & Technology*, **37**, 1016–1024.

21 Laville, N., Ait-Aissa, S., Gomez, E., Casellas, C. and Porcher, J.M. (2004) Effect of human pharmaceuticals on cytotoxicity, EROD activity and ROS production in fish hepatoctes. *Toxicology*, **196**, 41–55.

22 Kolpin, D.W., Furlon, E.T., Meyer, M.T., Thurmann, E.M., Zaugg, S.D., Barber, L.B. and Buxton, H.T. (2002) Pharmaceuticals, hormones, and other organic wastewater contaminants in U.S. streams, 1999–2000: a national reconnaissance. *Environmental Science & Technology*, **26**, 1202–1211.

23 Prados, M., Paillard, H. and Roche, P. (1995) Hydroxyl radical oxidation processes for the removal of triazine from natural water. *Ozone: Science & Engineering*, **17**, 183–194.

24 Adams, C.D. and Randtke, S.J. (1992a) Ozonation byproducts of atrazine in synthetic and natural waters. *Environmental Science & Technology*, **26**, 2218–2227.

25 Bellamy, W.D., Hickman, T., Mueller, P.A. and Ziemba, N. (1991) Treatment of VOC-contaminated groundwater by hydrogen peroxide and ozone oxidation. *Research Journal Water Pollution Control Federation*, **63**, 120–127.

26 Duguet, J.P., Bruchet, A. and Mallevialle, J. (1990) Application of combined ozone–hydrogen peroxide for the removal of aromatic compounds from a groundwater. *Ozone: Science & Engineering*, **12**, 281–294.

27 Duguet, J.P., Bernazeau, F. and Mallevialle, J. (1990) Research note: removal of atrazine by ozone and ozone-hydrogen peroxide combination in surface water. *Ozone: Science & Engineering*, **12**, 195–197.

28 Arslan-Alaton, I. and Caglayan, A.E. (2005) Ozonation of procaine penicillin G formulation effluent, Part I: process optimization and kinetics. *Chemosphere*, **59**, 31–39.

29 Westerhoff, P., Nalinakumari, B. and Pei, P. (2006) Kinetics of MIB and geosmin oxidation during ozonation. *Ozone: Science & Engineering*, **28**, 277–286.

30 Paillard, H., Brunet, R. and Doré, M. (1987) Application of oxidation by a combined ozone/ultraviolet radiation system to the treatment of natural water, *Ozone: Science & Engineering*, **9**, 391–418.

31 Glaze, W.H., Kang, J.-W. and Chapin, D.H. (1987) The chemistry of water treatment processes involving ozone, hydrogen peroxide and ultraviolet radiation. *Ozone: Science & Engineering*, **9**, 335–352.

32 Aieta, E.M., Reagan, K.M., Lang, J.S., Mc Reynolds, L., Kang, J.-K. and Glaze, W.H. (1988) Advanced oxidation processes for treatment of groundwater contaminated with TCE and PCE. *American Water Works Association*, **5**, 64–72.

33 Elovitz, M.S., von Gunten, U. and Kaiser, H.-P. (2000) Hydroxyl radical/ozone ratios during ozonation processes. II. The effect of temperature, pH, alkalinity, and DOM properties. *Ozone: Science & Engineering*, **22**, 123–150.

34 Adams, C.D., Randtke, S.J., Thurmann, E.M. and Hulsey, R.A. (1990) Occurrence and treatment of atrazine and its degradation products in drinking water. Proceedings of the Annual Conference of American Water Works Association, 18th–21st June, Cincinnati, Ohio, USA, pp. 871–885.

35 Heil, C., Schullerer, S. and Brauch, H.-J. (1991) Untersuchung zur oxidativen Behandlung PBSM-haltiger Wässer mit Ozon. *Vom Wasser*, **77**, 47–55.

36 Gilbert, E. (1991) Kombination von Ozon/Wasserstoffperoxid zur Elimination von Chloressigsäure. *Vom Wasser*, **77**, 263–275.

37 Baus, C., Sacher, F. and Brauch, H.-J. (2005) Efficiency of ozonation and AOP for Methyl-tet-Butylether (MTBE) removal in waterworks. *Ozone: Science & Engineering*, **1**, 27–35.

38 Behar, D., Czapski, G. and Duchovny, I. (1970) Carbonate radical in flash photolysis and pulse radiolysis of aqueous carbonate solutions. *Journal of Physical Chemistry*, **74**, 2206–2210.

39 Mizuno, T., Tsuno, H. and Yamada, H. (2007) A simple model to predict formation of bromate ion and hypobromous acid/hypobromite ion through hydroxyl radical pathway during ozonation. *Ozone: Science & Engineering*, **29**, 3–11.

40 Hoigné, J. and Bader, H. (1977) Beeinflussung der Oxidationswirkung von Ozon und OH-Radikalen durch Carbonat. *Vom Wasser*, **48**, 283–304.

41 Duguet, J.P., Bruchet, A. and Mallevialle, J. (1989) Geosmin and 2-methylisoborneol removal using ozone or ozone/hydrogen peroxide coupling, ozone in water treatment. Proceedings of the 9th Ozone World Congress, Vol. 1, pp. 709–719.

42 Adams, C.D. and Randtke, S.J. (1992) Removal of atrazine from drinking water by ozonation. *Journal of American Water Works Association*, **84**, 91–102.

43 Masten, S.J. and Hoigné, J. (1992) Comparison of ozone and hydroxyl radical-induced oxidation of chlorinated hydrocarbons in water. *Ozone: Science & Engineering*, **14**, 197–214.

44 Legrini, O., Oliveros, E. and Braun, A.M. (1993) Photochemical processes for water treatment. *Chemical Reviews*, **93**, 671–698.

45 Forni, L., Bahnemann, D. and Hart, E.J. (1982) Mechanism of the hydroxide ion initiated decomposition of ozone in aqueous solution. *Journal of Physical Chemistry*, **86**, 255–259.

46 Glaze, W.H., Schep, R., Chauncey, W., Ruth, E.C., Zarnoch, J.J., Aieta, E.M., Tate, C.H. and Mc Guire, M.J. (1990) Evaluating oxidants for the removal of model taste and odor compounds from a municipal water supply. *Journal of American Water Works Association*, **5**, 79–83.

47 Boncz, M.A., Bruning, H., Rulkens, W.H., Zuilhof, H. and Sudhölter, E.J.R. (2005) The effect of salts on ozone oxidation processes. *Ozone: Science & Engineering*, **27** (4), 287–292.

48 Xiong, F. and Legube, B. (1991) Enhancement of radical chain reactions of ozone in water in the presence of an aquatic fulvic acid. *Ozone: Science & Engineering*, **13**, 349–361.

49 Peyton, G.R. and Glaze, W.H. (1987) *Mechanism of Photolytic Ozonation*, ACS Symposium-Series 327, American Chemical Society, Washington DC, pp. 76–87.

50 Xiong, F. and Graham, N.J.D. (1992) Research note: removal of atrazine through ozonation in the presence of humic substances. *Ozone: Science & Engineering*, **14**, 283–301.

51 Schulten, H.R. and Schnitzler, M. (1993) A state-of-the-art structural concept for humic substances. *Naturwissenschaften*, **80**, 23–30.

52 Buxton, G.V., Greenstock, C.L., Helman, W.P. and Ross, A.B. (1988) Critical review of rate constants for reactions of hydrated electrons, hydrogen atoms and hydroxyl radicals ($°OH/°O^-$) in aqueous solutions. *Journal of Physical Chemistry Reference Data*, **17**, 513–886.

53 De Laat, J., Berger, P., Poinot, T., Karpel Vel Leitner, N. and Doré, M. (1995) Modeling the oxidation of organic compounds by H_2O_2/UV. Estimation of kinetic parameters. Proceedings of the 12th Ozone World Congress, Lille, France, pp. 373–384.

54 Grasso, D. and Weber, W.J. (1989) Mathematical interpretation of aqueous-phase ozone decomposition rates. *Journal of Environmental Engineering*, **115**, 541–559.

55 Hoigné, J. (1998) Chemistry of aqueous ozone, and transformation of pollutants by ozonation and advanced oxidation processes, in *The Handbook of Environmental Chemistry Quality and Treatment of Drinking Water* (ed. J. Hubrec), Springer Verlag, Berlin.

56 Kelly, K.E. (1992) Investigation of Ozone Induced PCE Decomposition in Natural Waters. Department of Environmental Sciences & Engineering, University of North Carolina, Chapel Hill, NC.

57 Kuo, C.H., Li, K.Y., Wen, C.P. and Weeks, J.L. Jr. (1977) Absorption and decomposition of ozone in aqueous solutions. *Water*, **73**, 230–241.

58 Kuosa, M., Haarlo, H. and Kallas, J. (2005) Axial dispersion model for estimation of ozone self-decomposition. *Ozone: Science & Engineering*, **27**, 409–417.

59 Liao, C.-H. and Gurol, M.D. (1995) Chemical oxidation by photolytic decomposition of hydrogen peroxide. *Enviromental Science & Technology*, **29**, 3007–3014.

60 Mizuno, T., Tsuno, H. and Yamada, H. (2007) Development of ozone self-decomposition model for engineering design. *Ozone: Science & Engineering*, **29**, 55–63.

61 Sotelo, J.L., Beltrán, F.J., Benítez, F.J. and Beltrán-Heredia, J. (1987) Ozone decomposition in water – kinetic study. *Industrial Engineering and Chemical Research*, **26**, 39–43.

8
Modeling of Ozonation Processes

Models in general are a mathematical representation of a conceptual picture. They are important tools that help us understand the system being investigated. Model development is closely interconnected with experimental design. In order to develop a good model we have to know what variables influence the oxidation process and to possess data to analyze that show how they affect it. On the other hand, to design experiments, we already have to have a conceptual picture of the process in order to decide which variables are important and should be varied. The two steps should be carried out iteratively. The result – a good model – will allow us to continue to study the influence of important parameters on the oxidation process without having to make more experiments. Such a model, for instance, can be used to predict results for other compounds and/or at various operating conditions. This ability can be used to help design and optimize treatment plants.

Our conceptual picture of the ozonation process generally includes the physical processes and chemical processes shown in Figure 8.1. Mass balances and rate equations for the oxidants and their reactants are the basic tools for the mathematical description. The type of model to be used depends on how complex the reality is that we are trying to describe and the desired application of the model. As Levenspiel pointed out "the requirement for a good engineering model is that it be the closest representation of reality that can be treated without too many mathematical complexities. It is of little use to select a model which closely mirrors reality but is so complicated that we cannot do anything with it" ([1], p. 359). In cases where the complete theoretical description of the system is not desirable or achievable, experiments are used to calculate coefficients to adjust the theory to the observations; this procedure is called semi-empirical modeling.

The accepted degree of complexity is increasing with the advances made in computing capacity and modeling software. However, the increased complexity needs to be balanced against our ability to understand the model and the limits of its applicability. For example, a simple model for ozonation using lumped parameters to describe the reactants may be preferable to a more complicated one that considers many individual constituents by using scores of coupled equations to describe reaction rates and equilibrium. The computational complexity is not necessarily the problem, but rather the measurement of concentrations and rate constants for the individual constituents. The decision on how complex to make

Ozonation of Water and Waste Water. 2nd Ed.   Ch. Gottschalk, J.A. Libra, and A. Saupe
Copyright © 2010 WILEY-VCH Verlag GmbH & Co. KGaA, Weinheim
ISBN: 978-3-527-31962-6

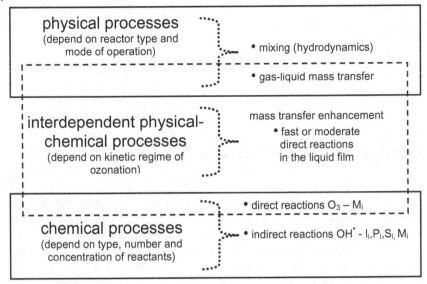

Figure 8.1 Physical and chemical processes important in ozonation modeling.

the model must be based on how the model will be used and what data is available or can be obtained. Model development, therefore, begins with determining the purpose of the model. Which questions must be answered by the model? Which outputs will answer the questions?

The next step in modeling is to identify all the physical and chemical processes that influence that output. These have been discussed in detail in previous chapters and are summarized in Figure 8.1. The physical processes include hydrodynamics (i.e., gas- and liquid-phase mixing) and mass transfer of ozone gas into the liquid phase, which depend on the reactor type and mode of operation. The chemical processes include, ideally, all direct and/or indirect reactions of ozone with water constituents, which depend on the type, number and concentrations of the reactants. And of course these processes cannot be seen separately. The kinetic regime of the reactions, how fast they take place, determines the degree with which the two processes influence each other. For instance, fast or moderate direct reactions of ozone with a pollutant M_i in the liquid film can enhance mass transfer, causing an increased flow rate of ozone into the system, which in turn increases the oxidation rate (see Section 6.2.1), making it essential that models take this interdependence into account.

The next problem is to determine which of the physical and chemical processes must be considered for the particular application at hand. Is a mathematical expression necessary to describe the influence of each process on the results, or can some of the processes be regarded as constant and considered in the boundary conditions for the model? Which set of variables have to be included in the model? The answers to these questions usually differ for drinking-water and waste-water applications because of the differences in reactant concentrations and often depend on the type and mode of the reaction system.

The need to consider the physical process of mixing is especially basic to all models – the mass balance must be chosen to describe the system hydrodynamics. For example, lab-scale systems used for experiments are often semibatch stirred-tank reactors or bubble columns that can be assumed to be completely mixed. This simplifies the description of the system hydrodynamics in the models immensely. Concentration gradients within the reactor do not have to be considered and the integral mass balance for a CFSTR can be used (Section 8.1). Mass balances for other reactor types can be found in Chapter 4.

Consideration of the other physical process, mass transfer, is often dependent on the application. It does not have to be considered explicitly in a model if the reaction is not mass-transfer limited, that is, if the reaction rate does not depend on the mass-transfer rate. This is often the case in drinking-water applications, but rare in waste-water applications. The independence from mass transfer must be verified for the system or the effect of mass transfer on the reaction must be considered in the model. This is discussed further in Section 8.1.

The models for the chemical processes also are different depending on the application. In drinking-water ozonation the chemical processes are well known. They comprise the complex hydroxyl radical-chain mechanism with reactions of initiation, propagation and termination (see Section 2.1.1 and Table 8.1). The main problem is to describe the influence of the initiators (I_i), promoters (P_i) and scavengers (S_i) on the OH° concentration, depending on the water matrix. There is a wealth of literature and complicated equations available to describe their interdependencies. Examples are given in Section 8.2 of various approaches that have lead to a sufficient – but not complete – description of experimental results in drinking water. In general, waste water has a similar or even more complex matrix and few – predominately direct – reactions are known that may develop in series and /or parallel. Therefore, simplified chemical models are generally used in this area, often showing limited validity when the boundary conditions change (Section 8.3).

This chapter can only present an overview of the modeling activity in ozonation. It is meant to familiarize the experimenter with the methods and literature on modeling, so that they can begin to use them to design their experiments and evaluate their results. Additional information can be found in a number of sources. A general introduction to modeling is provided by Hendricks for drinking water and by Dochain and Vanrolleghem [2, 3] especially for dynamical modeling in waste-water treatment, while Beltran [4] offers much information specifically on kinetic modeling of ozonation.

The discussion of modeling is divided into separate sections for each application after a short introduction to the general modeling problem and some basic modeling tools (Section 8.1). The division into two sections according to application area (Sections 8.2 and 8.3), however, is not based on clear criteria. The important difference is not the application *per se*, but rather the type of reactions that take place in the application, which in turn depends on the relative concentrations of the target compounds and other ozone-consuming compounds and dissolved ozone. The differences are first illustrated in the following section. Based on this understanding, the reader will be better able to choose which type of model to use depending on the questions to be answered and on the composition of the water.

8.1
Ozone Modeling

The general questions we are seeking to answer in our ozonation experiments and with a model are of course: Can an ozonation process be used to reach our treatment goal? and which is the most economical configuration? This must be broken down into more specific questions such as: Can ozonation remove the target compound to the required degree? Does the oxidation proceed via the direct reaction or via the hydroxyl radical? How much ozone and/ or other oxidant do we need to add to achieve the desired degree of removal of the target compound? What operating conditions will achieve the highest removal at the lowest cost?

The first specific question, whether the target compound can be oxidized by ozone via direct or indirect reactions must be answered on the basis of experimental data, either from the literature or from lab experiments. Experiments are necessary to answer the other questions, too; however, the experimental effort can be reduced if we can develop a model to describe our system. If we can build a model to predict how the various constituents in the water affect the oxidation rate of the target compound and its by-products, the number of experiments necessary to design and optimize a reactor system can be reduced substantially.

For instance, in drinking-water applications, models have been developed to describe the influence of the water constituents (e.g., pH, alkalinity). The effect of operating conditions, for example, changes in the ozone dose rate can be investigated with an appropriate model. The affect of the addition of H_2O_2 can be modeled by manipulating the $OH°$ concentration. If the model adequately describes the hydrodynamics in the reactor, the affect of the flow pattern and/or the position of the H_2O_2 dosing point, residence (or contact) time in the reactor can be studied.

8.1.1
General Description of the Ozone Modeling Problem

The success of the model depends on whether we understand and can quantify the complexity of the system. As we mentioned above, this is where we start to have problems in the waste-water area. To illustrate the difficulties we will first describe the time courses that we commonly find in most ozonation experiments, highlighting the typical differences between the two application areas.

Generally in both areas, experiments are made in semibatch lab-scale STRs or BCs where the liquid phase is assumed to be perfectly mixed. The concentrations of the target compound M and its oxidation intermediates change over time. The time courses for the concentrations shown in Figure 8.2 are typical for both areas.

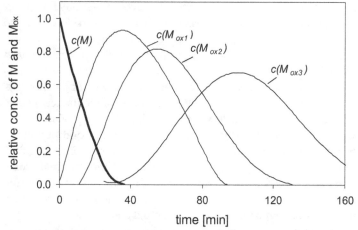

Figure 8.2 Change in the concentrations of the target compound M and its intermediates (M_{oxi}) as oxidation proceeds.

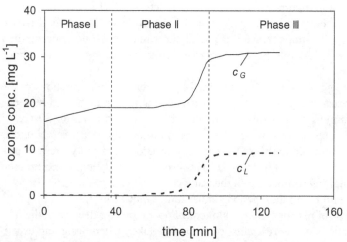

Figure 8.3 Change in the gas and liquid O_3 concentrations as ozonation of a high concentration of target compound M proceeds.

As oxidation proceeds, M disappears and intermediates M_{oxi} are formed, which can be further oxidized.

The differences between the applications are due to the type of the oxidation reactions that occur and their kinetic regimes as the target compound disappears. These differences can be illustrated by following the time courses of the ozone concentrations in the gas and liquid phases through the various stages in the semibatch ozonation of the target compound M at a relatively high concentration (Figure 8.3) and matching them to the kinetic regime.

Phase I: Initially a fast reaction of ozone with a high concentration of M (and/or other ozone-consuming compounds) occurs with no measurable dissolved-ozone concentration in the bulk liquid. It is assumed that no radical reactions occur in the film or bulk liquid. Oxidation takes place via direct reaction almost completely in the liquid film at the surface of the bubbles. The reaction is mass-transfer limited (fast or instantaneous regime). An enhanced mass-transfer rate results in a low gas-phase effluent concentration compared to the influent ozone gas-phase concentration. By-products less reactive with ozone are produced.

Phase II: As the concentration of the target compound and its oxidation intermediates decrease, there is a transition from the fast to moderate kinetic regime. The reaction rates decrease, less ozone is consumed, and the dissolved-ozone concentration increases. There is less enhancement of the mass-transfer rate and it decreases. The gas-phase ozone concentration rises. Oxidation still takes place mainly via direct reaction, but predominantly in the bulk liquid.

Phase III: As oxidation proceeds, there is a transition from the moderate to slow kinetic regime. The oxidation process is limited only by the reaction rate. The indirect reactions begin to dominate. The rate of ozone decomposition increases and radical-chain processes are initiated and promoted. Hydroxyl radical scavengers can play an important role. The by-products produced are not reactive with ozone and less and less organic material reactive with the hydroxyl radical being present. The mass-transfer rate and the dissolved-ozone concentration reach steady state. At this point the mass-transfer rate is equal to the ozone-decomposition rate.

The kinetic regime is mainly determined by the relative concentrations of the reactants. In waste-water applications in batch operation, with high concentrations of ozone-consuming compounds, usually all three phases occur over time. In batch drinking-water applications, where the concentration of the target compound is relatively low compared to that of the oxidants, reactions and concentrations profiles follow those described in *Phase III*. However, a relatively high concentration of ozone-consuming compounds in the water can cause changes in ozone concentrations in the initial reaction period similar to those in *Phase II*.

Since the type of reactor system and mode of operation also determine the relative concentrations of the reactants, care must be taken when designing experiments and models to be used for predicting results in various reactor types. For instance, CFSTRs are usually operated at steady state and designed to be completely mixed. The concentration is then constant over time and space. Consequently, the kinetic regime remains constant, too. It depends on the desired effluent concentration, which, for completely mixed reactors, is also the concentration in the reactor. Since this is very low in most cases, dissolved ozone is often present in the reactor and the kinetic regime is slow. Indirect reactions often play a role in the removal mechanism when the concentrations of ozone-consuming compounds in the reactor are low. Therefore, the oxidation reactions and products can be completely different in semibatch and CFSTR experiments.

Thus, we see that it is not the application *per se* that determines the model, but rather model choice depends on the reaction system and the relative concentrations of the reactants. All ozone-consuming compounds should be considered, not just the target compound. The difficulties in scaling up from lab-scale to full-scale also become more apparent. The large dependence of the oxidation rate on the physical processes in waste-water applications makes them much more dependent on scale. Scaling up from small to big often causes changes in the mixing patterns, which in turn affects the mass transfer. Even when great care is taken to keep the geometry and energy density of the reactor similar, the physical processes are invariably affected.

Sections 8.1.2 and 8.1.3 summarize the basic modeling tools for both application areas: the chemical model (reaction mechanisms) and mathematical models (mass balances and rate equations), while Sections 8.2 and 8.3 look more closely at the common methods of the modeling process for drinking-water and waste-water applications, presenting examples from the literature.

8.1.2
Chemical Model of Ozonation

The choice of the chemical model, that is, which reaction mechanisms are to be considered in the model, depends on what answers one wants. A reaction scheme of the chemical processes involved in ozonation would include, ideally, all direct and/or indirect reactions of ozone with the target compound, its oxidation products and with the water constituents, that is, initiators (I_i), promoters (P_i) and scavengers (S_i) that were discussed in Chapter 2 and are shown schematically in Figure 8.4. The pertinent indirect reactions can be chosen from the complex reaction scheme of the chain reaction (see Table 8.2) or simplified to the overall reaction for ozone decomposition initiated by the hydroxide ion (8.3).

The model would consider that these reactions with ozone and/or the hydroxyl radical can occur in series or in parallel, and would follow the oxidation steps of reaction intermediates until stable end products were reached. The reaction scheme would consist of the following types of reactions:

$$M + z_i O_3 \rightarrow M_{ox} \qquad (8.1)$$

$$M_{ox} + z_i O_3 \rightarrow \text{Products} \qquad (8.2)$$

$$OH^- + 3\,O_3 + H^+ \rightarrow 2\,OH^\circ + 4\,O_2 \qquad (8.3)$$

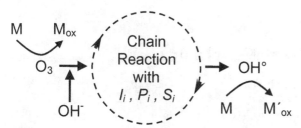

Figure 8.4 Schematic overview of direct and indirect reactions in water ozonation.

$$M + z_i OH° \rightarrow M'_{ox} \tag{8.4}$$

$$M'_{ox} + z_i OH° \rightarrow Products \tag{8.5}$$

However, in reality, we must reduce the complexity of the system and concentrate on those compounds and reactions that are of the most importance for answering our questions. Often in drinking water only the disappearance of the target compound is of interest, while in waste water partial mineralization may be the goal. The main oxidizing species may also differ according to the application. Therefore, a reaction scheme must be developed that considers the most relevant reactions. Drinking-water applications may only consider Equations 8.3–8.5, while waste-water applications may only include Equations 8.1 and 8.2. Examples of reaction schemes for each application can be found in Sections 8.2 and 8.3.

The next step is to develop a mathematical description of this reaction scheme for the reaction system used. Rate equations using concentrations and rate constants are needed to calculate the reaction rates for each reaction considered in the chemical model. These are used in mass balances on the reaction system to describe the appearance and disappearance of the reactants and their products. The mathematical equations are discussed in the following section.

8.1.3
Mathematical Model of Ozonation

The mathematical models in ozonation are comprised of mass balances and rate equations for the oxidants and their reactants. These have been presented in detail in previous Chapters 4 and 7, but are summarized here for reference for the following discussion. The balance for a CFSTR is used to illustrate the method. See Chapter 4 for other reactor types.

8.1.3.1 Mass Balances

The mass balance for the absorption of ozone in a CFSTR under the assumption that the gas and liquid phases are ideally mixed ($c_L = c_{Le}$, $c_G = c_{Ge}$) and considering the enhancement factor E for mass transfer, are as follows:

gas phase:

$$V_G \cdot \frac{dc_G}{dt} = Q_G(c_{Go} - c_G) - E \cdot K_L a \cdot V_L(c_L^* - c_L) - \sum r_{Gi} \cdot V_G \tag{8.6}$$

liquid phase:

$$V_L \cdot \frac{dc_L}{dt} = Q_L(c_{Lo} - c_L) + E \cdot K_L a \cdot V_L(c_L^* - c_L) - \sum r_{Li} \cdot V_L \tag{8.7}$$

total material balance at steady state:

$$Q_G(c_{Go} - c_G) - \sum r_{Gi} \cdot V_G = Q_L(c_{Lo} - c_L) - \sum r_{Li} \cdot V_L \tag{8.8}$$

where $\Sigma r_{Li} \cdot V_L$ and $\Sigma r_{Gi} \cdot V_G$ represent the sum of all ozone-consuming reaction rates in the liquid and gas phases. The direct reactions with the target compound and its intermediate oxidation products, as well with all other oxidizable organic and inorganic compounds are considered, in addition to the ozone-decomposition reactions with the water matrix. Usually, the ozone-consumption rates in the gas phase $\Sigma r_{Gi} \cdot V_G$ are neglected for nonvolatile compounds; however, the gas system should be checked that measurable ozone decomposition is not occurring (see Section 5.6).

Note: In the following section we will use $r(O_3)$ to represent the liquid phase ozone-consuming reaction rate instead of r_L to be more complementary to the symbol for the target reaction rate $r(M)$.

When the chemical reactions increase the rate of mass transfer above that found due to mere physical ozone absorption, the mass-transfer enhancement E must be evaluated. Fast or moderate reactions of ozone with a target compound M or other ozone-consuming compounds can result in $E > 1$. If only slow direct reactions of ozone with the reactants occur in the bulk liquid, no enhancement takes place ($E = 1$) and the experimental effort as well as the mathematical description becomes less complicated.

A general rule is that in drinking water with few ozone-consuming compounds, micropollutant removal usually proceeds in the slow kinetic regime without mass-transfer enhancement, whereas in waters with high amounts of ozone-consuming compounds, the fast or moderate kinetic regime usually applies and $E > 1$. However, it is important to remember that in nonsteady-state operation, all three regimes can occur in succession and E can change over time. This is discussed further in Section 8.3.

The model can include relationships between the mass-transfer rate and the operating conditions if such changes are to be modeled (see Section 6.2 for details), or the mass-transfer coefficient and equilibrium concentration can be given as constants.

The mass balance for the target compound M in the liquid phase, assuming that compound M is nonvolatile and is not stripped, is:

$$V_L \cdot \frac{dc(M)}{dt} = Q_L(c(M)_o - c(M)) - r(M) \cdot V_L \tag{8.9}$$

In the case of a semibatch process with $Q_L = 0$, Equation 8.9 reduces to:

$$\frac{dc(M)}{dt} = -r(M) \tag{8.10}$$

The mass balance for the intermediate considers the generation rate as well as the oxidation rate. For example, for the generation rate $r(M)$ and the oxidation rate $r(M_{ox})$, the mass balance becomes:

$$V_L \cdot \frac{dc(M_{ox})}{dt} = Q_L(0 - c(M_{ox})) + r(M) \cdot V_L - r(M_{ox}) \cdot V_L \tag{8.11}$$

8.1.3.2 Rate Equations

In order to solve the mass balances, we need kinetic models to describe the rates of ozone consumption and the reaction rates of the target compound and its intermediates. The rate of the direct ozonation of a compound is usually determined by the concentration of each reactant, so a second-order rate equation is used. The oxidation rate of the compound and the accompanying ozone-consumption rate are related to each other by the stoichiometric coefficient z_i for the reaction (see 8.1):

$$r(O_3)_i = z_i \cdot r(M_i) = z_i \cdot k_i \cdot c(M_i) \cdot c_L \tag{8.12}$$

For the ozone-consumption rate, we can define an overall rate $r(O_3)$ that takes the sum of all ozone reactions into consideration. The direct reactions of ozone with the target compound and its intermediate oxidation products, as well as with all other oxidizable organic and inorganic compounds are considered, in addition to the ozone decomposition reactions with the water matrix.

$$r(O_3) = \sum r(O_3)_i = \sum z_{O3i} r(M_i) + k_d \cdot c_L \tag{8.13}$$

The parameters that influence the ozone-consumption rate are the concentrations and the rate constants, shown schematically in Figure 8.5. The dissolved-ozone concentration is determined by how fast ozone is transferred into solution through mass transfer and how fast it reacts in solution with the target compound and intermediates as well as with compounds in the water matrix to produce OH° radicals.

The equation to describe the oxidation rate of the target compound M should take into account that direct and indirect reactions can occur at the same time. Therefore, the oxidation rate is usually described as the sum of the two reaction

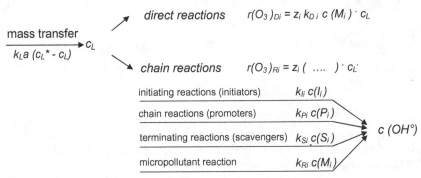

Figure 8.5 Parameters affecting the dissolved-ozone concentration c_L.

rates. A second-order rate equation is commonly accepted for each reaction pathway, so that:

$$r(M) = [k_D c_L + k_R c(OH°)] \cdot c(M) \qquad (8.14)$$

M target compound
k_D reaction rate constant for the direct reaction
k_R reaction rate constant for the indirect (radical) reaction

We see then that for each reaction considered in the chemical model, rate constants and concentrations of the reactants are needed to calculate the reaction rates. Rate constants can be determined experimentally or taken from the literature (see Chapter 7). Such a rate equation can be written for each compound of interest, for example, for each intermediate M_{ox} and/or each compound involved in the indirect chain reaction mechanism.

8.1.3.3 Solving the Model

The combination of the mass balances with the appropriate rate equations produces a mathematical description of the system. The resulting model is usually a set of ordinary differential equations. The method for its solution depends on the assumptions and boundary conditions chosen for the reaction system. Various simplifications can be made to reduce the modeling complexity. Which simplifications are chosen depends in part on the kinetic regime of the reaction discussed in Section 8.1.2. Some of the most common simplifications are:

1) **The contribution of indirect reactions is negligible** ($k_D \geq k_d$). This simplification is most often used for the fast to moderate kinetic regime in *Phase I*. No dissolved ozone to initiate the chain mechanism is available in the liquid bulk. However, in the moderate kinetic regime there are compounds that react directly with dissolved ozone, that is, in the liquid bulk, but their reaction rate constants k_D would be higher than that for ozone decomposition by the hydroxide ion ($k_d = 70\,M^{-1}s^{-1}$) [5].

2) **The contribution of the direct reaction is negligible** ($k_D \ll k_d$). This simplification is often used for reactions in *Phase III* because the reaction rate constants for the direct reaction of ozone with many organic micropollutants are low ($k_D < 1\,M^{-1}s^{-1}$). In contrast, ozone decompositions comparatively fast (reaction with OH^-: $k_d = k_1 = 70\,M^{-1}s^{-1}$; reaction with hydrogen peroxide, resp. HO_2^-: $k_9 = 2.8 \times 10^6\,M^{-1}s^{-1}$) forming hydroxyl radicals, which in turn react extremely fast with almost all types of organic micropollutants ($k_R = 10^6-10^{11}\,M^{-1}s^{-1}$). In such cases, only indirect reactions need be modeled.

3) **The concentrations for ozone and the radicals involved in the indirect reaction are assumed to be at steady state, that is, the change in the concentrations over time is negligible.** The assumption of a steady-state ozone concentration for the direct reaction is common for reactions in *Phase III*. It is based on the relatively large concentration of ozone compared to that of the micropollutants.

This is usually the case in drinking-water applications even for batch experiments. Measurements can be made to confirm the assumption. The assumption of a steady-state concentration for the hydroxyl radical is based on experimental results. Several authors have shown that the indirect reaction of OH° with organic compounds is pseudo-first order due to the steady-state concentration of the hydroxyl-radicals (e.g., [6, 7]). Further assumptions are that the concentrations of the intermediates, for example, $O_2^{°-}, O_3^{°-}, HO_3^{°}$ and organic radicals, are also at steady state [8].

If the steady-state simplification for the concentrations of ozone and OH° is used, the reaction rate can be simplified to a pseudo-first-order relationship:

$$r(M) = k'c(M) \tag{8.15}$$

with:

k' reaction rate coefficient, pseudo-first order

The reaction rate coefficient k' is dependent on the direct (k_D) and indirect (k_R) oxidation rate constants as well as the ozone and OH° steady-state concentrations:

$$k' = \underbrace{k_D c_{LSS}}_{\text{direct}} + \underbrace{k_R c(OH°)_{ss}}_{\text{indirect}} \tag{8.16}$$

where the subscript ss denotes steady-state conditions

The model and method of solving it usually depends on the application. In drinking water the oxidation of the intermediates is often not modeled. Simplifications 2 and 3 are commonly used. The resulting system of equations can often be solved analytically.

In waste-water applications, the oxidation of the target compound as well as major intermediate compounds has been of interest. If the model includes mass balances for multiple reactions with nonlinear rate expressions (i.e., not first-order reaction rates) the equations must be solved with numerical methods. For example, these can be solved using a high-quality ordinary differential equation (ODE) solver, available in commercial software such as MATLAB, Maple, etc. or open source software such as Octave, R, etc.

8.1.4
Summary

Although the above chemical and mathematical models are used in both drinking-water and waste-water applications, the following discussion of modeling is divided into separate sections for each application. The different emphasis in each section comes from the differences presented in Section 8.1.1. They can be summarized as:

In drinking-water oxidation, the relative concentration of the oxidant to target compound is very high so that

- the reaction rate is usually limiting;
- the indirect reaction is very important;
- whereas the direct reaction is often small to negligible;
- the concentration of O_3 and $OH°$ are assumed to be at steady state [6], as well as that of the intermediates like $O_2^{°-}$, $O_3^{°-}$, $HO_3^°$ and organic radicals [8].

In waste-water oxidation, the relative concentration of the oxidant to target compound is very low so that

- mass transfer is often limiting;
- the direct reaction is very important and partly occurs in the liquid film;
- whereas the indirect reaction is often small to negligible;
- the concentration of dissolved ozone c_L is often zero, with the elimination rate of M as a function of the mass-transfer rate.

This leads to largely differing areas of interest for the two fields. Models to determine the concentration of the $OH°$ as a function of the water matrix are needed in drinking-water applications, while models that incorporate the influence of the hydrodynamics and mass transfer on the reaction rate are necessary for waste-water applications.

8.2
Modeling of Drinking-Water Oxidation

Models in drinking-water applications are often developed to describe the influence of the water constituents on the oxidation rate of the target compound. The effect of operating conditions, for example, the ozone dose rate or the addition of H_2O_2 can also be investigated by adding additional equations to describe the effect on the mass transfer and the $OH°$ production from H_2O_2. If the model adequately describes the hydrodynamics in the reactor, the effect of the flow pattern and/or the position of the H_2O_2 dosing point, residence (or contact) time in the reactor can be studied. The use of models can reduce the number of experiments necessary to design and optimize a reactor system substantially. Models that adequately describe the processes taking place in the system for one compound can often be applied to predict the results with other compounds.

8.2.1
Chemical and Mathematical Models

Often, only the disappearance of the target compound is of interest, so that intermediate reactions are neglected in the chemical model. Furthermore, direct reactions do not play a significant role in the oxidation of the target compound. Instead, the indirect reactions and the chain reactions involved in the production of radicals must be described by the model. Therefore, Equations 8.3 and 8.4 and some subset of reactions from Table 8.2 are usually considered in drinking-water applications.

The mathematical models used to describe the oxidation rates of the target compound are based on material balances on the reaction system and reaction rate equations. The general equations for drinking-water ozonation are first summarized in Table 8.1 and then the steps in model development are explained. Since experiments are often carried out semibatch in STRs or BCs where the liquid phase is assumed to be perfectly mixed, the liquid-phase balances can be simplified to Equations 8.18 and 8.19. Furthermore, in most cases the reactions

Table 8.1 Mathematical model for drinking-water ozonation (slow kinetic regime).

Mass balances
  Ozone gas phase

$$V_G \frac{dc_G}{dt} = Q_G (c_{Go} - c_{Ge}) - k_L a (c_L^* - c_L) V_L \qquad (8.17)$$

Liquid phase
- Ozone

$$\frac{dc_L}{dt} = k_L a (c_L^* - c_L) - \left(\sum r(O_3)_i + k_d \cdot c_L \right) \qquad (8.18)$$

- Target compound

$$\frac{dc(M)}{dt} = -r(M) \qquad (8.19)$$

Reaction rates
- Ozone
  Oxidation

$$r(O_3)_i = z_i \cdot r(M_i) = z_i \cdot k_i \cdot c(M_i) \cdot c_L \qquad (8.20)$$

  Decomposition

$$r_d = k_d c_L^n \qquad (8.21)$$

- Target compound

$$r(M) = [k_D c_L + k_R c(OH^\circ)] \cdot c(M) \approx k_R c(OH^\circ) \cdot c(M) \qquad (8.22)$$

Mass-transfer rate

$$k_L a = f(Q_G, \text{energy input}) \qquad (8.23)$$

Optional: mass balance and rate equation for H_2O_2

$$V_L \frac{dc(H_2O_2)}{dt} = Q_L c(H_2O_2)_o - r(H_2O_2) \cdot V_L \qquad (8.24)$$

$$r(H_2O_2) = k_{H2O2} c(H_2O_2)^n \qquad (8.25)$$

8.2 Modeling of Drinking-Water Oxidation

develop in the slow kinetic regime, so that no mass-transfer enhancement occurs ($E = 1$) and the balances for ozone can be written without considering E (8.17 and 8.18).

The rate equations are needed to solve the mass balances. The ozone-consumption rate for each compound considered is related to the reaction rate by the stoichiometric coefficient z_i for the reaction (8.20). The rate of ozone decomposition must also be quantified (8.21) and is discussed below. Generally the rate of disappearance of the target compound of interest (M) can be described by the sum of the direct and indirect reaction rate, however, as pointed out in Section 8.1.3, if the target compound reacts rather slow with ozone the assumption that direct reactions are negligible can be used (8.22).

To describe the effect of operating conditions in the system, an equation is usually required for the mass-transfer coefficient as a function of the important operating conditions. Often semiempirical correlations are used (see Section 6.2). It is also possible to include the effect of the addition of H_2O_2 to the reaction system by adding a mass balance for H_2O_2 to the model and an appropriate reaction rate equation (8.24 and 8.25). This of course affects the production of $OH°$ and must be considered in the determination of the hydroxyl radical concentration.

The above general equations define our system, but they must be put together to answer the questions asked. For example, to describe the effect of the water matrix on the disappearance rate of the compound, the next step is to combine the target compound mass balance with the rate equation. Using the simplification of only indirect reactions, it becomes:

$$\frac{dc(M)}{dt} = -k_R c(OH°) \cdot c(M) \tag{8.26}$$

This second-order relationship contains the indirect rate constant and two unknowns: the concentration of hydroxyl radicals and the concentration of target compound. Values for rate constants can be found in the literature for a wide variety of substances or can be measured (see Chapter 7). At least a second equation is needed for our model to calculate the OH radical concentration as a function of the water matrix. It is to determine this OH radical concentration that a wide variety of methods with varying degrees of complexity have been developed. These are found in Section 8.2.2.

Before moving on to hydroxyl radical determination, though, it is important to recall one of the simplifications introduced in Section 8.1.3: the assumption of steady state for the concentrations of ozone and the radicals involved in the indirect reaction. As we will see below, all methods to determine the $OH°$ concentration require this assumption. Moreover, almost all methods require knowledge of the steady-state dissolved-ozone concentration due to its influence on the decomposition rate and, thus, on the generation rate of $OH°$. Therefore, we will first illustrate the general method to develop an additional equation for the dissolved-ozone concentration from the liquid-phase ozone balance.

The dissolved-ozone concentration is strongly influenced by the ozone-decomposition rate, which in turn is dependent on the water matrix, with its initiators (I_i), promoters (P_i) and scavengers (S_i) as well as on the organic reactants (Figure 8.4). This can be seen in the liquid phase ozone mass balance. A simplification for the ozone-consumption rate is to lump the individual rates in the chain reaction and the direct reactions with the ozone consuming compounds in the water matrix together in the ozone-decomposition rate r_d.

The ozone-decomposition rate has been studied by several authors. The dependence of the rate on the dissolved-ozone concentration varied from zero to second order. The variation is ascribed to differences in the water matrices and analytical methods. Several authors have developed empirical or semiempirical relationships as a function of the water constituents (see Section 7.2 and Table 7.1). However, to illustrate the method, a simple function will be used here with $n = 1$:

$$r_d = k_d \cdot c_L \qquad (8.27)$$

The liquid-phase ozone mass balance (8.18) can be simplified for steady-state concentrations of dissolved ozone c_{LSS} to:

$$0 = k_L a (c_{LSS}^* - c_{LSS}) - k_d \cdot c_{LSS} \qquad (8.28)$$

Rearranging gives the following equation for the dissolved steady-state ozone concentration:

$$c_{LSS} = \frac{k_L a (c_{LSS}^*)}{k_d + k_L a} \qquad (8.29)$$

Methods for measuring $k_L a$ and k_d for the water matrix used can be found in Section 6.3.

The model for semibatch then reduces to the mass balance for the target compound in conjunction with methods for determining the steady-state concentrations of dissolved ozone and the hydroxyl radicals. It can usually be solved analytically. The complexity usually lies in the determination of the OH° concentration, as will be seen below.

8.2.2
Methods to Determine the Hydroxyl-Radical Concentration

In order to predict the degradation rate of a target compound, knowledge of the concentration of the oxidizing species, that is, ozone and/or OH° is necessary. While the ozone concentration can be measured directly; the OH° concentration is not directly accessible due to the low steady-state concentration. A diversity of methods has been developed for determining the OH° concentration as a function of the water matrix and operating conditions. In particular, in natural waters the target compound and the other matter present in the water compete for the available $OH°$ that are produced by the decomposition of ozone. Therefore, a large number of compounds can be involved in the reaction and therefore must be considered in the methods.

Depending on the amount of kinetic information available for the initiating and scavenging reactions, the $OH°$ concentration can be assessed using rate equations at different levels of complexity. Alternatively, methods to indirectly measure the concentration have been developed. This section provides examples of the various types of methods and the underlying chemical models that have been developed to determine the $OH°$ concentration. These are described as well as the modeling results achieved with each method.

The first example illustrates the use of indirect measurements, where experimental data and literature values are combined to determine the $OH°$ concentration. Then, a general model based on the important known reaction mechanisms is presented, followed by models with varying degrees of complexities and simplifications. Further help in choosing an appropriate model can be found in the discussion by Peyton [8]. In his "Guidelines for the Selection of a Chemical Model for Advanced Oxidation Processes", he gives an overview of the different levels of model complexity and a general idea when to use which level. Glaze et al. [9] provide further examples for the calculation of the $OH°$ concentration for various AOPs, including ozonation, ozone with hydrogen peroxide and the hydrogen peroxide / ultraviolet combination.

8.2.2.1 Indirect Measurement

Various methods have been developed to indirectly measure the $OH°$ concentration via the direct measurement of the oxidation rate of a model compound. The so-called R_{CT} concept proposed by Elovitz and von Gunten [10] is based on the measurement of the oxidation of a model compound that only reacts with the hydroxyl radical ($OH°$), for example, para-chlorobenzoic acid (pCBA, with $k_D \leq 0.15 \, \text{l mol}^{-1}\text{s}^{-1}$ and $k_R = 5 \times 10^9 \, \text{l mol}^{-1}\text{s}^{-1}$). This measurement is performed in a *homogeneous* batch system. An ozone stock solution is added to the water, for example, any type of natural water, in which the model compound is dissolved. Then, the disappearance of both, the model compound as well as ozone, are monitored as a function of time. If, furthermore, the condition holds that the model compound pCBA is at a very low concentration, then its presence does not contribute much to the overall scavenging of $OH°$ ($k_D \, c(pCBA) \ll \Sigma \, k_{Mi} c(M_i)$ $c(OH°)$). Thus the total rate of $OH°$ scavenging approximately equals $\Sigma \, k_{Mi} c(M_i)$ $c(OH°)$ and, in such a batch system or under the conditions of plug flow, the rate of pCBA disappearance, being only due to radical reactions, can be expressed as follows

$$\frac{-dc(pCBA)}{dt} = -k_{R,pCBA} c(pCBA) c(OH°) \tag{8.30}$$

Rearranging and integrating Equation 8.30 yields

$$\ln\left(\frac{c(\text{pCBA})}{c(\text{pCBA})_o}\right) = -k_{R,\text{pCBA}} \int c(\text{OH}°) dt \qquad (8.31)$$

where the term $\int c(\text{OH}°) dt$ represents the time-integrated concentration of hydroxyl radicals, which is synonymous with the OH°-exposure or the OH°-ct. Thus, the relative decrease of the model compound at any time interval $t_o - t$ is an indirect measurement of the OH°-exposure ($\int c(\text{OH}°) dt$).

And since the hydroxyl radicals are generated from ozone during ozone decomposition one can relate the OH°-exposure (or OH°-ct) to the – experimentally measured – O_3-exposure (or O_3-ct). Elovitz and von Gunten [10] named this ratio of exposures the R_{CT}-value and defined it as

$$R_{CT} = \frac{\int c(\text{OH}°) dt}{\int c(O_3) dt} \qquad (8.32)$$

From their measurements they found, that the R_{CT}-value was a constant for a specific water, which also means that R_{CT} is independent of the reaction time and therefore

$$R_{CT} = \frac{c(\text{OH}°)}{c(O_3)} \qquad (8.33)$$

also holds. This seems plausible if one considers that the underlying chemical model "*only*" involves the complex radical-chain mechanism depending on the initiators, promoters and scavengers contained in the individual water (see Table 8.4), but with negligible contribution of the radical and direct reactions of the added model compound (pCBA).

Substituting Equation 8.32 into Equation 8.31 gives

$$\ln\left(\frac{c(\text{pCBA})}{c(\text{pCBA})_o}\right) = -k_{R,\text{pCBA}} R_{CT} \int c(O_3) dt \qquad (8.34)$$

and shows that R_{CT} can be calculated from the experimentally measured decrease in the concentrations of pCBA and O_3. A graphical determination of R_{CT} is possible by plotting the left-hand side of Equation 8.34 versus the ozone exposure ($\int c(O_3) dt$).

Summing up, the R_{CT}-value characterizes the ozone reactivity in any natural or prepared water and can help to model the disappearance of any other oxidizable compound, for example, by competition kinetics (see Section 7.2.1; i.e. Equation 7.23) if the boundary conditions applied to pCBA still hold.

However, the work with natural waters from various sources showed that direct reactions of ozone are often observed within the first 60–120 s in such an experiment (e.g., with fast-reacting nitrites). This means that the graph generated from Equation 8.34 would show two regions with constant, but different R_{CT}-values.

8.2 Modeling of Drinking-Water Oxidation

A similar concept, the oxidation-competition value Ω, was developed earlier by Hoigné and Bader [11]. It also defines the amount of ozone needed to reach a certain degree of elimination of a given pollutant in a natural water, when this is only due to radical reactions.

As in the case of the determination of R_{CT}, a short initial period with ozone consumption due to fast direct reactions has to be considered in the appropriate experiments. To account for this, the rate constant for the direct oxidation has to be included into the mathematical expression for the disappearance of the compound (M_i) under study:

$$\ln\left(\frac{c(M_i)}{c(M_i)_o}\right) = -(k_{Di} + k_{Ri}R_{CT})\int c(O_3)dt \qquad (8.35)$$

Thus, Ω is also a global parameter that characterizes the ability of natural water to form hydroxyl radicals (OH°). Compared with the oxidation-competition value Ω the R_{CT}-concept is a more universal approach, since it is independent of the reaction time, while Ω refers to a specific reaction time in which exactly 67% of the model compound (e.g., pCBA) is removed.

For a more detailed description of both concepts the reader is referred to the original literature or to the comprehensive book of Beltrán [4]. The latter also includes their mathematical treatment in other than batch systems, for example, in a CSTR.

Another possibility to calculate the OH° concentration is directly from an experimentally determined reaction rate coefficient k', and from Equation 8.36:

$$c(OH°)_{SS} = \frac{k' - k_D c_{LSS}}{k_R} \qquad (8.36)$$

The reaction rate coefficient k' can be determined from the experimental observation of the elimination of a model compound over time (Chapter 7). It can be combined with literature values for k_D and k_R, and the experimentally determined ozone concentration to calculate the amount of OH-radicals that were present during the oxidation process. And if indirect oxidation dominates, which is often the case, the term for the direct oxidation ($k_D c_{LSS}$) can be neglected.

It is possible to use this OH° concentration to predict k' for the oxidation of other compounds under the same conditions. Von Gunten et al. [6] calculated the actual concentration of OH° using this general and easy way for the ozonation of surface water at neutral pH in a two-stage pilot plant. Atrazine was used as the model compound ($k_D = 6$–$24\,M^{-1}s^{-1}$), ozone decomposition was assumed to be of first order and the reactors completely mixed. Based on this model they were able to precisely predict the formation of bromate BrO_3^- by oxidation of bromide Br^- for a full-scale water-treatment plant. Bromate is a disinfection by-product (DBP) from the ozonation of bromide-containing waters and of concern because of its carcinogenic effects in animal experiments (see also Chapter 3).

8.2.2.2 Complete Radical-Chain-Reaction Mechanism

An equation for the steady-state concentration of OH° can be developed from the steady-state mass balance on the hydroxyl radical considering the important indirect reaction mechanisms for production and consumption of the OH°. This can be done for ozonation alone or for O_3/H_2O_2 systems. Table 8.2 summarizes the reaction mechanisms that have been considered in the following equations. The derivation of the equations can be found in [12].

for O_3:

$$c(OH°)_{SS} = \frac{c_{LSS}\left\{2k_1 10^{pH-14} + \sum_{i=1}^{m} k_{Ii} c(I_i)\right\}}{k_R c(M) + \sum_{i=1}^{n} k_{Pi} c(P_i) + \sum_{i=1}^{o} k_{Si} c(S_i)} \quad (8.37)$$

Table 8.2 Hydroxyl radical (indirect) reactions initiated by OH^- or H_2O_2.

Reaction	Reaction rate constant[a]
Initiating reaction (initiators)	
$O_3 + OH^- \rightarrow O_2^- + HO_2°$	$k_1 = 70\, l\,mol^{-1}\,s^{-1}$
$H_2O_2 \leftrightarrow HO_2^- + H^+$	$pK_a = 11.8$
$HO_2^- + O_3 \rightarrow HO_2° + O_3^-$	$k_9 = 2.8 \times 10^6\, l\,mol^{-1}\,s^{-1}$
$O_3 + I \rightarrow products$	k_I
Chain reaction (promoters)	
$O_3 + O_2^- \rightarrow O_3^- + O_2$	$k_2 = 1.6 \times 10^9\, l\,mol^{-1}\,s^{-1}$
$HO_3° \rightarrow OH° + O_2$	$k_3 = 1.4 \times 10^8\,s^{-1}$
$HO_4° \rightarrow O_2 + HO_2°$	$k_5 = 2.8 \times 10^4\,s^{-1}$
$OH° + O_3 \rightarrow HO_4°$	$k_4 = 2.0 \times 10^9\, l\,mol^{-1}\,s^{-1}$
$OH° + P \rightarrow products$	k_P
Scavenging reaction (scavengers)	
$OH° + HO_2° \rightarrow O_2 + H_2O$	$k_6 = 3.7 \times 10^{10}\, l\,mol^{-1}\,s^{-1}$
$OH° + HO_2^- \rightarrow HO_2° + OH^-$	$k_{10} = 7.5 \times 10^9\, l\,mol^{-1}\,s^{-1}$
$OH° + H_2O_2 \rightarrow HO_2° + H_2O$	$k_{11} = 2.7 \times 10^7\, l\,mol^{-1}\,s^{-1}$
$OH° + CO_3^{2-} \rightarrow OH^- + CO_3^-$	$k_7 = 4.2 \times 10^8\, l\,mol^{-1}\,s^{-1}$
$OH° + HCO_3^- \rightarrow OH^- + HCO_3°$	$k_8 = 1.5 \times 10^7\, l\,mol^{-1}\,s^{-1}$
$HCO_3^- \leftrightarrow CO_3^{2-} + H^+$	$pK_a = 10.25$
$OH° + HPO_4^{2-} \rightarrow OH^- + HPO_4^-$	$k_{12} = 2.2 \times 10^6\, l\,mol^{-1}\,s^{-1}$
$OH° + H_2PO_4^- \rightarrow OH^- + H_2PO_4°$	$k_{13} < 10^5\, l\,mol^{-1}\,s^{-1}$
$H_2PO_4^- \leftrightarrow HPO_4^{2-} + H^+$	$pK_a = 7.2$
$OH° + M \rightarrow products$	k_R
$OH° + S \rightarrow products$	k_S
Net equation for the generation of OH°	
$2\,O_3 + H_2O_2 \rightarrow 2\,OH° + 3\,O_2$	–
$3\,O_3 + OH^- + H^+ \rightarrow 2\,OH° + 4\,O_2$	–

a Reaction rate constants taken from literature as examples.

for O_3/H_2O_2:

$$c(OH°)_{ss} = \frac{c_{LSS}\left\{2k_1 10^{pH-14} + 2k_9 10^{pH-11.6} c(H_2O_2)_{ss} + \sum_{i=1}^{m} k_{Ii} c(I_i)\right\}}{k_R c(M) + \sum_{i=1}^{n} k_{Pi} c(P_i) + \sum_{i=1}^{o} k_{Si} c(S_i)} \quad (8.38)$$

The numerator contains all hydroxyl radical forming reactions and all initiating reactions are summarized ($\Sigma k_{Ii} c(I_i)$). The denominator contains all hydroxyl-radical-consuming reactions. The second term includes all reactions with promoters ($\Sigma k_{Pi} c(P_i)$), the third the reactions with scavengers ($\Sigma k_{Si} c(S_i)$). Similarly, the steady-state concentrations of ozone and hydrogen peroxide can be calculated from the liquid-phase mass balances. This model has been successfully used to describe the O_3/H_2O_2 treatment of methyl-*tert*-butyl ether (MTBE) in spiked tap water and in contaminated groundwater [13] as well as of mineral oil-contaminated waste waters [14].

The terms for the initiators $\Sigma k_{Ii} c(I_i)$ and scavengers $\Sigma k_{Si} c(S_i)$ are those with the biggest uncertainty or error. The values of these terms vary depending on the water matrix. Generally, there is not enough kinetic information at hand to use such a complicated model, so that various simplifications have been developed. The following examples illustrate the various approaches that can be taken.

8.2.2.3 Semiempirical Method Based on Observable Parameters

In this approach, the OH-radical concentration is calculated with a semiempirical formula that takes the main influencing – and observable – parameters into account, that is, pH, TOC, UV absorbance at 254 nm (SAC_{254}), inorganic carbon, alkalinity and concentration of the micropollutant M [15]. The OH° concentration is, therefore, calculated as a function of:

$$c(OH°) = f(c_L, pH, c(TOC), c(H_2O_2), SAC, c(M), c(HCO_3^-), c(Alk))$$

The model was developed considering the radical reactions with initiators, promoters and scavengers similar to the method in Section 8.2.2.2 and then modified to reflect the results of 75 experiments carried out in natural waters.

$$c(OH°)_{ss} = \frac{c_{LSS}\{2k_1 10^{pH-14} + 3.16 \times 10^{-7} 10^{0.42 pH} c(TOC) + 2k_9 10^{pH-11.6} c(H_2O_2)\}}{k_R c(M) + c(HCO_3^-)(k_7 + k_8 10^{pH-10.25})} \quad (8.39)$$

And, furthermore, based on 56 experiments carried out in natural waters the ozone decomposition term was correlated with observable parameters:

$$r_d = w c_L \quad (8.40)$$

where:

$$\log 10\, w = -3.93 + 0.24\, pH + 0.7537 \log SAC_{254} + 1.08 \log c(TOC) - 0.19 \log c(Alk) \quad (8.41)$$

$c(Alk)$: alkalinity in $mg\, l^{-1}\, CaCO_3$

Laplanche et al. [15] developed the above equations as part of their model to predict the removal of micropollutants by ozonation in a bubble column, with or without hydrogen peroxide addition. They modeled the bubble column as an n-CSTR in order to investigate the effect of different dosing points for the H_2O_2 addition. The reader is referred to the original paper for the complete model. It successfully described the results for three different experimental setups taken from the literature with ozonated spiked tap water or raw water of similar composition. In each case a good correlation between the experimental and calculated data was obtained.

8.2.2.4 Semiempirical Method Based on Observed Hydroxyl Radical Initiating Rate

The development of the next method started with the attempt to ignore all initiators by setting $\Sigma k_{Ii}\, c(I_i)$ to zero in Equations 8.37 [16]. $k_R\, c(M) + \Sigma\, k_{Pi}\, c(Pi)$ were assumed to be constant and expressed as $k_R\, c(M)_o$. The results showed that the initiator term was not negligible. The fitting of the experimental data with a hydroxyl-radical initiating rate β was necessary. Thus, the OH° concentration was calculated as a function of:

$$c(OH°) = f(c_L, pH, \beta, c(M), c(HCO_3^-))$$

Beltrán et al. studied the oxidation of atrazine in distilled water with and without scavengers (carbonate, *tert*-butanol) under different conditions of ozone partial pressure, pH, temperature and concentration of the scavengers [16]. The oxidation was modeled considering the above-mentioned reaction mechanisms (see Table 8.2) as well as the molar balances for atrazine and ozone. In a first approach they neglected the initiators and assumed that the oxidation term of atrazine and the intermediates are constant

$$k_R c(M) + \sum_{i=1}^{n} k_{Pi} c(P_i) = k_R c(M)_o \qquad (8.42)$$

with the explanation that atrazine and the intermediates have nearly the same rate constants due to the unselective character of OH° and the similar structure of the intermediates, and the concentration of all micropollutants is constant because of the negligible mineralization.

By comparing the calculated values for the OH° concentration according to Equations 8.37 and 8.38 in connection with Equation 8.42 to the experimental data, it was shown that the initiation term was not negligible and the simplified model not appropriate. Thus, the authors defined a hydroxyl-radical initiating rate β that includes all possible initiating reactions, calculated by the OH° concentration difference between the experiment and the model.

$$c(OH°)_{ss} = \frac{c_{LSS} 2k_1 10^{pH-14} + \beta}{k_R c(M) + \sum k_{Si} c(S_i)} \qquad (8.43)$$

β: hydroxyl radical initiating rate

The scavenger term $k_{Si}\,c(S_i)$ was calculated from known values. If only carbonates are present, it is calculated from

$$\sum_{i=1}^{2} k_{Si} c(S_i) = k_8 c(\text{HCO}_3^-) + k_7 c(\text{CO}_3^{2-}) \tag{8.44}$$

and based on the assumption that the formed carbonate radicals $\text{HCO}_3^\circ/\text{CO}_3^{\circ-}$ do not promote the radical chain reaction. Note, however, that this is not always the case (for further information see [17]).

With the help of this correction term the atrazine concentration could be calculated with a precision of ±15%. The ß-term showed no trend, no general considerations were possible. The prediction of the dissolved-ozone concentration was only possible in the presence of scavengers. The effect of ozone consumption due to direct reactions with the intermediates was estimated.

The same method was used for the combination of ozone with ultraviolet radiation [18]. Here, the previous model was extended by terms considering the degradation rate due to direct photolysis. Again, the model was able to predict the experimentally observed oxidation time course of atrazine as well as the liquid-ozone concentration with a precision of ±15%. However, the estimated concentration of hydrogen peroxide, which is formed during the AOP, was much too high. A plausible reason may be that decomposition of hydrogen peroxide by reaction with OH-radicals was not considered in the model.

In general, the model from Beltrán et al. uses ß as a kind of correction factor. Without experimental data from which ß can be calculated, no prediction is possible.

8.2.2.5 Empirical Selectivity for Scavengers

The following method was developed by Glaze and Kang [19–21] for the ozone/hydrogen peroxide process in a semibatch reactor. High organic background was not taken into account, nor any initiators other than OH⁻ and H_2O_2. In the model, the scavenger effect of $\text{HCO}_3^-/\text{CO}_3^{2-}$ is reduced by a selectivity term S_{PER}. The experimental effect of $\text{HCO}_3^-/\text{CO}_3^{2-}$ was found to concur with the model prediction. The OH° concentration was calculated as a function of:

$$c(\text{OH}^\circ) = f(c_L, \text{pH}, c(H_2O_2), c(M), c(\text{HCO}_3^-), S_{PER})$$

where the selectivity $S_{PER} = f(c(H_2O_2), c(\text{CO}_3^{\circ-}), \text{other reactions with } \text{CO}_3^{\circ-})$

A part of the OH-radicals destroyed by the reaction with the carbonate species will be regenerated by the following reactions:

$$\text{CO}_3^{\circ-} + H_2O_2 \rightarrow HO_2^\circ + \text{HCO}_3^- \quad k_{14} = 8 \times 10^5 \text{ l mol}^{-1} \text{ s}^{-1} \tag{8.45}$$

$$\text{HCO}_3^\circ + HO_2^- \rightarrow HO_2^\circ + \text{HCO}_3^- \quad k_{15} = 4.3 \times 10^5 \text{ l mol}^{-1} \text{ s}^{-1} \tag{8.46}$$

With the help of a term called selectivity S_{PER}, the part of $CO_3^{\circ-}$ that reacts with hydrogen peroxide as a fraction of all reactions with $CO_3^{\circ-}$ is included in the model [20]:

$$S_{PER} = \frac{k_{14}c(CO_3^{\circ-})c(H_2O_2)}{k_{14}c(CO_3^{\circ-})c(H_2O_2) + \sum_{j=1}^{n} k_j c(CO_3^{\circ-})c(I_j)} \quad (8.47)$$

S_{PER}: selectivity

That means that these reactions increase the steady-state concentration of OH°. The scavenger term of carbonate must be reduced by the selectivity term. The following expression is used in the model:

$$c(OH^\circ)_{SS} = \frac{c_{LSS}\{2k_1 10^{pH-14} + 2k_9 10^{pH-11.8} c(H_2O_2)_{SS}\}}{k_R c(M) + (1 - S_{PER})\{k_8 c(HCO_3^-) + k_7 c(CO_3^{2-})\}} \quad (8.48)$$

To determine the reaction rate Glaze and Kang divided the ozone/hydrogen peroxide system into three regions depending on different relationships concerning the molar transfer rate $F(O_3)$ of ozone and the molar feed rate $F(H_2O_2)$ of hydrogen peroxide.

Region 1: $F(O_3) > 2 F(H_2O_2)$: Ozone is measurable in the liquid, hydrogen peroxide is consumed as fast as it is added.

Region 2: $F(O_3) = 2 F(H_2O_2)$: Stoichiometric point where two moles of O_3 and one mole of H_2O_2 are necessary to produce two moles of OH°

Region 3: $F(O_3) < 2 F(H_2O_2)$: Hydrogen peroxide is measurable in the liquid, ozone is consumed as fast as it is added. ($S_{PER} = 1$)

For each region certain assumptions were possible. For further details the reader is referred to the original articles [19–21].

The model was tested with distilled water containing carbonate. As a tracer substance PCE was used to predict the rate constant with OH° [20]. Good results were found in the ozone transfer-limited region 3, in region 1 and 2 the results were poorer. In region 3 it was also possible to calculate the hydrogen-peroxide concentration in very good agreement with the measured values.

Using this approach of a selectivity term S_{PER} Sunder and Hempel [22] successfully modeled the oxidation of small concentrations of tri- and perchloroethylene ($c(M)_o$ = 300–1300 µg l^{-1}) by ozone and hydrogen peroxide in a synthetic ground water (pH = 7.5–8.5; $c(S_i)$ = 1–3 mmol CO_3^{2-} l^{-1}). In this study an innovative reaction system was used; the oxidation was performed in a tube reactor and mass transfer of gaseous ozone to pure water was realized in a separate contactor being located in front of the tube reactor. In this way a *homogeneous* reaction was achieved. Since the two model compounds react very slowly with molecular ozone ($k_D < 0.1 1 \text{mol}^{-1}\text{s}^{-1}$), nearly the complete oxidation was due to the action of hydroxyl radicals, which were produced from the two oxidants (O_3/H_2O_2). With $S_{PER} = 0.2$

Table 8.3 Overview of the discussed chemical models.

System / Author	Matrix	Promoter incorporated	Scavenger incorporated	Specials	Results
O_3/H_2O_2 [15]	Tap and raw water	OH^-, H_2O_2	HCO_3^-, CO_3^{2-}, TOC	Semiempirical	Good correlation
O_3, O_3/UV [16]	Distilled water with/ without scavengers	OH^-, UV	HCO_3^-, CO_3^{2-}, tert-butanol	β: hydroxyl radical initiating term	Prediction was only possible with β included
O_3/H_2O_2 [19–21]	Distilled water with carbonate	OH^-, H_2O_2	HCO_3^-, CO_3^{2-}	S_{PER}: selectivity	Only good results in Region 3
O_3/H_2O_2 [22]	Synthetic ground water	OH^-, H_2O_2	HCO_3^-, CO_3^{2-}	S_{PER}: selectivity	$S_{PER<} = 0.2$ excellent correlation

the experimental results could be excellently modeled and S_{PER} did not vary significantly with the variation of the initial oxidant concentrations and the pH.

8.2.2.6 Summary of Chemical and Mathematical Models for Drinking Water

Table 8.3 gives an overview of the discussed models.

One of the unsolved questions regarding the chemical-kinetic model is the influence of scavengers and initiators especially with water containing organic matrix. Further investigations are necessary to answer the question: to what amount does the organic matrix promote or terminate the chain reaction. Here, models could act as a tool to understand this process in more detail.

8.2.3
Models Including Physical Processes

Modeling of micropollutant removal during drinking- or ground-water ozonation is often found to be a process without mass-transfer enhancement, meaning that all reactions develop in the liquid bulk (slow kinetic regime, $E \leq 1$) [23]. However, in an attempt to develop a more general model, the work of Marinas et al. [23] was extended by the incorporation of fast reactions through the introduction of mass-transfer enhancement into the model [24]. The mass transfer and the resulting liquid-ozone concentration are calculated via mass balances.

Most models are based on the assumption of completely mixed liquid. Examples for more complex hydrodynamic systems useful in scale-up are given by Marinas et al. [23] as well as Laplanche et al. [15]. They achieved good predictions of the ozone (and hydroxyl radical) residual concentrations in pilot-scale bubble columns.

The importance of correct assumptions on reactor hydrodynamics became evident in modeling studies on full-scale applications of drinking-water disinfection with ozone [25, 26]. In both cases a n-CFSTR-in-series model was successfully used to describe the reactor operation. Using computational fluid dynamics (CFD) it was shown that the ct-value could be doubled by improving the plug-flow behavior in a full-scale ozone contactor [25]. This was experimentally verified. As pointed out above, for successful modeling in such systems all relevant reactions and the corresponding reaction rate constants should be known.

8.3
Modeling of Waste-Water Oxidation

In contrast to drinking-water applications, mathematical models in waste-water applications generally have to describe both physical and chemical processes. The basic steps are the same though. The appropriate mass balance to describe the reactor mixing and mass transfer must be chosen as well as rate equations to describe the important reactions. The added complexity is that the interactions of direct chemical reaction and mass transfer can enhance the transfer rate, which in turn speeds up the oxidation rate. This comes about if the oxidation rate is dominated by the rate of the direct reaction of ozone with the contaminants. Since the kinetic regime of the reactions, how fast they take place, determines the degree with which the two processes influence each other, much of the effort in modeling waste-water ozonation is devoted to determining the kinetic regime. This, however, can change over time and space in the reaction system depending on the reactor type and mode of operation, and makes the task of modeling even more challenging than in drinking-water ozonation.

Although this transition in the kinetic regime and the accompanying changes in the mass-transfer rate is a general phenomenon in semibatch experiments with waters containing relatively high concentrations of ozone-consuming compounds, these changes have long been neglected in waste-water studies. This was due to the fact that dissolved ozone was often not measured, and probably because of the lack of an adequate model for the moderate kinetic regime, which was, however, recently developed by Benbelkacem and Debellefontaine [27] and is discussed in the following section.

In addition, the complexity of the chemical reactions in waste waters containing multiple compounds offers major hurdles. Many direct reactions with ozone may occur in series or in parallel and indirect reactions may also play a role. Not surprisingly, modeling of waste-water ozonation has rather seldom been performed. Until today, with the exception of a handful of studies, virtually no attempt has been made in scale-up, neither from lab- to pilot- nor to full-scale.

The following section starts with an example of waste-water modeling that considers the mass transfer and continues with examples from the literature concentrating on the attempts that have been made to handle the complexity of the chemical processes as well as model mass-transfer enhancement.

8.3.1
Chemical and Mathematical Models

Modeling of waste-water ozonation is most often carried out for semibatch processes with the aim to predict the disappearance of the initial target compound(s), the concentration profiles of the (main) intermediates as well as those of liquid ozone and ozone in the reactor off-gas. The chemical models applied in the lab-scale studies on waste-water modeling discussed below can be roughly categorized by their reference to waters initially containing *low (1 to 3), intermediate (approx. 10)* or *high (>10)* numbers of pollutants. Since the chemical complexity can sky-rocket as oxidation proceeds, the reduction of complexity in a chemical model is often a prerequisite for any success. Even if initially only one compound is present, the number of oxidation products that subsequently undergo direct and indirect reactions with ozone can be quite large before mineralization occurs. Multiple pathways are usually viable. It is the complexity of the chemical processes, the large number of waste-water constituents, and the interdependency of mass transfer and chemical reaction that makes waste-water ozonation modeling so difficult. And thus in almost every case the models can be considered as semiempirical.

The mathematical models used in this area are based on the mass-balance equations of the appropriate hydrodynamic model for the gas and the liquid phases, including terms to describe the mass-transfer rate and combined with the reaction rate equations to describe the oxidation rates of the compounds of interest. Generally, however, only direct reactions of ozone are considered. For the direct reactions of ozone with an individual compound a second-order irreversible reaction is generally assumed, even for the reactions developing in the liquid-side film. Studies with a *low* number of compounds initially present are usually made in synthetic waste waters that often contain only one (methanol [28], phenol [29], 2-hydroxypyridine [30], 3-methylpyridine [31], maleic acid [5]) or two compounds (2,4- and 2,6-dichlorophenol [32]). In some, the oxidation products and their subsequent direct reactions with ozone are incorporated into the chemical model (e.g., [5, 29, 30, 31]), whereas in others, the primary goal is to model solely the initial compounds' disappearance along with the changes in the liquid- and gas-phase ozone concentrations (e.g., [32]). Few study the reaction to mineralization (e.g., [5, 28, 31, 33]).

In most cases the hydrodynamic conditions are assumed to be completely mixed for the liquid phase, while the gas phase has been variously modeled as plug flow, completely mixed, or as consisting of a number of n-CFSTR in series. It is noteworthy that in earlier studies on waste-water ozonation modeling an assessment of the regime of ozone absorption was often not performed and the accompanying mass-transfer enhancement was not determined explicitly. However, sometimes an influence of mass transfer on the pollutant removal rates was postulated from the differences between the experimental observations and the initial model results. So that in some cases to adjust the model to the experimental observations the influence of mass transfer was included by incorporating a time-dependent value of $k_L a$ into the model.

The work of Gurol and Singer [29] was an early milestone in this regard. The ozonation of phenol ($c(M)_o$ = 5–50 mg l^{-1}) in a small bubble column (gas-wash bottle, $V_L = 0.5$ l) could be modeled well if the effect of phenol on the mass-transfer rate was considered by introducing $k_L a = fc(M)$ into the model. To illustrate the procedure for modeling, the chemical model and the mathematical equations for this case are presented in detail here.

Phenol reacts fast with ozone and produces a variety of oxidation products. The production and further oxidation of three main intermediates, catechol CA, hydrochinone HY, muconic acid MUA, to stable end products (STA, e.g., glyoxal, glyoxylic acid, oxalic acid, formic acid) are followed in the model (Table 8.4). The chemical model considers only direct reactions.

The general mathematical model is summarized in Table 8.5. It includes the nonsteady-state mass balances for ozone in the gas and liquid phases for semibatch operation (8.17 and 8.49), where ozone decomposition is disregarded in both phases, as well as the mass balance for the original organic compound (8.19). The mass balance for the oxidation products (8.50) takes the generation and oxidation rates into account.

In general, the ozone-consumption rate considers the sum of all ozone-consuming reactions. In this case, the reactions with the target compound and

Table 8.4 Chemical model (direct reactions) in a semibatch study on the ozonation of phenol (Gurol and Singer [29]).

No.	Reaction	Stoichiometric factor	Direct reaction rate constant (k_D)
	Phenol (Ph)		$k_{Ph} = 400$ l mol^{-1} s^{-1}
1	Ph + O$_3$ → CA (catechol)	$z_1 = 1$	$k_1 = 0.8$ l mol^{-1} s^{-1}
2	Ph + O$_3$ → HY (hydrochinone)	$z_2 = 1$	$k_2 = 6.0$ l mol^{-1} s^{-1}
3	Ph + O$_3$ → MUA (muconic acid)	$z_3 = 1$	$k_3 = 88$ l mol^{-1} s^{-1}
4	Ph + 3 O$_3$ → STA	$z_4 = 3$	$k_4 = 305$ l mol^{-1} s^{-1} = $k_{Ph} - (k_1 + k_2 + k_3)$
5	CA + 3 O$_3$ → STA	$z_{CA} = 3$	$k_{CA} = 1.0 \times 10^3$ l mol^{-1} s^{-1}
6	HY + 3 O$_3$ → STA	$z_{HC} = 3$	$k_{HY} = 0.72 \times 10^3$ l mol^{-1} s^{-1}
7	MUA + 2 O$_3$ → STA	$z_{MUA} = 2$	$k_{MUA} = 2.2 \times 10^3$ l mol^{-1} s^{-1}

STA, stable end products, for example, glyoxal, glyoxylic acid, oxalic acid, formic acid.

Table 8.5 Mathematical model (direct reactions) for a semibatch study on the ozonation of phenol (Gurol and Singer [29]).

Mass balances
Ozone gas phase

$$V_G \frac{dc_G}{dt} = Q_G(c_{Go} - c_{Ge}) - k_L a(c_L^* - c_L)V_L \tag{8.17}$$

Liquid phase
Ozone (ozone decomposition neglected)

$$\frac{dc_L}{dt} = k_L a(c_L^* - c_L) - \sum r(O_3)_i \tag{8.49}$$

Target compound

$$\frac{dc(M)}{dt} = -r(M) \tag{8.19}$$

and intermediates

$$\frac{dc(M_{ox})}{dt} = r(M) - r(M_{ox}) \tag{8.50}$$

Reaction rates (only direct reaction considered)
Ozone

$$r(O_3)_i = z_i \cdot r(M_i) = z_i \cdot k_i \cdot c(M_i) \cdot c_L \tag{8.20}$$

Target compound

$$r(M_i) = k_i \cdot c(M_i) \cdot c_L \tag{8.51}$$

and intermediates

$$r(M_{oxi}) = k_{oxi} \cdot c(M_{oxi}) \cdot c_L \tag{8.52}$$

Mass-transfer rate

$$k_L a(O_3) = f(c(Ph)): \text{decreasing from 0.042 to 0.016 s}^{-1} \text{ with decreasing } c(Ph) \tag{8.53}$$

intermediates are taken into account, while ozone decomposition is neglected. The ozone-consumption rate for each compound is related to the reaction rate by the stoichiometric coefficient z_i given in Table 8.4. Here, the phenol reaction rate was modeled considering the four parallel oxidation reactions

$$r_{Ph} = k_{Ph} \cdot c(Ph) \cdot c_L \tag{8.54}$$

where $k_{Ph} = k_1 + k_2 + k_3 + k_4$, so that:

$$r_{Ph} = k_1 \cdot c(Ph) \cdot c_L + k_2 \cdot c(Ph) \cdot c_L + k_3 \cdot c(Ph) \cdot c_L + k_4 \cdot c(Ph) \cdot c_L \tag{8.55}$$

Furthermore, the further oxidation of the three intermediates shown in Table 8.4 was modeled.

An additional equation is needed for the mass-transfer coefficient in the model as a function of the important operating conditions. In this case, the mass-transfer rate was found to be affected by the presence of phenol in the water, which can cause smaller bubbles to form. Its effect changes as its concentration decreases due to oxidation. The change in $k_L a(O_3)$ values due to different concentrations of phenol was experimentally assessed, indirectly using mass-transfer experiments with oxygen (O_2) in combination with Equation 6.26 by using a factor of $k_L a(O_3) = 0.833\, k_L a(O_2)$. The empirical *alpha* factor (Equation 6.25), commonly used to account for changes in the physical mass-transfer rate due to interfacial area, etc., was used to adjust for the changes in mass transfer. A correction for the mass-transfer coefficient was developed as a function of the phenol concentration in the model: $k_L a(O_3)$ decreased nonlinearly from an almost constant value of $0.042\,s^{-1}$ at phenol concentrations between 300 and $50\,mg\,l^{-1}$ to approximately $0.016\,s^{-1}$ at phenol concentrations between 50 and $15\,mg\,l^{-1}$.

The next step is to combine the mass balances with the rate equations to obtain a mathematical description of the change in the concentrations over time. Substituting the reaction rates into the mass balances results in the following equations:

Ozone

$$\frac{dc_L}{dt} = k_L a(c_L^* - c_L) - (k_1 + k_2 + k_3 + 3k_4) \cdot c(\text{Ph}) \cdot c_L - 3k_{CA} \cdot c(\text{CA}) \cdot c_L$$
$$- 3k_{HY} \cdot c(\text{HY}) \cdot c_L - 2k_{MUA} \cdot c(\text{MUA}) \cdot c_L \tag{8.56}$$

Target and intermediate compounds

$$\frac{dc(\text{Ph})}{dt} = -r_{\text{Ph}}$$
$$= -k_1 \cdot c(\text{Ph}) \cdot c_L - k_2 \cdot c(\text{Ph}) \cdot c_L - k_3 \cdot c(\text{Ph}) \cdot c_L - k_4 \cdot c(\text{Ph}) \cdot c_L \tag{8.57}$$

$$\frac{dc(\text{CA})}{dt} = k_1 \cdot c(\text{Ph}) \cdot c_L - k_{CA} \cdot c(\text{CA}) \cdot c_L \tag{8.58}$$

$$\frac{dc(\text{HY})}{dt} = k_2 \cdot c(\text{Ph}) \cdot c_L - k_{HY} \cdot c(\text{HY}) \cdot c_L \tag{8.59}$$

$$\frac{dc(\text{MUA})}{dt} = k_3 \cdot c(\text{Ph}) \cdot c_L - k_{MUA} \cdot c(\text{MUA}) \cdot c_L \tag{8.60}$$

To find the solution to the problem, the differential equations were solved simultaneously with numerical methods. The disappearance of phenol and the concentration profiles of three intermediate products were described well when the appropriate $k_L a(O_3)$ values were used.

Although phenol can increase the physical mass transfer by causing smaller bubbles to form, mass-transfer enhancement due to reactions in the liquid-side film was probably also taking place in the initial phase of this fast reaction. In such a case, the increased mass transfer should be accounted for by the introduction of the enhancement factor into the mass-transfer rate equation (see Section 6.2.1 for more detail). Of course, inclusion of E in the model requires that its value can be determined. In a nonsteady-state system, it also changes over time as the reaction rates slow down and the enhancement of mass transfer declines. The location of the reactions moves from within the liquid film, back to the bulk liquid (see Section 8.1.1). It would be possible to calculate the mass-transfer rate directly from the gas-phase mass balance using the appropriate ozone measurements in the gas and liquid phases over time and Equation 8.6. However, the model of mass-transfer enhancement using the enhancement factor alone cannot be used to determine how much of the compound is reacting in the film.

In order to model reactions with transitions in the location of the reaction from the film to bulk, Benbelkacem and Debellefontaine [5, 27] proposed a model based on the film theory that can distinguish between reaction in the liquid-side film and bulk liquid. They developed their model for reactions in the moderate kinetic regime where mass transfer is enhanced due to reactions taking place in the liquid-side film besides in the bulk liquid. When, as oxidation proceeds, the reactions transition from moderately fast to slow, there are changes in the location of the reactions and mass-transfer enhancement over time. The mathematical model takes these changes into account.

The model was validated using the moderately fast oxidation of a simple symmetrical organic acid maleic acid and its oxidation products in a semibatch process ($c(M)_0 = 441\,mg\,l^{-1}$; $V_L = 8.5\,l$) [5]. The chemical model describes the oxidation of maleic acid to formic acid, glyoxylic acid and oxalic acid (Table 8.6). An assumption in the model development was that only the reaction with maleic acid was moderately fast and occurred in the film, while the oxidation of the products was slow and takes place only in the bulk liquid. The rate constants calculated with the model agreed with this assumption and are given in Table 8.6.

Table 8.6 Chemical model (direct reactions) for a semibatch ozonation of maleic acid (Benbelkacem et al. [5]).

No.	Reaction	Stoichiometric factor	Direct reaction rate constant (k_D)
1	MA + O_3 → z_{FA} FA + z_{GA} GA + CO_2 + HO_2		$k_{MA} = 12 \times 10^3\,l\,mol^{-1}\,s^{-1}$
2	FA + O_3 → CO_2 + HO_2	$z_{FA} = 0.9$	$k_{FA} = 0.5 \times 10^3\,l\,mol^{-1}\,s^{-1}$
3	GA + O_3 → z_{OA} OA + CO_2 + HO_2	$z_{GA} = 1$	$k_{GA} = 0.09 \times 10^3\,l\,mol^{-1}\,s^{-1}$
4	OA + O_3 → CO_2 + HO_2	$z_{OA} = 0.7$	$k_{OA} = 0.71\,l\,mol^{-1}\,s^{-1}$

The mathematical model uses mass balances on the gas and liquid ozone phases considering mass transfer and consumption rates, as well as mass balances for the oxidation of the four substances similar to those in Table 8.5 (Table 8.7). In this case, the enhancement factor E is used in the ozone balances (8.61 and 8.62) to describe the increase in mass transfer due to reaction. In addition, the ozone decomposition is also taken into account.

The major difference to the previous model comes in the description of the reaction rates. The rate equations have to consider the reaction taking place in the film as well as in the bulk liquid. This has been done by introducing a novel parameter D, the "depletion factor", into the model. The concept is explained in detail in Section 6.2.1. In short, by looking not only at the enhanced mass-transfer rate into the film, but also at the depleted rate coming out of the film, the amount consumed in the film can be calculated. Therefore in the model, the term $(E-D)$ is introduced into the rate equation (8.64) in order to account for the maleic acid reactions in the film. The enhancement as well as the depletion

Table 8.7 Mathematical model for the moderate kinetic regime in the semibatch ozonation of maleic acid (Benbelkacem et al. [5]).

Mass balances
  Ozone gas phase

$$V_G \frac{dc_G}{dt} = Q_G(c_{Go} - c_{Ge}) - E k_L a (c_L^* - c_L) V_L \quad (8.61)$$

  Liquid phase
    Ozone

$$\frac{dc_L}{dt} = E k_L a (c_L^* - c_L) - \left(\sum_{i=1}^{n} r(O_3)_i + k_d \cdot c_L \right) \quad (8.62)$$

  Target compound

$$\frac{dc(M)}{dt} = -r(M) \quad (8.19)$$

  And intermediates

$$\frac{dc(M_{ox})}{dt} = r(M) - r(M_{ox}) \quad (8.50)$$

Reaction rates (only direct reactions considered)
  Ozone

$$r(O_3)_i = r(M_i) \quad (8.63)$$

  Target compound

$$r(O_3)_{MA} = r(MA) = [(E-D) k_L a (c_L^* - c_L) + k_{MA} \cdot c(MA) \cdot c_L] \quad (8.64)$$

  and intermediates

$$r(O_3)_{oxi} = r(M_{oxi}) = -k_i \cdot c(M_{oxi}) \cdot c_L \quad (8.65)$$

8.3 Modeling of Waste-Water Oxidation

factors are then calculated as time-dependent parameters. Since the stoichiometric coefficient z_i for ozone is one in each reaction above (Table 8.6), the ozone-consumption rates and reaction rates are equal (8.63). The other reactions are assumed to take place only in the bulk liquid and follow the general second-order rate equation (8.65).

No additional equation is necessary to describe the changes in the mass-transfer rate, since the gas-phase ozone balance is used to find the mass-transfer rate. The values for $E \cdot k_L a$ can be calculated directly using the gas-phase mass balance from the appropriate ozone measurements in the gas and liquid phases over time in conjunction with experimentally determined physical mass-transfer coefficients for the operating conditions.

In the following step the mass balances are combined with the rate equations to obtain a mathematical description of the change in the concentrations over time. Substituting the reaction rates into the liquid-phase mass balances results in the following equations:

Ozone

$$\frac{dc_L}{dt} = E k_L a (c_L^* - c_L) - \left[(E-D) k_L a (c_L^* - c_L) + k_{MA} \cdot c(MA) \cdot c_L \right]$$
$$- k_{FA} \cdot c(FA) \cdot c_L - k_{GA} \cdot c(GA) \cdot c_L - k_{OA} \cdot c(OA) \cdot c_L - k_d \cdot c_L \quad (8.66)$$

Target compounds and intermediates

The stoichiometric coefficient z here relates the moles of the oxidation products to one mole maleic acid.

$$\frac{dc(MA)}{dt} = -r(MA) = -[(E-D) k_L a (c_L^* - c_L) + k_{MA} \cdot c(MA) \cdot c_L] \quad (8.67)$$

$$\frac{dc(FA)}{dt} = z_{FA} [(E-D) k_L a (c_L^* - c_L) + k_{MA} \cdot c(MA) \cdot c_L] - k_{FA} \cdot c(FA) \cdot c_L \quad (8.68)$$

$$\frac{dc(GA)}{dt} = z_{GA} [(E-D) k_L a (c_L^* - c_L) + k_{MA} \cdot c(MA) \cdot c_L] - k_{GA} \cdot c(GA) \cdot c_L \quad (8.69)$$

$$\frac{dc(OA)}{dt} = z_{OA} k_{GA} \cdot c(GA) \cdot c_L - k_{OA} \cdot c(OA) \cdot c_L \quad (8.70)$$

The mass balances were solved numerically using a Runge–Kutta method in Matlab software. The adjustment of the mass-transfer rate to describe experimental results occurs via the change in the term $(E\text{-}D)$ over time, instead of adjusting the physical mass-transfer coefficient. This requires a separate subroutine to solve the equations used to calculate the terms E and D using a trial and error method (see [5, 27] for details).

The results of the model show that only approximately one third of the initial reaction of maleic acid took place within the film. Although the dissolved-ozone concentration was almost zero, a large portion (approximately 68%) of the reaction of maleic acid with ozone did occur in the bulk liquid at the beginning of the experiment. Furthermore, the oxidation rate in all three phases of the kinetic regime (see Section 8.1.1) could be described with the model.

Trying to model the disappearance of 2-hydroxypyridine (HPYR) during ozonation in a semibatch STR ($c(M)_o$ = 0.85–5.15 mmol l^{-1}; n_{STR} = 380 min^{-1}, $k_La(O_3) \approx 0.02\,s^{-1}$ in bidistilled water), the mechanism of the direct reactions of ozone with the initial compound as well as with the most important organic intermediates, the ozone decomposition and the action of hydrogen peroxide were included in the chemical model [30]. In an attempt to improve this model, Tufano et al. [34] postulated that k_La varied with reaction time. As the reaction time increased, the concentration of HPYR decreased, changing the value of k_La, which is comparable with the effect of phenol in the work of Gurol and Singer [29]. Therefore k_La was used as an adjustable parameter in the model $k_La(O_3) = f(t)$. In an experiment HYPR of $c(M)_o$ = 1.22 mmol l^{-1} was completely removed within approximately 6 min. Assuming that $k_La(O_3)$ decreased from an initial value of 0.082 s^{-1} to 0.054 s^{-1} within the first 8 min of the experiment the time courses of the concentrations of HPYR, five oxidation products and the reactor off-gas c_{Ge} could be modeled well over a reaction time of 20 min. The mathematical model included ozone in the liquid, the gas bubbles and in the gas freeboard.

An interesting attempt to reduce the complexity of the chemical model that incorporated direct ozone as well as hydroxyl radical reactions was applied by Carini et al. [31] who lumped the oxidation products of 3-methylpyridine ozonation into two groups, "reaction intermediates (DTOC)" and "stable end products (STA)". In the physical model complete mixing was assumed for the liquid phase and a model of 1 to 30 CSTRs in series was used for the gas phase. Although various modifications of the chemical model were considered, the simulation did not closely match with the experimental observations. In particular, the ozone gas and liquid concentrations were overestimated by the model, probably due to an inadequate consideration of the kinetic regime of ozonation.

Reports on modeling waste-water ozonation with an *intermediate* number of compounds initially present are extremely rare. An ambitious example is found in the work of Stockinger [33]. In a lab-scale semibatch bubble column (V_L = 2.4 l) a synthetic waste water containing 11 nitro- and chloroaromatic compounds was ozonated. The chemical model comprised direct and indirect reactions with the contaminants in the liquid bulk for which k_D and k_R values were available from the literature. Reactions occurring in the liquid-side film between ozone and the contaminants were also considered. In the physical model complete mixing was assumed for the liquid phase and a model of four CSTRs in series was used for the gas phase. However, no close match between the measured and the calculated concentration profiles was obtained. Due to the high complexity of the chemical and hydrodynamic models, the limit of the available computing capacity was reached, which limited the possibilities of fitting the model.

Real waste waters normally contain a high number of individual substances. Due to this lumped parameters, such as COD or DOC, are used to quantify treatment success. This approach has been used for the ozonation of two complex industrial waste waters from a distillery and a tomato-processing company [35] as well as for domestic waste water [36]. In the chemical model COD was used as a global parameter for all reactions of ozone with organic compounds, and in [36] the model considered only the concentration of the ozone-reacting compounds among the overall COD, named COD*, that is, not the refractory part of the COD. In general, an irreversible second-order reaction is assumed to exist between ozone and the waste water:

$$\text{COD} + z_{\text{COD}}\text{O}_3 \xrightarrow{k'} \text{Products} \tag{8.71}$$

The physical model included reactor hydrodynamics of the reactors used, for example, in the STR both phases were modeled as completely mixed, while for the BC the liquid was assumed to be completely mixed and plug flow in the gas phase. Also, the interdependency of physical and chemical processes was assessed. Mass-transfer enhancement was observed in the case of distillery waste-water ozonation, where the initial kinetic regime of ozone absorption was moderate to fast, depending on the amount of ozone applied. In the case of tomato waste-water ozonation, the reaction developed in the slow kinetic regime. However, a comparison of the model calculations for COD, and the ozone concentrations c_{Ge} and c_{L} with the experimental measurements did not result in a close match, especially in the case of distillery waste-water ozonation and also for c_{Ge} in tomato waste-water ozonation. Furthermore, it has to be noted that little removal of COD was achieved experimentally: 10 and 35% in the lab-scale studies with distillery and tomato waste water, respectively, and only 5% in pilot-scale with tomato waste water. The model is interesting in its approach; however, further experimental verification with larger removal efficiencies would be necessary in view of the practical relevance of the model. Moreover, it remains questionable whether it is possible to lump all reactions in such complex waste waters together.

In the study on domestic waste-water ozonation modeling two remarkable results were made. Firstly, it was shown that the kinetic coefficient k' of the global reaction in Equation 8.71 depended nonlinearly on the remaining (oxidizable) COD* concentration and was furthermore dependent on the pH value. Principally, it decreased to zero with decreasing concentrations of COD* and, interestingly, the minimal concentrations were higher for higher pH levels (approx. 25 mg l^{-1} at pH 2 and 4; 30 mg l^{-1} at pH 7.5 and 37 mg l^{-1} at pH 9). Secondly, it was stated that the chemical model was incomplete, since obviously radical reactions also occurred. This was derived from a considerable underestimation of the ozone dose (or the c_{Go}) being necessary for a given amount of COD removal in the model compared with the experimental results. Therefore, it was concluded that radical reactions should also be included into the chemical model.

In that case, the methods for estimating the hydroxyl radical concentration developed for drinking-water applications can be applied in principle. However, the composition of the organic fraction is much more complex as well as present in higher concentrations in waste water. Furthermore, it is usually unknown whether a compound acts as a scavenger or initiator, let alone the value of its reaction rate constant. Therefore, modeling success may be difficult to achieve. Nevertheless, Andreozzi et al. [14] have successfully modeled COD removal from waste waters when AOPs (O_3/H_2O_2 and O_3/UV) were applied and Safarzadeh-Amiri [13] modeled the degradation of MTBE in spiked tap water and contaminated groundwater considering both direct and radical reactions.

8.3.2
Empirical Models

The works of Whitlow and Roth [37] as well as of Beltrán et al. [38] employed an empirical approach for modeling. This procedure uses a global rate law of nth order for the observed disappearance of all target contaminants, which is for semibatch processes

$$\frac{dc(M)}{dt} = -k'c(M)^n \tag{8.72}$$

and for continuous processes at steady-state

$$k' = \frac{Q_G(c(M)_o - c(M)_e)}{V_L c(M)^n} \tag{8.73}$$

and correlates the determined global rate coefficient k' with "significant observable parameters of the system" [37]:

$$k' = f(F(O_3), c(M)_o, pH, T, k_L a) \tag{8.74}$$

Thus, all individual reactions of ozone with organics are lumped together and ozone decomposition is globally incorporated by the effect of varying the pH. The influences of mass transfer as well as mass-transfer enhancement are also lumped in observable parameters such as the ozone dose rate ($F(O_3) = Q_G \cdot c_{G0} \cdot V_L^{-1}$) or the initial substrate concentration $c(M)_o$. As a result, the coefficients calculated, for example, by multilinear regression, are appropriate for the boundary conditions of the individual system but normally cannot be used in another system or if changes are made. Even if only one of the parameter changes, k' can be affected.

Therefore, the simplification to pseudo-order coefficients should be avoided or used only conditionally. Even in steady-state experiments, although all concentrations remain constant, it is important to remember that the pseudo-order coefficient k' is also dependent on the concentrations of the oxidants. For example if the steady-state dissolved-ozone concentration changes due to changes in the operating conditions, k' also changes. The same is true for the $OH°$ concentration.

8.3.3
Summary

Overall, there are serious difficulties and limitations in modeling waste-water ozonation since it generally suffers from incomplete information or high complexity of the model. Nevertheless, Table 8.8 gives an overview of some important works in the field of waste-water modeling. By comparing the "completeness" of the individual approaches the reader can get an idea of what should be done in his or her own research work. Due to the complexity of the matter it is strongly recommended to study at least the following works in detail Tufano et al. [34], Benbelkacem et al. [5], Cheng et al. [32], and to consider furthermore the novel approach to modeling the moderate kinetic regime in Benbelkacem and Debellefontaine [27] as well as the comprehensive work of Beltrán [4], which both outline in detail the mathematical aspects of modeling.

8.4
Final Comments on Modeling

Modeling tries to approximate real life with a mathematical expression for the purpose of trying out hypotheses and investigating behavior, with the goal of predicting and extrapolating the behavior under different conditions. We need a concept of how real life functions and then we build our model to describe it – we build in expressions for the various things that are important to us, ignoring those that are not relevant to the task at hand. Because of the complexity of ozonation, models are often only valid over a small range and under certain conditions. These have to be checked before using a model.

The major difficulty in modeling drinking-water applications with significant contribution from indirect reactions to the oxidation rate is the determination of the OH° concentration. Models with various simplifications to describe the influence of scavengers and initiators on the complex reaction mechanisms have been developed. The search for general models of indirect reactions is complicated by the fact that a multitude of compounds, both inorganic and organic carbon, can act as either scavengers or initiators.

To model the effect of these compounds on the oxidation rate, rate constants are necessary. The mechanism and rate constants for inorganic carbon are well known. However, the case is otherwise for organic carbon. Due to the multitude of possible organic compounds, the lumped parameter DOC is used to describe them. The results of the drinking-water modeling efforts reported above have shown that all DOC is not alike, not even in drinking water. Carrying this further to waste-water applications we realize that the complexity makes calculation of OH° concentrations from individual parameters almost impossible and that the validity of models using lumped parameters, that is, DOC, COD or pH to describe indirect reaction rates is severely restricted.

Table 8.8 Examples of semibatch ozonation waste-water modeling.

System	Source							
	Chang and Chian 1981 [28]	Tufano et al. 1994 [34]	Carini et al. 2001 [31]	Benbelkacem et al., 2003 [5]	Cheng et al., 2003 [32]	Stockinger 1995 [33]	Beltrán et al. 1995 [35]	Beltrán et al. 2001 [36]
Type of WW	Synthetic	Synthetic	Synthetic	Synthetic	Synthetic	Synthetic	Industrial	Domestic
WW complexity	Low	Low	Low	Low	Low	Intermediate	High	High
Contaminants	Methanol	2-HPYR	3-MEP	Maleic acid	2,4- & 2,6-DCP; mixture (50/50)	Nitro-, chloro-aromatics (Σ11)	COD_o (mg l^{-1}) A: 6400 B: <100	COD_o = 285 mg l^{-1} COD*
Oxidation products	FA, FAD	5 products	DTOC, STA	FA, GA, OA EP	n.a.	Various	COD	n.a.
Mineralization	Yes / (TOC)	No	Yes / (TOC)	Yes	No	Yes	No	No
Type of reactor	BC	STR	BC	BC	STR	BC (TIS: 1|4)	STR	BC
$k_L a$ (10^{-2} s^{-1})#	4.2–11.1	~1.0–2.0	0.23	1.14–1.69	1.134–1.605	0.159	0.543	1.08
Initial regime E (t = 0) !	Slow $E \approx 1$	Moderate E = 1.2–1.5	n.a n.a.	Moderate E = 1.27	Fast E = 20–30 $E_{mix} \approx 80$	Moderate at pH 2; fast at pH 12; E(t = 0) n.a.	A: fast to moderate B: slow	Fast E = n.a.
pH / T in °C	9 / 25	5 + TBA / 20	Uncontrolled 7,5 → 3 / n.a.	7 / 20	(5) 7 / n.a.	2 and 12 / 20	A: 8–8.5 / 17 B: 7–7.5 / 18	7.5 / 20
Simulation results	Good: TOC, c_{Ge}, c_L	Good: 2-HPYR and 3 products, c_{Ge}, c_L	Good: 3-MEP, pH; poor: c_{Ge}, c_L, DTOC, STA	Excellent for all compounds	Good: 2,4-DCP at pH 7, c_{Ge}; poor: 2,4-DCP at pH 5	Poor for all compounds and c_L	B: good: c_{Ge}, c_L B: poor: COD (±30%)	Good: COD poor: c_{Ge}, l (ozone dose)
Specific remarks		In model $k_L a$ = f(t): 0.082 to 0.054 s$^{-1\#}$	OH° considered	Depletion factor	E = f(t) modeled	OH° considered simulation result	A: distillery B: tomato	E = f(t)

AA, acetic acid; FA, formic acid; FAD, formaldehyde; GA, glyoxylic acid; OA, oxalic acid; PA, pyruvic acid; EP, end products = CO_2 + H_2O; STA, sum of all identified stable carbonaceous end products (FA, AA, OA, PA); DTOC, reactive carbonaceous compounds (DTOC = TOC – 3-MEP – STA); COD*, actual COD – recalcitrant COD; n.a, not available; TIS: l|gas, number of tanks in series in hydrodynamic model; $k_L a(O_3)$ #, measured in pure water; ## decreases with c(M) over time.

Nonetheless, the long-term goal is to develop models with a large range of validity. Existing models must be checked, expanded and possibly discarded. New or expanded ones will be developed as our understanding increases, most likely leading to increased complexity. As pointed out in the introduction, however, it is important to keep the models usable.

References

1 Levenspiel, O. (1999) *Chemical Reaction Engineering*, John Wiley & Sons, Inc., New York.
2 Hendricks, D.W. (2006) *Water Treatment Unit Processes: Physical and Chemical*, CRC Press, Boca Raton.
3 Dochain, D. and Vanrolleghem, P. (2001) *Dynamical Modelling & Estimation in Wastewater Treatment Processes*, IWA Publishing, London.
4 Beltrán, F.J. (2004) *Ozone Reaction Kinetics for Water and Wastewater Systems*, Lewis Publisher, Boca Raton, London, New York, Washington D.C.
5 Benbelkacem, H., Cano, H., Mathe, S. and Debellefontaine, H. (2003) Maleic acid ozonation: reactor modeling and rate constants determination. *Ozone Science & Engineering*, **25**, 13–24.
6 Von Gunten, S., Hoigné, J. and Bruchet, A. (1995) Oxidation in ozonation processes. Application of reaction kinetics in water treatment. Proceedings of the 12th Ozone World Congress, Lille, France, pp. 17–25.
7 Yao C.C.D. and Haag, W.R. (1992) Rate constants for reaction of hydroxyl radicals with several drinking water contaminants. *Environmental Science & Technology*, **26**, 1005–1012.
8 Peyton, G.R. (1992) Guidelines for the selection of a chemical model for advanced oxidation processes, water pollution. *Research Journal Canada*, **27** (1), 43–56.
9 Glaze, W.H., Beltrán, F., Tuhkanen, T. and Kang, J.-W. (1992) Chemical models of advanced oxidation processes. *Water Pollution Research Journal Canada*, **27** (1), 23–42.
10 Elovitz, M.S. and von Gunten, U. (1999) Hydroxyl radical/ozone ratios during ozonation processes. I. CT concept. *Ozone Science & Engineering*, **21**, 239–260.
11 Hoigné, J. and Bader, H. (1979) Ozonation of water: oxidation-competition values of different types of water used in Switzerland. *Ozone Science & Engineering*, **1**, 357–372.
12 Gottschalk, C. (1997) *Oxidation organischer Mikroverunreinigungen in natürlichen und synthetischen Wässern mit Ozon und Ozon/Wasserstoffperoxid*, Shaker Verlag, Aachen.
13 Safarzadeh-Amiri, A. (2001) O_3/H_2O_2 treatment of methyl-*tert*-butyl ether (MTBE) in contaminated waters. *Water Research*, **35**, 3706–3714.
14 Andreozzi, R., Caprio, V., Insola, A., Marotta, R. and Sanchirico, R. (2000) Advanced oxidation processes for the treatment of mineral oil-contaminated wastewaters. *Water Research*, **34**, 620–628.
15 Laplanche, A., Orta de Velasquez, M.T., Boisdon, V., Martin, N. and Martin, G. (1993) Modelisation of micropollutant removal in drinking water treatment by ozonation or advanced oxidation processes (O3/H2O2), ozone in water and wastewater treatment. Proceedings of the 11th Ozone World Congress, San Francisco, pp. 17–90.
16 Beltrán, F.J., García-Araya, J.F. and Acedo, B. (1994) Advanced oxidation of atrazine in water–I Ozonation. *Water Research*, **28**, 2153–2164.
17 Chen, S. and Hoffmann, M.Z. (1975) Reactivity of the carbonate radical towards aromatic compounds in aqueous solutions. *Journal of Physical Chemistry*, **79**, 1911–1912.
18 Beltrán, F.J., García-Araya, J.F. and Acedo, B. (1994) Advanced oxidation of atrazine in water–II Ozonation combined with

ultraviolet radiation. *Water Research*, **28**, 2165–2174.

19 Glaze, W.H. and Kang, J.-W. (1988) Advanced oxidation processes for treatment of groundwater with TCE and PCE: laboratory studies. *Journal of the American Waterworks Association*, **5**, 57–63.

20 Glaze, W.H. and Kang, J.-W. (1989) Advanced oxidation processes. Description of a kinetic model for the oxidation of hazardous materials in aqueous media with ozone and hydrogen peroxide in a semibatch reactor. *Industrial Engineering Chemical Research*, **28**, 1573–1580.

21 Glaze, W.H. and Kang, J.-W. (1989) Advanced oxidation processes. Test of a kinetic model for the oxidation of organic compounds with ozone and hydrogen peroxide in a semibatch reactor. *Industrial Engineering Chemical Research*, **28**, 1581–1587.

22 Sunder, M. and Hempel, D.C. (1996) Reaktionskinetische Beschreibung der Oxidation von Perchlorethylen mit Ozon und Wasserstoffperoxid in einem Rohrreaktor. *Chemie Ingenieur Technik*, **68**, 151–155.

23 Marinas, B.J., Liang, S. and Aieta, E.M. (1993) Modeling hydrodynamics and ozone residual distribution in a pilot-scale ozone bubble-diffusor contactor. *Journal of the American Waterworks Association*, **85** (3), 90–99.

24 Huang, W.H., Chang, C.Y., Chiu, C.Y., Lee, S.J., Yu, Y.H., Liou, H.T., Ku, Y. and Chen, J.N. (1998) A refined model for ozone mass transfer in a bubble column. *Journal Environmental Science & Health*, **A33**, 441–460.

25 Cockx, A., Do-Quang, Z., Liné, A. and Roustan, M. (1999) Use of computational fluid dynamics for simulating hydrodynamics and mass transfer in industrial ozonation towers. *Chemical Engineering Science*, **54**, 5085–5090.

26 Smeets, P.W.M.H., van der Helm, A.W.C., Dullemont, Y.J., Rietveld, L.C., van Dijk, J.C. and Medema, G.J. (2006) Inactivation of Escherichia coli by ozone under bench-scale plug-flow and full-scale hydraulic conditions. *Water Research*, **40**, 3239–3248.

27 Benbelkacem, H. and Debellefontaine, H. (2003) Modeling of a gas-liquid reactor in batch conditions. Study of the intermediate regime when part of the reaction occurs within the film and part within the bulk. *Chemical Engineering and Processing*, **41**, 723–732.

28 Chang, B.-J. and Chian, E.S.K. (1981) A model study of ozone-sparged vessels for the removal of organics from water. *Water Research*, **15**, 929–936.

29 Gurol, M.D. and Singer, P.C. (1983) Dynamics of the ozonation of phenol–II, Mathematical Model. *Water Research*, **16**, 1173–1181.

30 Andreozzi, R., Caprio, V., D'Amore, M.G., Insola, A. and Tufano, V. (1991) Analysis of complex reaction networks in gas-liquid systems, the ozonation of 2-Hydroxypyridine in aqueous solutions. *Industrial Engineering and Chemical Research*, **30**, 2098–2104.

31 Carini, D., von Gunten, U., Dunn, I.J. and Morbidelli, M. (2001) Modeling ozonation as pre-treatment step for the biological batch degradation of industrial wastewater containing 3-Methyl-Pyridine. *Ozone Science & Engineering*, **23**, 359–368.

32 Cheng, J., Yang, Z.R., Chen, H.Q., Kuo, C.H. and Zappi, M.E. (2003) Simultaneous prediction of chemical mass-transfer coefficients and rates for removal of organic pollutants in ozone absorption in an agitated semibatch reactor. *Separation and Purification Technology*, **31**, 97–104.

33 Stockinger, H. (1995) Removal of biorefractory pollutants in wastewater by combined ozonation-biotreatment. Dissertation ETH No 11 063. Zürich.

34 Tufano, V., Andreozzi, R., Caprio, V., D'Amore, M.G. and Insola, A. (1994) Optimal operating conditions for lab-scale ozonation reators. *Ozone Science & Engineering*, **16**, 181–195.

35 Beltrán, F.J., Encinar, J.M. and Garcia-Araya, J.F. (1995) Modeling industrial wastewater ozonation in bubble contactors: scale-up from bench to pilot plant. Proceedings of the 12th Ozone World Congress, May 15–18, International Ozone Association, Lille, France, Vol. 1, pp. 369–380.

36 Beltrán, F.J., Encinar, J.M., Garcia-Araya, J.F. and Álvarez, P.M. (2001) Domestic

wastewater ozonation: a kinetic model approach ozone. *Science & Engineering*, **23**, 219–228.

37 Whitlow, J.E. and Roth, J.A. (1988) Heterogeneous ozonation kinetics of pollutants in wastewater. *Environmental Progress*, **7**, 52–57.

38 Beltrán, F.J., Encinar, J.M. and Garcia-Araya, J.F. (1990) Ozonation of o-Cresol in Aqueous Solutions. *Water Research*, **24**, 1309–1316.

9
Application of Ozone in Combined Processes

The previous chapters in Part B have dealt with the basics of the ozonation process. As seen in the discussion of full-scale ozonation applications (Chapter 9), ozone is rarely used alone. The combination of ozone with other water-treatment processes can often greatly increase effectiveness and cost efficiency of ozonation, or the addition of ozonation to an existing production process can increase efficiency in achieving production goals. This chapter deals with such process combinations.

Process combinations make sense that utilize ozone's effectiveness in:

- disinfection;
- oxidation of inorganic compounds;
- oxidation of organic compounds, including taste, odor, color removal; and
- particle removal.

Part of ozone's effectiveness in these four areas is derived from its production of OH°-radicals. Combined processes, that is, advanced oxidation processes, represent alternative techniques for catalyzing the production of these radicals and expands the range of compounds treatable with ozone. An overview of these processes is presented in Section 9.1.

Treatment combinations often introduce a third phase into the system, either solid or liquid, intentionally (e.g., adsorption onto a solid, absorption into a liquid, biodegradation) or because of contamination (soil, cutting oils). The effect a third phase can have on the ozonation step is discussed in Section 9.2.

The advantages of combining ozonation with biodegradation stem from the finding that many oxidation products of biorefractory pollutants are easily biodegradable. Combining chemical oxidation with a biological process can minimize the amount of oxidant needed and thus reduce operating costs (Section 9.3). An example where all four areas of ozone's effectiveness are utilized in combination with production processes is found in ozone applications in the semiconductor industry (Section 9.4).

9.1
Advanced Oxidation Processes

Advanced oxidation processes (AOPs) have been defined by Glaze *et al.* [1] as water-treatment processes that involve the generation of highly reactive radical intermediates, especially OH°. Even ozone alone at high pH-values (pH > 8) is one kind of AOP. Because ozone reacts with most organic contaminants in natural waters primarily through the nonselective indirect pathway, AOPs represent alternative techniques for catalyzing the production of these radicals, thereby accelerating the destruction of organic contaminants. Since the radicals are relatively nonselective in their mode of attack, they are capable of oxidizing all reduced materials and are not restricted to specific classes of contaminants as is the case with molecular ozone.

The versatility of AOPs is also enhanced by the fact that they offer different possible ways for OH° production. These possibilities are listed in Table 9.1. AOPs usually require at least a combination of two technologies to create hydroxyl radicals. The processes are classified depending on the type of combination. In general, AOPs can be divided into two groups:

- based on the combination of chemicals;
- based on catalytic reaction.

In addition to chemicals, radiation energy can also be used to produce OH° radicals. When UV-radiation from lamps or radiation from the sun is used, the terms photochemical and photocatalytic are used to refer to the process, and the light sources are referred to as UV or solar. Both chemical and photochemical AOPs are covered in Section 9.1.1.

The catalytic AOPs can be divided into homogeneous and heterogeneous catalysts, depending on the water solubility of the catalyst. Homogeneous catalytic processes in aqueous phase are mainly based on dissolved metal ions, for example, the Fenton reaction [5] using Fe^{2+}/Fe^{3+} salt and hydrogen peroxide with or without

Table 9.1 Advanced oxidation processes.

Processes	Combination	Chapter/Section
Chemical		
Chemical	O_3 / pH ↑	2
	O_3 / H_2O_2	9.1.1
Photochemical Processes	O_3 / UV	9.1.1
	H_2O_2 / UV	9.1.1
Catalytical		
Homogeneous (Photo)catalysis	Fe^{2+} or Fe^{3+} / H_2O_2 or O_3/ (UV or solar)	9.1.2
Heterogeneous (Photo)catalysis	TiO_2 / H_2O_2 or O_3 / (UV or solar)	9.1.2/9.2

radiation. Heterogeneous catalytic processes use solid catalysts that are either suspended or immobilized. Details of the catalytic ozonation are described in Section 9.1.2. Some aspects about the heterogeneous catalysts are covered in Section 9.2 on three-phase systems.

The research in these areas started more than 100 years ago. Some examples:

Fenton's (Fe^{2+} or Fe^{3+} / H_2O_2) have been known and used since 1890. Krystiakowski (cited in [2]) already studied the effect of radiation on H_2O_2 decomposition in 1900. Taube (1956) [3] studied the photoreaction of ozone in solution and the formation of H_2O_2 and O_3.

The following sections are meant to give an overview of the AOP processes. For a comprehensive treatment of the fundamentals and application of AOPs in combination with UV-radiation, we recommend consulting Oppenländer [4].

9.1.1
Chemical AOPs

The principles and goals of chemical AOPs are only briefly summarized in this section, since the mechanisms of these AOPs were already described in Chapter 2, the importance and the influence of various parameters affecting the oxidation process was discussed in Section 7.4. Some examples of existing processes and some general hints for the application of chemical AOP in investigations are then presented.

9.1.1.1 Principles and Goals
Ozone is one of the strongest oxidants in drinking- and waste-water treatment. However, due to the slow reaction rate constants and mostly incomplete mineralization with the direct reaction of ozone, treatment methods with an even stronger oxidant, the OH-radical, were developed such as:

- ozone / hydrogen peroxide;
- ozone / UV-radiation;
- UV-radiation / hydrogen peroxide.

By combining two oxidants, the oxidation potential increases and the treatment can be more successful than with a single oxidant (UV, H_2O_2 or O_3). Much theoretical and practical work has already been carried out in this field. Comprehensive reviews can be found in Peyton [6], Camel and Bermond [7].

9.1.1.2 Existing Processes

9.1.1.2.1 Ozone / Hydrogen Peroxide (O_3 / H_2O_2)
The oxidation potential of O_3 / H_2O_2 is based on the fact that the conjugate base of H_2O_2 can initiate ozone decay, which leads to the formation of OH° [8]. This combination is also called peroxone. Many studies have found that the addition of hydrogen peroxide enhanced the efficiency of oxidation of organic substances since the early studies of Brunet et al. [9] and Duguet et al. [10]. It can be used for treating groundwaters

and soils contaminated with various types of organics. A bench-scale study of TCE and PCE and an associated pilot-scale study of contaminated groundwater with these compounds in Los Angeles, showed O_3 / H_2O_2 could oxidize these highly chlorinated organic compounds [11].

The advantage of this process lies in the removal of compounds relatively non-reactive with ozone. If compounds already react very fast with ozone, the addition of hydrogen peroxide is nearly ineffective, which was shown by Brunet *et al.* [9] for benzaldehyde and phthalic acid, where the functional groups on the aromatic ring are relatively reactive toward molecular ozone [12]. In contrast, the oxidation of oxalic acid, which is often an end product in the case of molecular ozone reactions, was significantly accelerated with the addition of hydrogen peroxide.

An optimum dose ratio of H_2O_2 / O_3 has often been shown to be in a molar range of 0.5–1 depending on the presence of promoters and scavengers. Peroxide itself can act as a scavenger as well as an initiator, so searching for the optimum dose ratio is important. Enhancement of ozone mass transfer over that during ozonation alone can be expected in many cases.

9.1.1.2.2 **Ozone / UV Radiation (O_3 / UV)** The combination of O_3 / UV-radiation is very effective at producing $OH°$, because of the strong photolysis of ozone. Compared to the combination of hydrogen peroxide with UV-radiation, the oxidation potential is higher due to the higher extinction coefficient of O_3 (ε_{254nm} = 3300 mol l^{-1} cm^{-1}) versus that of H_2O_2 ($\varepsilon_{254\ nm}$ = 19 mol l^{-1} cm^{-1}). An additional removal mechanism in photochemical combinations is the direct photolysis of the target compound.

Prengle *et al.* [13] were the first to see the commercial potential of the O_3 / UV system in waste water. They showed that this combination enhances the oxidation of complexed cyanides, chlorinated solvents, pesticides and lumped parameters like COD and BOD (cited by [1]).

Energy costs, though, are especially of concern for this treatment combination. Photochemical processes require relatively high energy expenditures. Paillard *et al.* [14] summarized that O_3 / UV treatment is not suitable for the removal of aliphatic organohalogen compounds because less energy was required to remove them using air stripping. They found that an energy consumption of about 290–380 Wh m^{-3} was required for 90% removal for unsaturated organohalogen compounds, whereby no optimization was done and could reduce these numbers.

In other studies it was shown that ozone in combination with UV-radiation is capable of removing halogenated aromatic compounds and that the oxidation is faster than with ozone alone [15, 16]. The degradation of the herbicide alachlor was improved when ultraviolet radiation was added compared to ozone treatment alone [17, 18]. Direct photolysis of alachlor did contribute to the decomposition, however, the hydroxyl-radical mechanism was proposed as the major reaction pathway [18].

The optimum ozone-UV ratio varies depending on the water and its matrix, so no general guidelines can be given. However, in order to be effective, dissolved ozone must be present in the liquid. This is not always the case in waste-water applications, especially not in semibatch operation. If this combination is to be used to achieve a higher degree of mineralization than that possible with ozone alone, in semibatch operation the UV-radiation should only be started after dissolved ozone is measured.

9.1.1.2.3 Hydrogen Peroxide / UV-Radiation (H_2O_2 / UV)

The addition of hydrogen peroxide to UV-radiation also can be used to produce OH° to eliminate pollutants. Hydrogen peroxide is cheaper than ozone production, and the application is less complicated and requires fewer safety precautions than the more toxic ozone. All this allows hydrogen peroxide / UV-radiation to be easily included in a treatment scheme. It has its drawbacks though.

The photolysis of one mole of hydrogen peroxide leads to two mole OH°, making it seem that this combination is the ideal treatment when looking only at the theoretical yield of the oxidant (see Chapter 2). In practice, hydrogen peroxide is a poor absorber at 254 nm and the efficiency in producing OH° is low. Because of its low extinction coefficient ($\varepsilon_{254nm} = 19\,\text{mol}\,l^{-1}\,\text{cm}^{-1}$) a surplus of H_2O_2 is necessary or a longer UV-exposure time.

In the case of drinking-water treatment, high concentrations of hydrogen peroxide in the effluent of the treatment plant can be a problem. Often a limit exists, in Germany it is $0.1\,\text{mg}\,l^{-1}$. The treatment of pesticide-containing groundwater with hydrogen peroxide / UV was studied by the research group from Wabner and taken to pilot scale [19, 20]. The pesticides could be removed to under the required level ($0.1\,\text{ng}\,l^{-1}$) from $0.25\,\text{ng}\,l^{-1}$ for atrazine and $0.55\,\text{ng}\,l^{-1}$ for desethylatrazine, the problem concerning the remaining concentration of hydrogen peroxide was not solved.

If the combination is used as a pretreatment for a biological stage, however, the remaining hydrogen peroxide may not pose a problem due to adaptation of the bacteria. Even when very high concentrations of hydrogen peroxide were used ($>1\,\text{g}\,l^{-1}$), the bacterial stage became adapted to the concentrations [21].

9.1.1.2.4 Comparison

If properly used, chemical AOPs generally result in higher oxidation rates than ozone alone, but need to be evaluated for effectiveness, costs and possible side-effects [21]. However, for removing taste and odor problems in drinking water, ozone alone is sufficient without the addition of hydrogen peroxide or ultraviolet radiation.

For drinking-water treatment plants a comparison of the AOPs showed that the combination ozone / hydrogen peroxide is the most efficient and inexpensive combination followed by ozone / UV-radiation [21].

The mechanism in Chapter 2 showed that O_3 / UV and O_3 / H_2O_2 are similar, in the first one H_2O_2 is formed *in-situ*, while it is added in the second process. When a substance to be oxidized absorbs strongly in the UV region, the O_3 / UV process can be much more effective, which was reported by Peyton *et al.* [16] in the case of tetrachloroethylene. For some photolytically labile substances like pesticides the direct photolysis rate is so large that little is to be gained from using ozone. On the other hand, when the substances are not photolyzed directly, the use of UV combined with O_3 to generate hydrogen peroxide makes little sense.

An advantage of the O_3 / H_2O_2 process is that it does not require maintenance, such as cleaning or replacement of a UV lamp, and the power requirements are usually lower. Treatment plants that already use ozone as a treatment step can easily add hydrogen peroxide to increase the reaction rate.

9.1.1.3 Experimental Design

In addition to the general aspects for experimental design (see Chapter 4) the following aspects should be considered:

9.1.1.3.1 Define System

Water Due to the enormous influence of the water composition always use the same water that is to be treated later.

Oxidant Avoid treating water by AOP, which can easily be oxidized by ozone or UV itself. Use screening methods to find out if you are using the right dose rates.

If the reaction rates in the water are faster than the mass-transfer rate, a mass-transfer limitation can occur. Ozone could already be consumed before the second oxidant is added. Then, addition of the second oxidant will not improve oxidation efficiency and it is not a combined process. A pretreatment with ozone alone could be a possible solution before using combined oxidants. Even if ozone is not generally consumed before the addition of the second agent, water composition and flow rates often vary, so that the role of transfer limitation should always be considered.

Reactor Conceptually, the application of O_3 / UV and O_3 / H_2O_2 is quite simple. In the first case the water is treated by UV lamps while, or after, the ozone/oxygen gas stream is bubbled through the solution. In the second case, hydrogen peroxide is added while ozone is bubbling through the solution. In practice, however, it is necessary to pay attention to several details.

For the O_3 / UV process Prengle [13] has recommended the use of stirred photochemical tanks (STPR) to obtain better mass transfer. Simultaneous ozone contacting and irradiation was found to be more successful than in sequence due to the need for good ozone transfer to sustain the OH-radical reaction. A promising alternative to the STPR may be the use of static mixers in conjunction with a bubble-column-photochemical reactor and recirculation pump.

The adoption of treatments based on O_3/UV and H_2O_2/UV system requires the use of suitable UV sources and of appropriate photochemical reactors [23] where a good mixing is provided for effective absorption of the emitted light. Commercial AOP reactors have been developed for full-scale application by, for example, Chemviron and Wedeco.

For the AOP combinations with ultraviolet lamps, low-pressure mercury lamps are often used. Their major output is at 254 nm (85% of total intensity), which is important for the efficiency of the ozone photolysis. Mercury lamps, which have a quartz envelope without TiO_2 doping, also emit the 185 nm, line which produces ozone and aids in hydrogen-peroxide photolysis. For drinking-water applications where the hydrogen-peroxide concentration must be kept below the legal limits, lamps emitting a large fraction below 254 nm should be considered.

9.1.1.3.2 Select Analytical Methods In order to calculate the mass balances necessary for data evaluation, the concentration of the incoming and outgoing ozone gas stream as well as the liquid concentration of ozone and / or hydrogen peroxide must be measured. In the case of UV-radiation the number of photons should be measured by actinometry.

9.1.1.3.3 Determine Experimental Procedure See Section 4.2.

9.1.1.3.4 Evaluate Data and Assess Results See Section 4.1.

9.1.2
Catalytic Ozonation
Anja Kornmüller and Dieter Lompe

In catalytic ozonation, ozone is used in conjunction with dissolved (homogeneous) or solid (heterogeneous) catalysts to produce highly reactive radicals capable of oxidizing a large number of compounds. Although the catalytic reaction pathways do not always proceed via the formation of hydroxyl radicals, both homogeneous and heterogeneous catalytic systems have the production of radicals in common with the advanced oxidation processes (AOP), and are, therefore, presented in this section. On the other hand, reaction systems with heterogeneous catalysts usually have three phases (gas/ water / solid), so that the mass-transfer resistances and adsorption processes involved are topics discussed later in the section on three-phase systems Section 9.2. Furthermore, some of the applications mentioned in Section 9.2, for example, the regeneration of adsorbent having active surfaces can also be categorized as catalytic ozonation. But instead of splitting the discussion of catalysts up, the common principles and goals are presented together, while the reader is referred to the other sections for rounding out the topic.

In the last two decades there has been a significant increase in studies on catalytic ozonation featuring a wide range of catalysts and target compounds. While full-scale applications were already in use for certain catalysts in the beginning of the 1990s, these seem to have declined in recent years and current research is more focused on a return to small-scale investigations comparing various catalysts and elucidating the complex mechanisms involved.

9.1.2.1 Principles and Goals
In catalytic ozonation, activated carbon or metal species in the water, dissolved as ions, dispersed or fixed on support material, initiate a quantitatively enhanced

production of highly reactive radical species. Thus, catalytic ozonation can be regarded as an AOP. While hydroxyl radicals can play an important role, some studies describe the formation of other oxidizing species (e.g., organic or oxygen radicals) without the generation of hydroxyl radicals. The higher yield of radicals can be used to either achieve a higher degree of target compound removal with the same amount of ozone (or AOP-oxidants) or reduce the amount of oxidant required to reach the same degree of removal. The advantages of this "improved oxidation efficiency" over conventional AOPs has to be weighed against the drawbacks associated with the addition of a treatment step to separate or recover the catalysts.

Depending on how the catalyst is present in the reaction system, catalytic ozonation can be characterized as

- *homogeneous* catalytic ozonation where dissolved metal ions are the catalysts; or
- *heterogeneous* catalytic ozonation where activated carbon or metal oxides that are dispersed or fixed on support material act as catalysts.

In *homogeneous* catalytic ozonation ionic species of transition metals such as Fe(II), Mn(II), Co(II) are used as catalysts, which initiate the decomposition of ozone and result in the production of hydroxyl radicals through a direct or indirect reaction pathway (via $O_2^{\bullet-}$ and $O_3^{\bullet-}$). As mentioned above, some homogeneous catalysts can also form complexes with organic molecules without the decomposition of ozone to OH-radicals. For example, during ozone oxidation of oxalic acid at pH 6 in the presence of Co(II) as the catalyst, a Co(II)-oxalate complex is formed first. This complex is then oxidized by molecular ozone to Co(III)-oxalate, which under further attack by ozone, decomposes to an oxalate radical and the catalyst Co(II). Similarly, surface complexes with organic acids have been described for heterogeneous catalysts [24, 25]. Compared with heterogeneous catalytic ozonation, the homogeneous processes have the advantage that only gas–water mass-transfer resistance exists. Their significant drawback is that an additional treatment step has to be applied for the separation or recovery of the catalysts.

In *heterogeneous* catalytic ozonation, dispersed solid metal oxides such as TiO_2, Al_2O_3, Fe_2O_3, MnO_2, or metals and/or metal oxides fixed on supports, for example, TiO_2, Al_2O_3 or $Cu-Al_2O_3$, are applied as catalysts. Furthermore, granular or powdered activated carbon can also be used as catalyst. Similar to homogeneous catalysts, the main effect of the heterogeneous catalyst is that they enhance the decomposition of ozone and the generation of hydroxyl radicals. Target compound removal results from the complex interaction between the catalyst, ozone and the target compound(s), which involves various consecutive and/or parallel steps of mass transfer (gas–water, water–solid and gas–solid) and chemical reactions. In general, three different mechanisms occur in heterogeneous catalytic ozonation:

1. sorption of ozone to the catalyst surface, where the active radical species are formed, which then react with the target compounds (M) in the aqueous phase;
2. sorption of the target compound (M) to the catalyst surface followed by the reaction with dissolved (molecular) ozone; or

3. both, ozone (dissolved or gaseous) and target compound (M) sorb to the catalyst surface and react through direct or indirect pathways in the sorbed phase.

The reaction rate, selectivity and ozone consumption strongly depend on the catalyst surface properties, for example, specific surface area, surface active sites, surface charge, porosity and pore volume. Serious operational problems can occur by catalyst poisoning, for example, due to sorption of nontarget compounds. Also, desorption processes can negatively influence the oxidation efficiency.

The overall goal of catalytic ozonation is to outperform conventional ozonation or AOPs at removing dissolved compounds from water. Depending on the target compound and the properties of the catalyst, the specific goals of catalytic ozonation can be broken down into the following (with the first two being predominant):

- Improve oxidation efficiency to overcome the limitations in target compound removal and / or efficient ozone use compared with AOP and conventional ozonation.

- Achieve a higher degree of mineralization, especially with respect to the already highly oxidized end products of oxidation processes, such as carboxylic acids. This can reduce or prevent bacterial regrowth in the following systems, which is an important subgoal in drinking-water applications.

- Achieve a more selective oxidation of target compounds, especially by oxidation in the sorbed state.

- Reduce the influence of scavengers or promote hydroxyl radical reactions at unusual conditions, like low pH, by oxidation in the sorbed state.

If these goals are met, catalytic ozonation can result in less ozone consumption and better economic efficiency.

Examples of both homogeneous and heterogeneous catalytic ozonation are presented in Section 9.1.2.2. Emphasis is laid on the treatment goals, as well as on technical and/or operational advantages of the processes, not forgetting uncertainties that still exist. Finally, Section 9.1.2.3 provides useful advice for experimentation in catalytic ozonation.

9.1.2.2 Existing Processes and Current Research

Catalytic ozonation is mostly studied at the lab-scale with the aim of removing single compounds. Only a few full-scale applications exist. An overview of some recent literature investigations is given in Table 9.2. The majority have studied heterogeneous catalytic ozonation. Due to their high number and similarities only selected studies are discussed in detail below. For comprehensive overviews of studies on catalytic ozonation as well as on reactions kinetics the reader is referred to Kasprzyk-Hordern [25] and Beltrán [26].

9.1.2.2.1 Improved Oxidation Efficiency
Many authors have found that the presence of heterogeneous catalysts improves the oxidation efficiency of ozonation.

Table 9.2 Examples of catalytic ozonation: lab-scale studies and full-scale applications.

Type of reactor	Catalyst	Target compound	Ref.
Goal: improved oxidation efficiency			
Full-scale fixed-bed reactor (Ecoclear® process)	special grade AC	landfill leachate	[27]
lab-scale fixed-bed reactor (Ecoclear® process)	special grade AC	textile wastewater	[28]
Slurry batch-reactor, lab-scale	Al_2O_3, Fe_2O_3 / Al_2O_3, TiO_2 / Al_2O_3	oxalic acid, chloroethanol, chlorophenol	[29]
SBR in lab-scale, semibatch or continuous mode	11 catalysts, e.g., TiO_2, CoO, CuO and mixed metal oxides	p-chlorobenzoic acid (pCBA)	[24]
Column in lab-scale, batch mode	MnO_2 loaded on GAC or GAC	nitrobenzene	[30]
Bubble column, lab-scale, continuous mode	various *homogeneous* catalysts (e.g., Fe^{2+}, Mn^{2+}, Cu^{2+}) and *heterogeneous* (MnO_2, Ni_2O_3, Fe_2O_3)	m-dinitrobenzene (m-DNB)	[31]
Differential-flow reactor, lab-scale, semibatch mode	activated carbon	benzothiazole	[32]
Fluid bed reactor, lab-scale, semibatch mode	activated-carbon fiber	phenol, COD	[33]
Goal: achieve a higher degree of mineralization			
Dispersed lab scale reactor	TiO_2	oxalic acid	[34]
SBR in lab-scale, semibatch mode	CuO / Al_2O_3, Cu(II)	oxalic acid	[35]
Stirred-basket reactor in lab-scale, semibatch mode	homogeneous Fe(III); Fe_2O_3	oxalic acid	[36]

Table 9.2 Continued

Type of reactor	Catalyst	Target compound	Ref.
Slurry SBR, lab-scale, semibatch mode	activated carbon; various homogeneous (e.g., Cu) and heterogeneous (e.g., Ru / CeO$_2$) catalysts	pyruvic acid	[37, 38]
SBR, lab-scale, semibatch mode	various homogeneous catalysts, e.g., Mn^{2+}, Co^{2+}, Ti^{4+}	citric acid	[39]
Fixed-bed column, lab-scale, batch mode	granular supported metal	about 30 organic compounds commonly found in wastewater	[40]
Subgoal: prevent bacterial regrowth			
Agitated glass flask, lab-scale	TiO$_2$ fixed on alumina beads	synthetic fulvic acid	[41]
Iron-oxide-coated ceramic membranes (cut-off 5 kDa), lab-scale	sintered iron oxide nanoparticles	biodegradable organics, bacteria	[42]

Activated carbon is commonly used as the catalyst. Since the process involves both adsorption and oxidation, both processes have to be considered as removal mechanisms when evaluating the processes. For example, in a heterogeneous ozonation process that used activated carbon that was loaded by manganese dioxide the removal efficiency for nitrobenzene was 2.0–3.0 times higher than that achieved by ozonation alone, as well as 1.5–2 times higher compared with activated carbon alone [30]. Various forms of activated carbon have been investigated. Instead of granular activated carbon, activated-carbon fibers were used in a fluid bed reactor for the catalytic ozonation of phenolic waste water [33].

The first heterogeneous catalytic ozonation process that has gained practical importance in full-scale applications is the Ecoclear® process. In this process ozone gas and the polluted water are fed cocurrently into a fixed-bed reactor where special-grade activated carbons serve as the catalysts. To protect the activated carbon from being oxidized itself, it has to be preloaded with the pollutant(s) before the first contact with ozone. It has been used successfully for the treatment of biologically pretreated landfill leachates since 1992. In a few full-scale applications problems occurred due to precipitation of calcium-oxalate during long-term operation. Many industrial waste waters have been investigated in lab-scale or pilot-scale Ecoclear® systems. For example, waste water from chemical production containing substances like bisphenol-A, xylidine, dichloroethane in sulfuric acid, fluoroaromatics, nitroaromates, siloxanes, as well as other industrial process water including waste waters from textile and paper production.

Improved oxidation efficiency, meaning lower specific ozone consumption in catalytic ozonation with activated carbon compared with conventional noncatalytic ozonation, has been shown in many lab-scale tests with various contaminants. The comparison was made using glass beads instead of activated carbon in the same fixed-bed reactor [43]. Similar operational problems, though, can be found. Both conventional and catalytic ozonation processes can experience problems due to precipitation of calcium oxalate. Clogging was observed at low ozone doses when treating waste waters that contained organic compounds as well as sufficient hardness [43]. At low ozone doses clogging was observed due to the conversion of partly oxidized contaminants, for example, by polymerization or precipitation of calcium salts like calcium oxalate. If organic substances are mineralized calcium carbonate might precipitate in the presence of hardness. Clogging depends on temperature and chemical conditions and has to be taken into serious consideration.

The reaction mechanisms that occur in heterogeneous ozonation processes are complicated and still under investigation. Since the catalysts are present as solids, reactions at both the surface and in the aqueous phase can take place. In addition, the parameters in solution can also affect the reactions causing different mechanisms to occur. In activated-carbon fixed-bed reactors, Kaptijn [27] attributed the oxidation not to hydroxyl radicals but to highly active oxygen radicals, which were assumed to be formed from ozone on the surface of the carbon and to react with the adsorbed organic compounds. Other authors regard activated carbon as an initiator of the radical chain reaction in which ozone is transformed into hydroxyl radicals at or in the aqueous phase (e.g., [44]). Consistently, Beltrán et al. [45] explained an improved ozone-decomposition rate in the presence of activated carbon, as being due to decomposition taking place simultaneously as a catalytic heterogeneous surface reaction and a homogeneous aqueous phase reaction. In other work it was suggested that above the pH of the point of zero charge, dissociated acid groups present on the activated-carbon surface could be responsible for the observed increase in ozone decomposition [32]. It was shown that selective ozonation of organic compounds that easily adsorb on activated carbon (e.g., xylidine) are oxidized more thoroughly than dissolved compounds like acetate and other COD that were only slightly oxidized [46].

9.1.2.2.2 Achieve a Higher Degree of Mineralization

As mentioned above, in ozonation highly oxidized species such as oxalic acid can be formed that are not further oxidized by ozone and are therefore sometimes called ozonation end products. Due to the enhanced production of (hydroxyl) radicals, catalytic ozonation has a high potential to overcome this limitation and to enable further mineralization. In a study on the oxidation of oxalic acid using TiO_2 as solid catalyst, a higher degree of total organic carbon removal was achieved compared to the O_3/H_2O_2

process [34]. As an additional advantage, the catalytic oxidation was hardly affected by the presence of hydroxyl radical scavengers such as sodium bicarbonate. This was attributed to reactions occurring close to or on the catalyst surface. In comparing the effects of ozone, O_3/H_2O_2 and catalytic ozonation (TiO_2 fixed on alumina beads) on fulvic acid removal, the highest biodegradable dissolved organic carbon concentrations (BDOC) were achieved by O_3/H_2O_2, while catalytic ozonation resulted in a higher degree of mineralization and consequently a lower BDOC [41]. In a study on heterogeneous ozonation of approximately 30 organic compounds that are commonly found in waste water, a high degree of TOC removal of 90% was shown for carboxylic acids, phenolic compounds, amines and other compounds that are almost totally inert in ozonation alone [40].

The complex reaction mechanisms of transition-metal catalysts are also under investigation. Studies on the catalytic ozonation of oxalic acid using Fe(III) (homogeneous) and Fe_2O_3 (heterogeneous) as catalysts showed improved ozonation efficiency [45]. At a low pH value of approximately 2.5, the ozonation rate improved by 25% with Fe(III) and by 65% with Fe_2O_3 compared to those of the noncatalytic ozonation process. The oxidation was assumed to proceed via metal-oxalate complexes and not via hydroxyl radicals. In the case of chemical-reaction control, both adsorption of oxalic acid and surface reaction between nonadsorbed ozone and adsorbed oxalic acid can be the rate-controlling steps.

Similarly, the catalytic ozonation with copper (homogeneous Cu(II) and heterogeneous CuO/Al_2O_3 catalysts) at low pH improved the effectiveness of the ozonation of oxalic acid [35]. However, the addition of phosphate buffer decreased the reaction rate. Two effects may be responsible for this. While the known reaction of the phosphate ion as radical scavenger was assumed to be partially responsible, the adsorption of phosphate on the alumina surface was also suspected to inhibit the essential first initiation step in the radical chain mechanism.

9.1.2.2.3 Prevent Bacterial Regrowth

It is well known that ozonation can cause disinfection as well as partial oxidation of organic compounds and that both can result in the formation of easily biodegradable oxidation products such as organic acids and aldehydes that facilitate the regrowth of bacteria in the water supply lines after an ozonation step. Therefore, a high degree of mineralization due to catalytic ozonation would reduce the concentration of BDOC and limit bacterial regrowth. Analogously, catalytic ozonation could reduce or prevent biofouling in membrane reactors. Both aspects were combined in a study on drinking-water preparation from surface water. A treatment scheme that combined ozonation and ultrafiltration by iron-oxide-coated ceramic membranes was applied by Karnik et al. [42]. Besides the conventional ozonation, catalytic ozonation also takes place at the iron-oxide-coated membrane. The goal is the degradation of the sorbed or trapped ozonation by-products, for example, aldehydes, which would otherwise facilitate regrowth of bacteria. The catalytic decomposition of ozone to hydroxyl radicals at the iron-oxide surface is also assumed to be the cause for the inactivation of bacteria so that biofouling control is achieved.

In summary, the mechanisms of catalytic ozonation are complex and not fully understood yet. Furthermore, the great variety of catalysts and operating conditions add to this complexity.

9.1.2.3 Experimental Design

In this section only aspects of experimental design that are specific to catalytic ozonation are considered. The parameters relevant for experiments in general with AOPs and three-phase (gas / water / solid) systems are found in Sections 9.1.1.3 and 9.2.3.

9.1.2.3.1 **Define System** Due to the strong dependence of catalytic ozonation efficiency on the catalyst, a comprehensive characterization of its properties is a prerequisite. This should include important general properties such as chemical composition and purity. For heterogeneous catalysts, the specific properties of the solids should be determined, such as density, particle size, surface area, surface charge, active sites, porosity, pore-volume distribution as well as chemical and mechanical resistances.

In addition to adsorption, several other processes between the target compound and the catalyst can occur, such as complexation, ion exchange and ligand exchange, which complicate the interaction between the catalyst, the compound(s) and ozone. Therefore, besides considering the influence of the pH on the ozonation mechanisms, its influences on surface charge and the other processes mentioned, especially ion exchange, have to be taken into account. Likewise the temperature can influence the involved processes differently. While a rise in temperature increases the ozone decomposition and chemical reaction rates, it simultaneously decreases the dissolved-ozone concentration in water. More importantly though, in heterogeneous catalytic ozonation an increase in temperature may decrease the adsorption rate, but be aware that some compounds, for example, surfactants have an inverse temperature behavior in adsorption.

In the case of studies with synthetic (waste) waters the use of buffers should be carefully investigated, because some (e.g., carbonate and phosphate) are known for their high affinity to metal oxides, which can result in blockage of catalyst surface and changes in surface charge. Both can prevent the adsorption of the target compound. Furthermore, the reaction of calcium with organic and inorganic oxidation products might cause clogging. As an example, the polymerization or precipitation of partly oxidized contaminants can occur with calcium, for example, forming calcium oxalate. If ozonation achieves mineralization of organic substances the formation of carbon dioxide may enable the precipitation of calcium oxalate.

Finally, in heterogeneous ozonation, an ubiquitous supply of ozone to the catalyst is important. Otherwise, the extremely short lifetime of hydroxyl and other radicals cause a limited reaction zone at the solid catalyst without using the full capacity of the reactor.

9.1.2.3.2 Select Analytical Methods

Oxidants Due to the extremely short lifetime of radicals it is not possible to analyze them directly. In homogeneous catalytic ozonation, hydroxyl radicals can be quantified by using certain organic tracer compounds, which are known to react only with radicals but not with dissolved ozone. Due to the short lifetime of radicals and the possible sorption of the tracer compound on the catalyst, this approach is not suitable in heterogeneous systems. Therefore, only the overall ozone consumption can be determined by a mass balance of ozone in the gas phase.

9.1.2.3.3 Determine Experimental Procedure
The adsorption characteristic of the solid catalyst should be known. Since adsorption is often one step in the heterogeneous catalytic ozonation, adsorption properties are required for understanding the complex process that may comprise adsorption, desorption and oxidation as well as various interactions between these processes, which cannot be distinguished during catalytic ozonation. Therefore, it can be helpful to carry out adsorption experiments first to determine the equilibrium adsorption isotherm of the compound under conditions similar to the catalytic ozonation experiment to be performed later. Also, ozone oxidation of the compound in the aqueous phase should be assessed independently.

If the heterogeneous catalyst is not used as a fixed bed but dispersed in water an additional treatment step, e.g., filtration, has to be included to remove the catalyst particles from water (see also Section 9.2.3). For homogeneous catalysts, especially certain metals, and depending on the application discharge limits to the receiving water might have to be considered.

9.1.2.3.4 Evaluate Data and Assess Results
Try to close the compound mass balance as far as possible, since this is necessary in order to evaluate how well the system performs. At the end of a heterogeneous catalytic ozonation experiment the solid catalyst should be extracted with an appropriate solvent in order to determine if unoxidized target compound has remained adsorbed on the catalyst surface.

9.2
Three-Phase Systems
Anja Kornmüller

Commonly at least two phases are present in ozonation processes, the ozone gas and the water to be treated. A third phase such as a second fluid or a solid phase may also be present. Such three-phase systems may be found in drinking-water and waste-water applications, as well as in treatment processes for gaseous or solid wastes. The third phase can be intentionally introduced to improve the selectivity of the ozone reaction or the oxidation of residual compounds (e.g., solvents, or heterogeneous catalysts – which were discussed previously in Section 9.1.2), or the third phase can be a constituent already present in the water to be treated (e.g., bacteria and particles in untreated drinking water, or oil contamination in waste water). Two types of three-phase systems discussed in this section are gas / water / solvent and gas / water / solid systems.

Full-scale applications are common especially in drinking-water treatment for particle removal and disinfection, while in waste-water treatment sludge ozonation and the use of catalysts in AOP have been applied occasionally. However, most applications are limited to laboratory and pilot-plant scale. Even at lab-scale, ozonation is seldom studied in water–solvent systems and seems favorable only in special cases. In recent years, research areas for three-phase ozonation include soil treatment, oxidative regeneration of adsorbers and catalytic ozonation. In general, potential still exists for new developments and improvements in ozone applications for gas / water / solvent and gas / water / solid systems.

The principles and goals of ozone application in both types of three-phase systems are discussed in Section 9.2.1. Since mass transfer may decisively influence the oxidation efficiency and outcome in these complex systems, the additional mass-transfer resistances and their effects on mass transfer are explained in detail in this section. The gas / water / solvent system is used as an example for both types of system, leaving the reader to adapt the principles to the gas / water / solid systems by him/herself. This also applies to heterogeneous catalytic ozonation, which was discussed in the section on AOPs (Section 9.1.2). Examples of ozone application in both types of three-phase systems are presented subsequently in Section 9.2.2, with emphasis on their goals, as well as technical advantages and disadvantages, while the last Section 9.2.3 provides useful advice for experimentation in three-phase systems.

9.2.1
Principles and Goals

Three-phase systems contain a second fluid or solid phase in addition to the water and gaseous ozone / oxygen or ozone / air phase. They can be classified according to whether the solvent or solid phase is dispersed in water or segregated. Ozone can be directly or indirectly gassed to the reactor. If it is directly gassed, the chemical reaction in the liquids or in the liquid film around particles occurs simultaneously with the mass transfer. In indirectly gassed reactors, the ozone is absorbed in pure water or the second liquid phase prior to entering the reactor containing the compound to be treated (Table 9.3).

The primary target compounds of ozonation are often dissolved in the water phase, although there are applications in which they are present in the solvent, for example, highly lipophilic polycyclic aromatic hydrocarbons (PAHs) in dispersed oil droplets, or adsorbed on the solid, for example, in the regeneration of spent adsorbents or treatment of contaminated soils. In general, oxidation products tend to be more hydrophilic than the original compound and therefore have

Table 9.3 Three-phase systems classified by the state of the third phase in water.

State of third phase in: gas / water / solvent	gas / water / solid
• nondispersed phases / segregated phases (*fluorocarbons in membrane reactors*)	• nondispersed phases / segregated phases (fixed *bed* catalysts; regeneration of adsorbents; *in-situ* or *ex-situ* treatment of contaminated soil)
• solvent dispersed in water (oil droplets in water-oil emulsions, with or without emulsifier)	• particles dispersed in water (organic or *inorganic*, e.g., sludge from biotreatment processes; contaminated soil in reactor)

a higher affinity to the water phase. The tendency of a solute to partition between water and solvent or water and solid has to be considered carefully in three-phase systems.

9.2.1.1 Gas / Water / Solvent Systems

The most frequent goals of ozonation in gas / water / solvent systems are

- to make use of the increased solubility of both ozone and target compound(s) (M) in the solvent;
- to establish a high-rate selective direct reaction between ozone and target compound(s) (M);
- and thus to achieve the intended degree of target compound removal with less ozone.

In these systems two processes can occur: a chemical extraction of the organic target compound from the water into the solvent followed by ozonation in this phase or the diffusion of ozone from the solvent into water with subsequent reactions in the water phase.

The feasibility of a water / solvent-phase ozonation depends mainly on the properties of the solvent. The following properties of the solvent are favorable:

- low vapor pressure;
- nontoxic and immiscible with water;
- high ozone solubility;
- inertness against ozone;
- high affinity for the target compound(s).

A fast direct reaction between the target compound (M) and molecular ozone is a prerequisite for a selective oxidation. Nontarget compounds present in the water phase should not have a high reactivity to ozone since ozone might then be consumed to a large extent in the aqueous phase and might not be available for the oxidation of the target compound in the solvent phase.

The selectivity of the ozone reaction in pure solvent or water–solvent systems is known from early studies conducted by chemists for analytical and preparative purpose [47]. Solvents inert to ozonation (e.g., pentane, carbon tetrachloride) provide an opportunity to produce and study oxidation products of the ozonolysis, such as ozonides at low temperature [48]. The achievable selectivity depends much on the distribution of the compound between the gas, water and solvent phases, which should be checked by partition coefficients from the literature or experimentally determined.

Only in the last two decades have ozonation techniques been developed and studied that utilize specifically the higher ozone solubility, enhanced mass-transfer rates, higher reaction rates, etc., in water–solvent systems.

9.2.1.2 Gas / Water / Solid Systems

Ozonation in three-phase systems containing solids normally has one or more of the following goals:

- change in the solids (better sedimentation / filtration or reduction of solids mass);
- change in compounds adsorbed on the solids (transformation or mineralization);
- oxidative regeneration of spent adsorbents for reuse;
- improvement of oxidation efficiency due to catalytic production of radicals.

Ozone applications in gas / water / solid systems cover a wide range of media such as sludge, soils, adsorbents and catalysts. Disinfection of bacteria, which can be regarded as a three-phase system, is a well-described and established application. The preozonation for particle removal is discussed frequently, especially in the treatment of surface water, where different organic (e.g., bacteria, viruses, algae, suspended organic matter) and inorganic (e.g., silica, aluminum and iron oxides, clay) particles can be present.

9.2.2
Mass Transfer in Three-Phase Systems

Most of the parameters that influence the rates of mass transfer and chemical reaction, and therefore the efficiency of the system, have already been discussed in Chapter 6. However, in addition to the resistances to mass transfer found in gas / water systems, more resistances exist in three-phase systems. To illustrate the additional complexity, the resistances are explained using the ozonation of an oil / water-emulsion containing polycyclic aromatic hydrocarbons (PAHs) as an example. The three phases in such a system are: the gas phase (G) containing a mixture of ozone and oxygen, the liquid water phase (W) and a second liquid phase of dispersed oil consisting of dodecane (O). The target compounds PAHs are mainly dissolved in the dispersed oil phase due to their hydrophobic behavior and low water solubilities. Therefore, in order for a reaction between ozone and the PAH to occur, ozone has to be transferred from the gas bubbles to the oil droplets. This requires mass transfer at two interfaces – from the gas to the water phase and from the water to the oil phase. Direct contact between gas bubbles and oil droplets is neglected due to the difference in their sizes and inner pressure of very small oil droplets.

The development of the equations for mass transfer in a three-phase system follows the same steps used in a two phase system (Chapter 6). The ozone mass transfer flux N can be described using an overall mass-transfer coefficient gas / oil $K_{G/O}$ that includes mass-transfer resistance at both interfaces:

$$N = K_{G/O}(c_G - c_O^*) \tag{9.1}$$

Following the development of two-film theory (compare Section 6.1.4), the overall mass-transfer resistance is the sum of the resistances at each interface which consists of:

1. mass-transfer resistances in both films at the interface gas / water, $k_G a_{G/W}$ and $k_L a_{G/W}$;
2. mass-transfer resistances in both films at the interface water / oil, $k_L a_{W/O}$ and $k_O a_{W/O}$.

In addition to the

3. diffusional resistance inside the oil droplet, which might limit the oxidation reaction (see concentration profile c_O in Figure 9.1).

The first transfer resistances are those found in the ozonation of aqueous solutions, while the second and third resistances are due to the additional third phase. An experimental determination of the overall mass-transfer coefficient gas / oil $K_{G/O}$ is not possible, because the dissolved-ozone concentration c_O cannot be measured in the dispersed oil phase. This prevents the determination of the

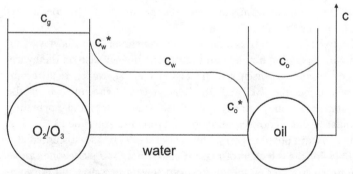

Figure 9.1 Schematic ozone mass transfer from an oxygen/ozone bubble via water to an oil droplet and possible concentration profiles [49] with c_G ozone concentration in the gas phase, c_W^* ozone equilibrium concentration in water at interface gas / water, c_W dissolved-ozone concentration in water, c_O^* ozone equilibrium concentration in oil at interface water / oil and c_O dissolved-ozone concentration in oil.

corresponding film mass-transfer coefficient $k_L a_{W/O}$. Hence, the influence of mass-transfer resistances can only be assessed theoretically and studied individually in experiments designed to selectively exclude the other resistances if possible.

To evaluate mass transfer at each interface, it is important to know which side is the controlling one. For ozone mass transfer at the gas/water interface, the relatively high Henry coefficient H and low solubility of ozone in water is decisive. With $H = c_G/c_W{}^* = 4.7$ (30 °C) a gas-side resistance can be neglected ($H \cdot k_G a_{G/W} \gg k_L a_{G/W}$) and the controlling resistance to ozone mass transfer at the gas/water interface is on the aqueous film side.

For the resistances in the ozone mass transfer water/oil, the ozone partition coefficient oil/water and the ratio of both mass-transfer coefficients in the films are important. The latter depends on the diffusion coefficients according to the film theory (see Equation 6.2). Both diffusion coefficients in this example were calculated using the modified Wilke–Chang equation [50]. The ozone diffusion coefficient is higher in dodecane ($D_O = 2.65 \times 10^{-9}\,\text{m}^2\,\text{s}^{-1}$) than in water ($D_W = 2.2 \times 10^{-9}\,\text{m}^2\,\text{s}^{-1}$, 30 °C). Based on the available oxygen partition coefficient dodecane/water [51], the ozone solubility in dodecane is 8.1 times higher than in water. Therefore, the condition $(c_O{}^*/c_W) \cdot (D_O/D_W) \gg 1$ is fulfilled and a controlling ozone mass-transfer resistance can principally be only on the aqueous film side at the water / oil interface.

Consequently, because ozone mass transfer at both interfaces is dominated by transport resistances on the respective aqueous film sides, the overall mass-transfer coefficient gas / oil $K_{G/O}$ can be defined as:

$$K_{G/O} = \frac{1}{\dfrac{1}{k_L a_{G/W}} + \dfrac{1}{k_L a_{W/O}}} \tag{9.2}$$

For the case where the second liquid is first saturated with ozone before it is used to extract the pollutants from water, two distinct mass-transfer coefficients can be defined, one for each step.

In general, solvents and solids as well as water constituents can have varying influences on $k_L a$, depending not only on their properties but also on the hydrodynamics of the system (as discussed in Section 6.2). Therefore, it is recommended to determine the $k_L a$ and the *alpha*-factor (ratio of $k_L a$ in the three-phase-system to the one in water) in the three-phase system under operating conditions similar to those corresponding to the ozonation experiments.

The addition of a third phase not only adds resistances to mass transfer but also more possibilities for the direction of mass transfer. For example, ozone can be transferred from gas to water or solvent, or from solvent to water, and the compound can be transferred from water to solvent or *vice versa*. The third phase also brings complications when considering the chemical reactions as discussed below.

The driving force for the mass transfer of the compound in the three-phase system can be determined with the solvent / water partition coefficient, just as the partition coefficient for gas / liquid phases, the Henry's Law constant, is used to determine the driving force for the mass transfer of ozone. A compound tends to diffuse from phase to phase until equilibrium is reached between all three phases. This tendency of a compound to partition between water and solvent can be estimated by the hydrophobicity of the solute. The octanol / water partition coefficient K_{ow} is a commonly measured parameter and can be used if the hydrophobicity of the solvent is comparable to that of octanol. How fast the diffusion or transfer will occur depends not only on the mass-transfer coefficient in addition to the driving force but also on the rate of the chemical reaction as well.

The mass-transfer rate of the solute to the solvent phase has to be considered compared to its reaction rate in the solvent. The system is controlled by the chemical reaction, if the oxidation of the solute in the solvent is slower than the mass-transfer rate water / solvent of ozone. This is reversed in the case of a diffusion-controlled or mass-transfer limited system. The reaction mechanism of the solute with ozone should be considered in order to utilize a selective ozonation at the water / solvent interface or in the solvent. Depending on the phobicity of the oxidation products, they might stay in the solvent phase and/or diffuse into the water phase until reaching equilibrium. Hence, these oxidation products might be further oxidized at the interface water / solvent and/or in the aqueous phase. In both cases the reaction of oxidation products might interfere and limit the ozone reaction of the compound in the solvent phase.

9.2.3
Existing Processes and Current Research

Generally, most of the existing processes are examples from lab-scale studies. Only a few full-scale applications are known, for example, the particle removal processes in drinking-water ozonation or the ozonation of effluents from the final biological stages of waste-water treatment plants. Below, various processes are discussed, categorized predominately according to the system, the type of dispersion of the third phase (solvent or solid) in the aqueous phase as well as the purpose of the treatment.

9.2.3.1 Gas / Water / Solvent Systems
In most reports on gas / water / solvent systems, ozonation was applied to model (waste) waters where the target substances were contained in the water phase, though examples of the treatment of pollutants contained in the solvent phase do also exist. The experiments were often conducted to study the working principles of such systems with their general goals (see Section 9.2.1). Also work on the development of special reactors types has been carried out. Table 9.4 gives an overview of some studies.

Nontoxic fluorinated hydrocarbons were some of the earliest and are even now the most often used solvents in lab-scale gas / water / solvent three-phase systems, mainly in the ozonation of chlorinated compounds [52–54, 56, 57]. Generally, the partition coefficient of the compound between water and the solvent can be

Table 9.4 Examples of investigated lab-scale gas / water / solvent systems.

Dispersion of solvent (third phase) in water / reactor	Solvent	Solute (M)	References
Nondispersed continuous or dispersed solvent phase / three step system of ozone saturator for solvent, STR for reaction and two-phase separator; emulsions in semicontinuous and continuous mode	fluorinated hydrocarbons (e.g., FC40 or FC77 from 3 M Co.)	phenols and chlorinated organic compounds, e.g., PCP, TCE; clofibric acid, diclofenac	[52–55]
Nondispersed, segregated phases / innovative hollow fiber membrane reactor	(FC43 or FC77)	toluene, phenol, acrylonitrile, TCE, nitrobenzene	[56, 57]
Solvent dispersed in water (with or without emulsifier) / standardized STR	oil droplets in oil / water-emulsions	2–5 ring PAHs	[49, 58, 59]
Ozone-loaded solvent phase used for contaminants extraction / sealed separating funnel	Volasil 245 (decamethylcyclopentasiloxane)	phenol, 2-chlorophenol, 2,3-dichlorophenol, 1,3-dichlorobenzene, o-nitrotoluene, nitrobenzene	[60]

decisive for where (in which phase) and what type of reactions (direct or indirect) occur.

For example, an inert fluorinated hydrocarbon phase (FC40; from 3M Co.) with a high ozone stability and solubility (saturation concentration of 120 mg l^{-1} at 25 °C, under applied experimental conditions) was first saturated with ozone and then contacted with an aqueous solution containing PCP or various chlorinated organic pollutants [53, 54]. In these experiments a high degree of pentachlorophenol (PCP) destruction (95%) was obtained independently of the partition of PCP in the two liquid phases and it was concluded that ozone mass transfer was not a limiting factor in this system. Compared to an aqueous system, the specific ozone consumption was lower by 1/25, and at pH 10.3 the pseudo-first-order reaction rate constant of PCP was three orders of magnitude larger [53]. For compounds with low partition coefficients between the fluorinated hydrocarbon and water, which are practically insoluble in the perfluorinated phase, the reaction may proceed in the aqueous phase predominantly by the radical mechanism [55].

The heat-exchange fluid Volasil 245, which has properties comparable to fluorocarbons, was loaded with ozone prior to being used to extract and destroy various organic compounds, for example, phenol, o-nitrotoluene, nitrobenzene etc. [60]. Contact with the ozone-loaded Volasil 245 was effective for contaminants that have a high solvent / water distribution coefficient and undergo direct reactions with molecular ozone. In some cases, resistances in the solvent / water interface were suspected to limit the extent of contaminant elimination from the water phase.

A different approach in reactor design using the advantages of water–solvent systems was developed by Guha et al. [56]. A membrane reactor made of microporous Teflon hollow fibers was used. The space around the outside of the hollow fibers was filled with an inert fluorocarbon phase. The hollow fibers were divided into two groups: ozonated air was passed through the first group, waste water through the second. The target compounds (e.g., toluene, nitrobenzene, etc.) and ozone diffused through the membranes to the outer fluorocarbon phase used as reaction medium. Hydrophilic oxidation products were extracted back into the aqueous effluent. Only 40–80% of the pollutant was transformed, probably due to the resistance in the aqueous phase film. Therefore, two reactors in series were suggested to increase the degree of removal. This membrane-based reactor was improved for the treatment of volatile organic compounds in air by Shanbhag et al. [57]. Compared to conventional treatment where the ozone gas is bubbled directly through fluorocarbons, losses by fluorocarbon volatilization could be avoided in membrane aeration systems, however, additional boundary-layer and membrane resistances existed.

The ozonation of highly condensed polycyclic aromatic hydrocarbons (PAH) in an oil / water emulsion was studied by Kornmüller et al. [49]. One emphasis was on describing the ozone mass transfer using the model for three-phase systems described in Section 9.2.2 above (see Figure 9.1). Due to the lipophilic behavior and low water solubility of the PAH studied, they were initially dissolved in the oil phase, which was then emulsified in the water phase. First, the ozone mass-transfer coefficient for gas / water $k_L a_{G/W}$ was studied experimentally and optimized by variation of gas flow rate, ozone inlet gas concentration and stirrer rotational speed. Subsequently, the second ozone mass-transfer coefficient for water / oil $k_L a_{W/O}$ was investigated individually by variation of the oil droplet size. Evaluation of the experimental results based on the theoretical model showed that mass-transfer limitations did not exist at the water / oil interface and were not caused by diffusional resistance inside the oil droplets. Therefore, a microkinetic description independent of mass transfer was possible using a first-order reaction with regard to the PAH concentration. The effects of pH variation and addition of scavengers indicated a selective direct reaction mechanism of PAH inside the oil droplets [58]. For example, in the ozonation of the five-ring condensed benzo(e) pyrene this was confirmed by the oxidation products being formed: a secondary ozonide and oxepinone (hydroxytriphenylenol[4,5-cde]oxepin-6(4H)-one), which are specific for an ozonolysis as known from ozonation in solvents [59].

9.2.3.2 Gas / Water / Solid Systems

The discussion of existing processes of gas / water / solid systems is grouped according to the treatment goals. Table 9.5 gives an overview of the various systems that are mostly studied in lab-scale or, in the case of sludge ozonation, sometimes also in full-scale. Heterogeneous catalytic ozonation, which can be regarded both as an AOP and a three-phase system, is described separately in Section 9.1.2.

9.2.3.3 Change in the Solids

The principle of sludge ozonation is based on applying ozone on biosolids from biotreatment processes to change the surface properties of the floc and/or destroy the cell-membrane structures. Ozonation to change the surface properties has been studied under aspects of bulking control, improving the settleability and dewaterability [61, 62, 68], whereas ozonation to destroy cell structures has been investigated to stabilize the biological phosphate removal by producing soluble organic substrates [63], and to reduce excess sludge production [65, 66, 69]. Of course, ozonation can bring multiple benefits. For example, in addition to controlling bulking, sludge ozonation can cause a simultaneous reduction in excess sludge [66].

Many studies have shown that sludge ozonation can significantly reduce the production of excess sludge in biological processes. If ozonation is combined with a biodegradation step to achieve mineralization of the oxidation products after the cell structure has been destroyed, it is called sludge disintegration. The disintegration of sludge makes use of the treatment scheme: floc disintegration due to cell destruction (suspended solids), solubilization to biodegradable organics and subsequent mineralization due to further oxidation of soluble organics by bacteria. Solubilization, that is, a transformation of digested sludge into soluble substances such as proteins, lipids and polysaccharides, was observed at a specific ozone consumption of $0.5 \, g \, g^{-1}$ (ozone per dry matter) [67]. With the same dose, a mass reduction of 70% due to mineralization and a volume reduction of 85% due to enhanced dewaterability of the municipal sludge were achieved when compared to an untreated control [68]. The reduction of suspended solids of 5 to 25% in the bioreactors after ozone treatment was proportional to the ozone dose between 0.02 and $0.09 \, g \, g^{-1}$ (ozone per total suspended solids of initial excess sludge) [71].

In full-scale applications, complete elimination of excess sludge was achieved in a pharmaceutical and a municipal waste-water plant [64, 65]. Likewise, zero excess sludge production was obtained during the winter season using pilot-scale ozone disintegration and subsequent biodegradation [70]. An overview of the results of various full-scale applications of sludge ozonation is presented in [72].

Results from sludge ozonation studies are difficult to generalize because of the complexity of sludge, the numerous parameters that influence the process and the different treatment schemes used. Often the experiments are focused only on the ozone dose or ozonation rate necessary to achieve the goal without adequate consideration of further influencing and process parameters that are important for process optimization. For instance, in combined processes such as sludge disintegration, changes in the operation of the biological process can also have as much or even more effect on the results as changes in the ozonation parameters.

Table 9.5 Examples of gas / water / solid systems grouped according to the treatment goal.

Dispersion of solid (third phase) in water / reactor	Solid	Compound (M) / adsorptive (M)	References
1. Change in solids: particle removal – improved settleability, dewatering of sludge, mineralization – zero excess sludge,			
Solid dispersed in water / shaker flasks, bubble columns, STRs, SBRs or full-scale waste-water treatment plants	Various biosolids		[61–68]; [70, 71]
2. Change in the compounds adsorbed on the solids – removal of contaminants, or change in the compounds associated with the solids – disinfection of soilborne pathogens,			
Solid with (very) little water / *in-situ* soil remediation or lab-scale soil columns	Contaminated soil; porous media (glass beads, sands, soils); quartz sand and soil	PAHs (Pyr, BaP); Phenanthrene, diesel-range organics; soilborne pathogens, e.g., *Fusarium oxysporum*	[73–75]
Solid dispersed in water / STR	Contaminated soil	NOM	[76]
3. Change in compounds adsorbed on solids – regeneration of spent adsorbents, enhancement of oxidation efficiency			
Nondispersed, segregated phases / fixed-bed reactors	Octadecyl silica gel (ODS)	PAH (BeP)	[77]
Solid dispersed in water / STR	AC (Filtra-sorb 400)	TCE	[78]
Solid dispersed in water / lab-scale magnetic-stirred reactor	Highly active granulated iron hydroxide (ß-FeOOH)	Fulvic acids	[79]
Nondispersed, segregated phases / adsorber columns, differential-flow reactor,	Highly active granulated iron hydroxide (ß-FeOOH); cucurbituril	Reactive dyes	[80, 81]
Solid dispersed in water / lab-scale magnetic-stirred reactor; nondispersed, segregated phases / basket reactor	Perfluorooctylalumina (PFOA)	Toluene, chlorobenzene, cumene, nitrobenzene; methyl tertiary butyl ether (MTBE), ethyl-tertiary butyl ether (ETBE)	[82, 83]

The applicability of ozonation for sludge disintegration is very site specific. Each country or region has different boundary conditions, such as legal regulations for sludge disposal (esp. organic dry matter content) and availability of alternative disposal options (agricultural use, incineration, landfill) as well as their specific costs. Therefore, the feasibility of each application must be evaluated individually. Stensel and Strand [69] compared the costs for sludge disintegration with ozone to conventional solids handling for both a municipal and an industrial waste-water treatment plant in the USA. They concluded that although ozonation was very effective at reducing excess sludge, the costs were higher than conventional handling and disposal in a landfill. However, they suggest that ozonation may be cost-effective when disposal options are more costly, such as incineration.

9.2.3.4 Change in Compounds Adsorbed on the Solids

Besides changing the properties of the solid itself as discussed above, frequently the goal is only the change in compounds adsorbed to solids without changing the properties of the solids. Such selective oxidations reactions are preferred in the remediation of contaminated soils, the disinfection of soil pathogens and the oxidative regeneration of spent sorbents.

9.2.3.5 Soil Ozonation

The ozonation of contaminated soil can be carried out both *in situ* with direct injection into the soil or *ex situ* under controlled conditions in a reactor as a three-phase system. The goal is to increase the biodegradability of residual, nonvolatile organic compounds like PAH in contaminated soils. The water is present either as a water film covering the soil particles in the nonsaturated water zone (*in-situ* treatment) or as the continuous phase in the reactor where the soil is suspended in water (mainly onsite treatment). For *in-situ* treatment, ozone can be supplied either gaseous to the unsaturated soil zone (comparable to conventional soil venting) or dissolved in ozonated water that is injected/applied to the soil. For *ex-situ* treatment, ozone is applied directly to the soil dispersed in water.

The ozonation of the radioactive (C^{14})-labeled PAHs pyrene (Pyr) and benzo(a) pyrene (BaP) was studied in silica and soil [73]. Considerable percentages of both PAH were oxidized to water soluble, probably biodegradable substances (20–30%), but 10% were found in nonextractable and 30% in bound residues in the soil organic matter. The toxicity and stability of bound residues is still unknown under long-term degradation conditions. The ozonation of the soil natural organic matter (NOM) led to a decrease in the humic acid fraction with a reduction in average molecular size [76]. Because of an increase in the building-block and low molecular acid fraction, which are easily biodegraded, a fast and high bacterial regrowth can be expected after ozonation.

In the ozonation of phenanthrene in unsaturated porous media, the water content, the soil organic matter and the metal oxides were the decisive factors in the fate and transport of gaseous ozone. Maximum removal was found in columns packed with baked sand, whereby the catalytic ozone decomposition by metal oxides enhanced the OH-radical formation and consequently the phenanthrene removal [74]. Based on the results of mathematic models, Kim et al. [84] determined a significant retardation of gas-phase ozone transport by ozone consumption due to reactions with soil organic matter and phenanthrene. Therefore, a long operation time of 156 h is assumed to be necessary for complete phenanthrene removal in a 5-m soil column.

Besides the above-described goal of removing the soil contaminants, ozonated water can be used as soil disinfectant for the control of soilborne pathogens, for example, *Fusarium oxysporum* [75], which themselves can be regarded as a solid phase but mainly occur attached to the soil matter.

In general, it is not easy to control the oxidation reaction during *in-situ* soil treatment. While in lab soil-column experiments the pH can be controlled, in an *in-situ* application it will always decrease due to the formation of organic acids. This will cause shifts in the oxidation mechanism toward the direct oxidation pathway and in the chemical equilibrium of the soil. Furthermore, ozone applications will result in changes in the soil chemical constituents, that is, the cation exchange layer and the humic fraction. The consequences of these changes are still mostly unknown. A special lag-phase and a selection of bacteria in regrowth might be caused by the ozonation. Therefore, specific site characterization and screening tests are prerequisites for *in-situ* ozonation treatment.

Special precautions should be taken in applying *in-situ* soil ozonation due to the present state of knowledge. Even for homogeneous aqueous systems the oxidation products of most compounds and their toxic effects are not fully known yet, let alone for the complex ecological system "soil". More work is necessary to evaluate possible effects on the ecological system "soil" as well as safety aspects for a full-scale *in-situ* application.

9.2.3.6 Regeneration of Adsorbents

Adsorption is often used to remove low concentrations of micropollutants from ground water as a treatment step for producing drinking water. Chemical oxidative regeneration can be used for the oxidation of these compounds sorbed on adsorbents to regain adsorption capacity, allowing direct reuse of the adsorbent. Such regeneration of spent adsorbent can provide advantages (no shipment to regeneration units or disposal of concentrates) over thermal or other methods. Furthermore, the use of adsorption to enrich the target compound on the solids followed by its ozonation in the sorbed state can be more economical than ozonation of the complete water stream. This is especially true if the sorption of the target compound occurs selectively and nontarget ozone-consuming compounds remain in the water phase.

Essential criteria for selecting an adsorbent to use with ozonation are that the adsorbent itself does not react with ozone and that the compound adsorbed can be oxidized on the adsorbent to achieve complete regeneration. If both conditions are not met, another combination of compound and adsorbent should be chosen. Several adsorbents were tested for their compatibility to regeneration by ozone [77]. Adsorber polymers like Wofatit reacted with ozone and are therefore not suitable for oxidative regeneration. A favorable adsorbent was made of octadecyl silica gel particles (ODS), which hinder an electrophilic attack of ozone on the adsorbent due to its chemical structure with alkane side chains. Six cycles of adsorption and oxidative regeneration by ozone dissolved in water could be obtained with ODS loaded with benzo(e)pyrene without a significant loss in adsorption capacity. A direct reaction of ozone with the adsorbed compound was indicated by the two main oxidation products found from the reaction of sorbed benzo(e)pyrene, which were similar to the one mentioned above in the ozonation of homogeneous oil / water-emulsions. In contrast, complete regeneration of a Filtrasorb-400 activated carbon saturated with trichloroethylene (TCE) was not achieved with ozone / hydrogen peroxide [78]. In the case of reactive dyes adsorbed on cucurbituril, the dye ozonation in water is more efficient regarding the ozone consumption than the two-stage adsorption-regeneration process [81].

Highly active granulated iron hydroxide (β–FeOOH) loaded with fulvic acids was regenerated using dissolved ozone in water [79]. The results indicated that an initial ozone concentration in the suspension higher than $8\,mg\,l^{-1}$, which is difficult to reach in lab-scale application, and an ozone dosage above $1.2\,g\,g^{-1}$ (ozone / adsorbed organic carbon) are necessary for a good regeneration efficiency. In the oxidative regeneration of β–FeOOH loaded with reactive dyes, gaseous and dissolved ozone in water were not efficient enough due to mass-transfer limitations, while a high regeneration capacity of 98.5% was achieved by the catalytic activation of hydrogen peroxide to OH-radicals that occurs directly at the highly active adsorbent surfaces [80].

Based on the above-mentioned work with fluorinated hydrocarbons in gas / water / solvent systems, Kasprzyk *et al.* [82, 83] studied the ozonation of toluene, chlorobenzene, MTBE etc. in the presence of perfluorooctylalumina (PFOA) and thereby used the capability of a monomolecular layer of nonpolar perfluorinated alkyl chains immobilized to solid alumina particles to dissolve simultaneously ozone and organic molecules. By favoring the direct reaction of molecular ozone a 24–43% higher degradation of compounds was achieved compared to ozonation alone.

9.2.4
Experimental Design

9.2.4.1 Define System
In both gas / water / solvent and gas / water / solid systems, it is important to keep dispersions and suspensions homogeneous during ozonation. A suitable type of reactor and effective mixing has to be provided depending on the properties of the system.

For gas / water / solvent systems, the vapor pressure of the solvent has to be checked before starting any experiment so that stripping of the solvent will be negligible. The solute should also be nonvolatile. The solvent should be nontoxic, immiscible with water and provide a high solubility for ozone. The inertness of the solvent against ozone and thus the stability of a three-phase system have to be guaranteed during the whole ozonation. If the solvent cannot be reused, it should be generally treatable and preferably biodegradable. An important safety aspect is that ozone (and pure oxygen) might be explosive on contact with highly reactive lipids, fats and oils.

Generally, the role of mass-transfer limitation should always be considered and investigated in the ozonation of three-phase systems. In the case of suspensions the k_La and the *alpha*-factor should be determined as discussed in Section 6.2. Kinetic modeling of the ozone reaction, which even in heterogeneous systems is often described as pseudo-first order, has to be viewed critically. Often the influences of the ozone mass transfers gas / water and water / solid were not examined. When studying the ozonation of compounds adsorbed onto or inside particles, a limitation of the reaction by the ozone diffusion through the interface water / solid or inside the particles can be expected in most cases. Using a model system with defined particle size and known compounds, the mass transfer water / solid can be studied and a kinetic model of the reactions might be developed, but real systems will be too complex for such an approach.

To achieve successful regeneration of spent adsorbents by ozonation of gas / water / solid systems, the adsorbents used have to be almost completely inert against ozone. Before considering the regeneration of an adsorbent with ozone, the ozone decomposition caused by the adsorbent alone should be tested. Activated carbon is not recommended as adsorbent if regeneration is to be carried out with ozone. Depending on the experimental conditions ozone reacts more or less strongly with its double bonds and / or is decomposed by the activated carbon. For example, activated carbon is often used as a gas phase ozone destructor in lab-scale setups, resulting in a consumption of the carbon by slow burning.

9.2.4.2 Select Analytical Methods

9.2.4.2.1 Compounds (M) Known compounds (M) and their oxidation products should be analyzed in both the water and solvent phases separately with methods that allow individual quantification if possible. While the oxidation progress can be described by overall parameters like DOC and COD in the aqueous phase, in the solvent phase these concentrations are normally dominated by that of the solvent itself. In gas / water / solid systems a method should be developed to measure a compound on the solid particles, for example, by extraction or dissolution of the particles, and to describe the oxidation in/at the solid phase.

9.2.4.2.2 Ozone Measuring the dissolved-ozone concentration is also complicated in three-phase systems. The photometric analysis of dissolved ozone by the indigo method can be disturbed by compounds or materials that scatter or absorb light [85], although a procedure for measuring the dissolved-ozone concentration in the presence of suspended material by the indigo method has been suggested [86]. However, a measurement in the presence of other compounds like oil might not be applicable. A correction based on the turbidity of homogeneous oil / water-emulsions did not provide reliable results with the indigo method [58]. Nevertheless, the method has been modified successfully for measuring dissolved ozone in the presence of a fluorocarbon phase [53].

When using an amperometric electrode as the measuring technique, interference from the solutes and solvent can occur. The solutes and solvent can adsorb to the semipermeable membrane of the electrode, therefore giving an additional resistance to the diffusion of ozone through this membrane to the electrolyte chamber. The use of an amperometric electrode is not recommended for water containing particles due to likely damage to the membrane. In these cases the ozone consumption can only be calculated from the ozone gas balance.

9.2.4.3 Determine Experimental Procedure

Homogeneous sampling is very important in both types of three phase-systems and can be studied by taking samples at different heights or locations in the reactor. The homogeneity has to be guaranteed also during analysis, for example, samples from water / solvent systems have to be stirred during TOC measurements. If a compound is analyzed in both water and solvent phase, a reproducible and efficient separation process of both phases has to be chosen. The common method for dispersed water / solid systems is filtration. Due to the high ozone solubility in solvents, after sampling further reaction of ozone has to be stopped.

When studying the regeneration of adsorbents, it is important to keep in mind that adsorption, desorption and oxidation processes occur simultaneously during oxidative regeneration. Due to the interaction of these processes, experiments need to be planned to evaluate these processes individually. It is advisable to first evaluate the aqueous reactions between the oxidants to be used for regeneration and the pollutant. The direct ozonation of pollutants in water might be more favorable than a complex two-stage process of adsorption followed by oxidative regeneration. However, if the pollutants are present in small concentrations in water, the enrichment by adsorption followed by ozonation in the sorbed state is often advantageous. Likewise, when the pH or scavenger concentrations are limiting the ozonation of a compound in water, a more selective ozonation might be achieved in the adsorbed state.

In the next step, it can be helpful to carry out separate conventional adsorption experiments with the compound and the adsorbent alone. These results can then be compared to the results from adsorption after the first oxidative regeneration. Repeated tests of pure adsorption compared to reloading of used adsorbent after oxidative regeneration reveal the influence of both adsorption and oxidative regeneration on the overall performance as well as the completeness of the regeneration capacity.

9.2.4.4 Evaluate Data and Assess Results

Prior to carrying out oxidation experiments, the mass balance of a compound should be examined under the conditions of aeration with oxygen or air, so that other elimination processes like stripping can be excluded. Stripping of a compound can be determined by absorption of the off-gas in water or an appropriate liquid in gas-wash bottles.

Mass balances of a compound have to cover all three phases. Before transformation of a compound by ozonation can be reported, the concentrations of the compound in all three phases have to be checked. A decrease in a compound concentration in one phase might not be caused by oxidation but by partitioning into another phase.

9.3
Chemical-Biological Processes (CBP)

Biological treatment is often the least expensive and most effective process for destroying organic pollutants. Many pollutants can be fully biodegraded (i.e., mineralized to CO_2 and water) by microorganisms in biological reactor systems. In contrast, many physical and chemical processes just concentrate the pollutants or transfer them from one medium to another, which then must be disposed of, leaving their ultimate fate in the environment unclear. Unfortunately, not all compounds are biodegradable. Combined chemical-biological processes (CBPs) are based on the finding that many of the products formed by oxidative chemical reactions with biorefractory pollutants are biodegradable.

For example, chemical oxidation processes such as ozonation or the advanced oxidation processes (AOPs, e.g., O_3/UV, O_3/H_2O_2, UV/ H_2O_2) can be used advantageously to reduce the molecular size of the original contaminants, making them more amenable to biodegradation. Ozonation is frequently combined with bioprocesses in both drinking-water and waste-water treatment schemes. Other chemical oxidation technologies that can be used in combination with biodegradation are Fenton's reagent, the photo-Fenton process, photolysis resp. photocatalysis and wet air oxidation (WAO) [87].

This section on chemical-biological processes starts with an introduction to the principles and goals of such combinations (Section 9.3.1). Examples of existing processes are then presented in Section 9.3.2. Since the CBPs already have a long history of successful practical application in drinking-water treatment, the emphasis in Section 9.3.2 is laid on their application in waste-water treatment, which has experienced intense research throughout the last decade. A succinct review of recent results can be found in Mantzavinos and Psillakis [87]. Last but not least Section 9.3.3 provides useful advice for experimentation with chemical-biological processes.

9.3.1
Principles and Goals

The combination of chemical and biological processes is called for, if one process alone cannot achieve the required treatment goals for drinking or waste water such as

- a decrease in the concentration of a specific toxic or biorefractory substance (M);
- removal of organic matter (measured as DOC or COD).

as completely as desired or as defined by the legislative requirements, or cannot achieve the goals economically. The coupling of the two types of processes tries to utilize the strengths of each process: biorefractory but easily oxidizable compounds (e.g., aromatics during ozonation) can be partially oxidized chemically, producing by-products (e.g., low molecular weight acids) that are difficult to ozonate further, but are easily (or at least more) biodegradable than the original compounds.

In combined processes, the treatment goal of each stage must be adapted to the overall goals. In the chemical step the predominant goal is to eliminate – but not to mineralize – the specific toxic or biorefractory substances. This means the oxidation must be carried out to the point that no original compounds are left and the by-products are less toxic and more biodegradable. The subsequent biological step is used to remove these by-products and of course mineralization is the ultimate goal because only this guarantees a complete removal of the substances of concern.

In order to judge whether the treatment goal has been achieved, analyses of individual compounds as well as lumped parameters are needed. The chemical methods DOC and COD are often supplemented with methods to measure the biodegradable matter based on oxygen demand, for example, the 5-day or ultimate BOD, or on organic carbon content such as assimilable organic carbon (AOC) or biodegradable DOC (BDOC). The latter two methods are also based on bioassays but are focused on different types of compounds and measurements. The AOC analysis was developed as an index of regrowth potential in drinking water, and is proportional to the density of test organisms that can grow in the solution [88]. It is not an absolute measure of carbon concentration. The BDOC is calculated as the difference between DOC analyses before and after a biodegradation step. Many methods for the biodegradation step have been suggested, from BOD flasks analogous to BOD measurements over 5–28 d [89] to continuous plug-flow bioreactors with analysis times of 2–3 h. Comparisons between AOC and BDOC show a large variability in correlation between the two methods, whereas stronger correlations exist between various methods for measuring BDOC [90] as well as between BDOC and BOD [89].

While DOC measurements can be used in most water applications to detect concentrations as low as $10\,\mu g\,l^{-1}$, BOD and COD are normally restricted to wastewater applications since the detection limits are 2 and $5\,mg\,l^{-1}$ O_2, respectively. Since BDOC is based on DOC measurements, it can also be used to measure low concentrations. In many waste-water applications, however, COD is still often chosen to monitor performance, mainly because most regulations still use COD as a controlling parameter. In oxidation reactions, however, the type of information gained from the DOC and COD measurements is quite different. DOC measures the amount of organic carbon present and is an indicator of the degree of mineralization, while COD measures the degree of oxidation achieved and reveals little about the state of mineralization. In chemical oxidation, a high COD removal does not necessarily correspond with a high DOC removal. Therefore, care has to be taken when comparing results from the literature, especially when the ozone-yield coefficient (the ratio of mass of ozone consumed to the mass of compound removed) based on different parameters is used.

The treatment goal of mineralization in CBPs is achieved by chemical oxidation as well as by biodegradation. The coupling of the two processes to attain a high degree of overall DOC removal with the lowest amount of oxidative agent is an optimization problem. The chemical stage must be operated on the fine line between achieving enough oxidization to make the compounds bioavailable, but not too much to keep expensive chemical mineralization at a minimum, as well as to avoid oxidizing biodegradable by-products.

An example of this balance can be seen in Figure 9.2, which shows a typical plot of the change in the DOC and BDOC concentration relative to the initial DOC concentration as a function of the specific ozone dose I^* in a batch system. As the specific ozone dose increases, the concentration of the DOC decreases continuously (Curve 1) indicating that mineralization is occurring. Initially there is no biodegradable DOC present in the mixture (Curve 2). As ozonation proceeds, the BDOC increases until a certain maximum is reached. Further ozonation oxidizes biodegradable matter.

In order to evaluate this two-stage process, it is advantageous to use Equation 4.9 from Section 4.2 to calculate the overall degree of removal $\eta(M)_\Sigma$ as the sum of removal achieved in each stage:

$$\eta(M)_\Sigma = \frac{c(M)_o - c(M)_i}{c(M)_o} + \frac{c(M)_i - c(M)_e}{c(M)_o} \quad (9.3)$$

This is plotted in Figure 9.2 as Curve 3. In this case, higher ozone doses I^* will actually lower the overall degree of removal $\eta(M)_\Sigma$ in the two-stage system.

The optimization problem to achieve a high degree of overall DOC removal with low ozone costs is practically solved by treatment schemes that combine oxidative

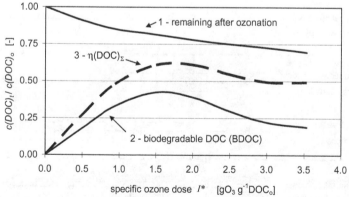

Figure 9.2 Typical changes in the relative concentrations of the total DOC remaining after ozonation (1), the biodegradable DOC (2), and overall degree of DOC removal $\eta(DOC)_\Sigma$ (3) as a function of the specific ozone dose I^* for the ozonation of model pollutant and subsequent biodegradation of its oxidation products in a batch chemical-biological system.

chemical and biological processes in series: single or multisequential stages as well as integrated systems with recycle streams between two or more stages are applied. In order to ease the discussion of CBP applications in the following section, the following four types of process schemes are defined: the (single-) sequential (CBP-type 1), the multisequential (CBP-type 2) and the integrated (CBP-type 3) chemical-biological process (Figure 9.3). The names are not standardized in the literature and the variety can cause confusion. For instance, the sequential chemical-biological process (CBP-type 1) is sometimes also called "integrated ozonation and biodegradation" in the literature.

If the water to be treated contains both biodegradable and biorefractory compounds, a biological pretreatment step prior to chemical oxidation is strongly recommended to remove the biodegradable compounds and reduce chemical consumption. Sometimes ozonation alone follows this biological pretreatment step (CBP-type 0). In the literature this process scheme is often referred to as post-ozonation, which is used as a final polishing step but not as a tool to further increase the bioavailability of the treated solution.

In general, though, a biological stage usually follows ozonation in both drinking and waste-water applications – at least to remove the easily biodegradable compounds to prevent bacterial regrowth, so that sequential configurations using CBP-type 1 are often found. In order to increase compound removal even further, effluent from the bio-stage can either be treated in a second oxidation stage (CBP-type 2) or recycled back to the original oxidation stage (CBP-type 3). The oxidation products are then transported into a biological stage where the biodegradable fraction is mineralized. This procedure can be repeated as often as the number of recycles or sequences until the required effluent conditions or threshold limits are met.

B 6 The Application of Ozone in Combined Processes

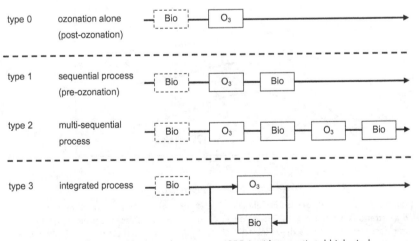

Figure 9.3 Types of chemical-biological processes (CBPs) with an optional biological pretreatment step.

Since the addition of each step involves increased investment and operating costs, one of the major concerns in the evaluation of alternative treatment schemes must be weighing the various costs against each other [91].

The next section gives an overview of existing processes and the research going on in the areas of combined ozonation and biodegradation of drinking water and waste water. The large difference in target-compound concentrations in the two applications leads to different treatment goals and especially different reactions in the treatment of the two water types. Indirect radical reactions usually dominate in drinking water, while direct reactions often play a large role in waste-water applications. Furthermore, the volume of water to be treated frequently differs greatly. Table 9.6 summarizes the main characteristics of the contaminants and the main goals of combined chemical-biological treatment in the two applications.

9.3.2
Existing Processes and Current Research

The application of ozone before a biological treatment process originates from studies in drinking-water treatment in the 1970s where ozonation units were used for the removal of organic trace compounds. Since ozonation alone did not–in every case–help to meet the low contaminant limits required by legislative regulations, an activated-carbon process was often installed after the ozonation step in order to eliminate remaining contaminants by adsorption [92]. It was observed that the resulting operation cycles of the activated-carbon filters were very long, enabling considerable cost savings compared with AC treatment alone. Detailed examination of the removal processes in a full-scale sequential chemical-biological application confirmed that biodegradation was responsible for the prolonged cycles [93]. Due to this observation, several hundred CBPs employing the sequential treatment scheme (CBP-type 1) were successfully put into operation in the field of drinking-water treatment since the late 1970s [94]. The more complex CBPs (type 2 and 3) play no role in this field of application. The specific goals as well as current research results in this area are dealt with in Section 9.3.2.1.

In contrast, all the four types of CBPs can be found in waste-water treatment. While post- and pre-ozonation (CBP-types 1 and 0) are characterized by rather simple installations and comparatively low capital costs, a high degree of mineralization can often only be achieved with a comparatively high amount of costly ozone being applied. Therefore, the more complex multisequential (CBP-type 2) or integrated (CBP-type 3) chemical-biological processes have been developed in which a high degree of overall DOC removal can be achieved with much less ozone due to increased biodegradation.

In such schemes attention has to be paid to reduce carryover of biomass into the ozonation stage, which is mostly done by using fixed-bed bioreactors or less frequently by applying a membrane separation process. The reverse is also true, ozone should not be carried over into the biological stage. Extensive research on these processes using model as well as real waste waters has shown that the degree

Table 9.6 Specific treatment goals in drinking-water and waste-water applications of chemical-biological processes as well as important operating conditions and pollutant characteristics.

Type	Drinking water (DW)	Waste water (WW)
Typical organic pollutants, concentration range and reactions		
Toxic or refractory target compounds (M)	• pesticides, herbicides, pharmaceuticals • substances influencing taste, odor and color • $\mu g\, l^{-1}$ to few $mg\, l^{-1}$	• various industrial contaminants • sometimes odorous and colored substances • $mg\, l^{-1}$ to approx. $1\, g\, l^{-1}$
DOC	• natural organic matter (NOM) • 1 to $20\, mg\, l^{-1}$	• various organics • 50 to $500\, mg\, l^{-1}$
Main reactions	• radical reactions	• direct reactions
Main treatment goals		
Chemical stage	• complete micropollutant transformation • maximization of BDOC-formation and avoidance of mineralization	• high degree of toxic and/or biorefractory substance elimination • optimization of BDOC-formation and minimized mineralization
Biological stage	• complete removal of "fast" and "slow" BDOC	• high degree of BDOC removal
CBP overall	• prevention of bacterial regrowth in distribution system • reduction of disinfection by-product formation potential (DBPFP)	• meeting effluent standards, e.g., high degree of overall DOC removal
Cost efficiency	• minimization of oxidant consumption	• minimization of oxidant consumption • optimized capital costs, i.e., optimized combination of CBP

of DOC removal depends much on the specific ozone absorption A^* or the specific ozone dose I^*. The main goals in the development of these processes, typical operating conditions, current research results and basic economical aspects are discussed in Section 9.3.2.2.

9.3.2.1 Drinking-Water Applications

Chemical stages in drinking-water treatment plants are required if organic substances like pesticides or herbicides from agricultural use or – as has lately been observed more frequently – pharmaceutical substances from human or veterinary medicine are present in the water source. Because of their low concentrations in the range of some nanograms up to a few milligrams per liter, they are commonly called micropollutants. Drinking-water standards normally restrict micropollutant concentrations to very low levels. Thus, comprehensive transformation of the micropollutants, if not complete mineralization, is the goal in drinking-water treatment.

Since drinking water is normally produced from ground water or surface waters, natural organic matter (NOM) is commonly present besides these micropollutants and can be oxidized as well. NOM is a complex mixture of organic materials (such as humic substances, hydrophilic acids, carbohydrates, amino acids, carboxylic acids etc.). Its removal is also an important goal in drinking-water treatment since NOM is a precursor of disinfection by-products (DBPs) when the water is subjected to postchlorination and, in addition, enhances biofilm regrowth in distribution networks. It is normally measured as dissolved organic carbon (DOC), though more advanced chromatographic measurements may be appropriate when trying to optimize for NOM removal [95]. The fraction of dissolved organic carbon that can be utilized by bacteria is sometimes referred to as biodegradable organic matter (BOM).

Sequential chemical-biological treatment (CBP-type 1) can achieve the three important goals in drinking-water treatment. The biodegradation of the oxidation products removes micropollutant by-products, and reduces the potential for bacterial regrowth in the drinking water distribution system as well as the disinfection by-product formation potential (DBPFP) of the treated water. Thus, chemical-biological pre-ozonation processes have an economical advantage, since mineralization of NOM and individual contaminants by chemical oxidation would require a great amount of expensive chemical agent.

The primary focus of the chemical treatment stage, therefore, is to partially oxidize the compounds by the action of ozone or hydroxyl radicals. In such applications, the hydroxyl radical reactions often dominate. This is due to the fact that dissolved ozone, triggered by the concentration of hydroxide ions in the water, decays fast to hydroxyl radicals (see Section 2.1.1) as well as to the rather low rates of the direct reaction of ozone with the micropollutants. The low direct rates result from comparatively small kinetic constants of the ozone direct reactions with many micropollutants as well as from their very low concentrations. However, since hydroxyl radicals react very fast with almost any organic substance, NOM is also partially oxidized in the chemical treatment stage. The formation of as much biodegradable DOC (BDOC) as possible from NOM along with the prevention of costly mineralization is another important goal in the chemical stage of CBPs in drinking-water treatment.

In the comparatively inexpensive biological stage, the BDOC produced from the NOM and the micropollutant is degraded. When activated carbon is used as filter material in the biological stage, a process that is commonly called the (ozone-enhanced) biological activated-carbon process ((OE)BAC), extended operation times of the activated carbon between two regeneration cycles are beneficial for the reduction of the operation costs. The OEBAC process was developed during the 1970s as pointed out above. It is the prototype of all CBPs and represents CBP-type 1 (see Figure 9.3). In drinking-water treatment it achieves good micropollutant removal at reasonable operating costs, which is why it has gained broad application.

9.3.2.1.1 Current Research Results
According to Yavich et al. [96], the control of biodegradable organic matter produced from ozonation is usually accomplished by rapid sand filtration with an empty bed contact time (EBCT) of 15–20 min in the United States of America. Even after long years of research and full-scale application, the studies on this process proceed (e.g., [96, 97, 98]) and the main question of interest is still prevention of bacterial regrowth in the distribution system.

In the study of Melin and Ødegaard [98] the effect of the biofilter loading rate on the removal rate of oxidation products of NOM was assessed. Expressed in terms of TOC, at loading rates up to $0.8\,mgCl^{-1}h^{-1}$, corresponding with EBCT of 20 min or more, the removal rates were at least 80% of the loading rates. At higher loading rates (tests were run up to $1.6\,mgC\,l^{-1}h^{-1}$) the removal rates decreased to approximately 50% of the loading rates. To prevent bacterial regrowth in the distribution system, it was concluded that the main concern in biofilter design should be minimizing the amount of biodegradable organic carbon in the filter effluent rather than obtaining the highest possible rates of removal.

Yavich et al. [96] applied ozonation and subsequent biodegradation to three different drinking waters each of which contained characteristic fractions of NOM. Based on biokinetic data the oxidation products of NOM ozonation were classified as "fast", "slow" or "non"-biodegradable organic carbon (BDOC). It was shown that the slowly biodegradable natural organic matter remains in the effluent and may cause bacterial regrowth in the distribution system. To overcome this problem, "stimulated" biodegradation, that is, the addition of a small amount of easily biodegradable carbon to the ozonated water, was successful in increasing removal in the biological stage.

9.3.2.2 Waste-Water Applications
The idea to use combined chemical-biological processes for the treatment of waste waters caught on around the mid-1980s. Increasing awareness about refractory organic substances in the effluents of municipal waste-water treatment plants (MWWTPs) was accompanied by the definition of a list of priority pollutants, many of which are poorly or scarcely biodegradable in normal activated-sludge processes [99, 100].

Thus, the requirements for waste-water treatment increased and existing technologies had to be applied in new combinations. In analogy to the OEBAC-processes in drinking-water treatment, ozonation combined with subsequent biodegradation under aerobic conditions was generally assumed to be a viable treatment option in such cases. Soon, this brought about a lot of research work dealing with the ozonation of individual priority pollutants and their subsequent biodegradation under aerobic conditions. The work of Gilbert [101] was an early milestone in this field. However, the regulatory priority soon became prevention, cleaning-up at the source of these waste waters and not at the end-of-the-pipe in the MWWTPs. Therefore, the treatment of segregated streams of specific and often highly concentrated waste waters (e.g., from pulp mills or distilleries or from processing cork, black-olives, textiles or pharmaceuticals; see also Section 9.3.2.2.2) came into the focus of the research work, though, such processes are still seldom found in industrial applications (see Section 3.3).

Considering the composition of pollutants in waste waters, the situation is generally comparable with the situation in drinking-water treatment. Mostly, one or more substances of specific interest have to be eliminated from the waste water, e.g., because of their toxicity, their color or their poor biodegradability. And besides such individual substances the waste waters normally contain a complex matrix of other organic–and inorganic–compounds, a situation principally similar to NOM in drinking waters. However, the concentration of the individual substances often ranges from some ten to a thousand milligrams per liter, thus being higher by a factor of 1000 compared with micropollutants in drinking water. In total, the complex matrix of dissolved organics often ranges between 50 to 500 mg l^{-1} DOC. Since many of such complex waste waters also contain some biodegradable substances, biological pretreatment prior to ozonation should generally be considered [87].

All in all, rather high concentrations of various oxidizable organic compounds are treated in the chemical stage of a CBP in waste-water treatment. In such systems with high concentrations, especially if ozonation is applied in a semi-batch chemical stage, direct reactions of ozone with the organic compounds often dominate and ozone mass transfer often limits the elimination rates (see also Chapter 6 and Section 8.3). It is important to take this into consideration when experimental results are to be assessed, for example, not only the specific ozone absorption A^* but also the specific ozone dose I^* should always be measured and reported.

A high degree of elimination of priority pollutants or a number of specific contaminants is normally the primary goal of the chemical treatment of waste water. As in the treatment of drinking water this means that they will be transformed, i.e., partially oxidized, by the action of ozone and/or hydroxyl radicals. The maximization of the formation of BDOC and minimization of mineralization are further important goals in the chemical stage of a CBP in waste-water treatment. Correspondingly, a high degree of biodegradation or biomineralization is aimed at in the biological stage. Last but not least the whole process will have to be designed to fulfil the legislative effluent standards for individual substances as well as the requirements for overall DOC (or COD) removal. On the whole this is a complex optimization problem.

9.3.2.2.1 Current Research Results
Meanwhile, a lot of research has been dedicated to the application of chemical-biological processes using ozone. Two questions are of major importance. First, how much ozone (or other oxidants) is necessary to almost quantitatively eliminate the specific compounds of interest, and secondly, how much ozone (or other oxidants) will be necessary to generate a high degree of biodegradable compounds without too much mineralization in the chemical stage at the same time. Since the simpler process types of post-ozonation (CBP-type 0) and pre-ozonation (CBP-type 1) have shown limited success in reducing the amount of ozone needed, the more complex multisequential and integrated processes, i.e., CBP-types 2 and 3, have been developed with this goal and have shown interesting potential in minimizing the ozone consumption.

9.3.2.2.2 Application of CBP-type 0 and CBP-type 1
As was already mentioned above, comprehensive research has been conducted on chemical oxidation and subsequent biological treatment of the oxidation products. Scott and Ollis [102] as well as Mantzavinos and Psillakis [87] have reported on the state-of-the-art work and the results. Most of the reported applications have been batch treatments in laboratory scale employing CBP-type 1 (O_3 + Bio). This research has covered a wide variety of organic compounds.

The early study of Gilbert [101] using CBP-type 1 on 28 substituted aromatic compounds showed that 100% elimination of the individual aromatic compound corresponded with 55–70% removal in COD and 30–40% in DOC due to ozonation and also resulted in good biodegradability. This was defined using the ratio BOD_5 to COD that remained after the chemical treatment step. The biodegradability was rated "good" for values of $0.4 \pm 0.1\,g\,BOD_5\,g^{-1}$ COD or higher.

Up to now, this approach (CBP-type 1) has often been used with the value of $BOD_5/COD = 0.4 \pm 0.1$ often quoted in the literature as an operating goal. However, operational goals and the dependency of removal efficiency on operating parameters must be determined for each specific application. It is important to keep in mind that the relationship between ozone dose and degree of removal is not linear, particularly in CBP (see Figure 9.2). Moreover, the maximum degree of overall mineralization in an optimized sequential chemical-biological process cannot be derived from it.

Research within the last decade has shown that the representation of the DOC removal, in either stage or for the whole process, as a function of the specific ozone dose I^* or the specific ozone absorption A^* provides a practical measure of how much ozone has to be produced and applied to attain the required degree of substrate removal. However, the specific ozone dose is system specific, meaning that it also depends on the mass-transfer rate and the hydrodynamics (e.g., batch or continuous mode of operation) of the ozonation system. Thus, optimizing the reaction conditions in the chemical stage offers the potential to minimize the ozone consumption as well as the operating costs.

Research on the lab-scale treatment of complex and often highly concentrated, not model but "real" waste waters (e.g., from cork processing, black-olives, pulp mill, distillery, textile processing) by means of the sequential chemical-biological process (CBP-type 1) has been intensified in the new millennium. The treatment in both stages was normally conducted in batch reactors, but a few studies also employed the sequencing batch-reactor technology (SBR) with integrated (CBP-type 3) [103] or continuous processes in both stages (CBP-type 1) [91]. A predominant treatment goal in all these works was to achieve a high degree of total substrate removal measured as COD or DOC (or TOC). In some cases color removal was another important goal. Since these complex waste waters often contain some biodegradable substances, ozonation alone after prior biodegradation (post-ozonation, CBP-type 0) has also been tested, mainly for pretreatment before discharge to a WWTP.

An overview of the main operating conditions and the results of substrate removal (measured as TOC or COD) is given in Table 9.7 for several studies on the treatment of such real waste waters employing CBP-types 0 or 1. It is important to note that the results of substrate removal in the ozonation stage were rather seldom reported as a function of I^* or A^*, although the necessary data for their calculation was available in the reports. Unfortunately, the final discharge concentrations to be met were also seldom addressed.

Combinations of CBP-type 0 are used when it is shown that biological (pre-treatment before chemical oxidation increases overall removal or significantly reduces ozone requirements. In the CBP-type 0 studies with complex waste waters in Table 9.7, an initial aerobic biological stage removed much of the organics contained in the model waste waters made-up of gallic acid, tannin or lignin (e.g., 20–50% TOC) [104]. Because of the high initial concentrations (TOC$_o$ of 1234, 608 and 620 mg l^{-1}, respectively) this was, however, not enough. To reach at least 80% overall DOC removal, specific ozone absorptions between $A^* = 4.0$ and 5.6 g O$_3$ g^{-1} TOC$_o$ had to be applied during post-ozonation and in the case of lignin η(TOC) > 70% could not be reached. In similar studies on cork processing and black-olive waste-water treatment biodegradation was able to remove 27% [106] and 86% [105] of the COD, respectively. In the latter case the reduction of the COD to a comparatively low concentration of approximately 100 mg l^{-1} still needed a rather high specific ozone dose of $I^* = 8.15$ g O$_3$ g^{-1} COD$_o$.

The applications of CBP-type 1 in Table 9.7 show that rather low specific ozone doses between $I^* = 0.16$–1.84 g O$_3$ g^{-1} COD$_o$ enabled overall COD reductions between 50 and 65%. These are much lower I^* than those used in the CBP-type 0 examples. This is due in part to the removal achieved in the biological stage, but also due to the fact that the overall removal is also less in these examples. Unfortunately, there was no discussion in the cited literature about whether this would fulfil the effluent requirements nor was it made evident how higher degrees of COD removal would depend on the specific ozone dose or the specific ozone absorption. This highlights the difficulties in comparing results between waste waters and between investigations, especially in combinations with biotreatments. Often, the amount of chemical oxidation required to achieve high reductions in the biological stage is very waste water and even reactor specific. In addition, treatment goals in the various studies vary, or indeed, are not stated. Generally, in order

308 | 9 Application of Ozone in Combined Processes

Table 9.7 Sequential chemical-biological treatment of complex waste waters (lab-scale batch processes) employing CBP-types 0 and 1.

Type of CBP	Origin of waste water	Initial substrate conc. c(M) in stage (B) or stage (C) (mg l^{-1})	Specific O$_3$ dose I* or O$_3$ absorption A* (g g^{-1})	Degree of substrate removal η(M) in each stage (C, B) and overall (CB or BC) (%)	Reference	Remarks
Bio + O$_3$ (type 0)	gallic acid[a]	TOC: 1234 (B) TOC: 438 (C)	A*: 5.6	B: [1] 49 C: 44 BC: 80	[104]	[1] degradation partly due to condensation
	tannin[a]	TOC: 608 (B) TOC: 586 (C)	A*: 4.0	B: 21 C: 79 BC: 80	[104]	
	lignin[a]	TOC: 620 (B) TOC: 440 (C)	A*: 5.0	B: 29 C: 58 BC:[2]70	[104]	[2] 80% TOC removal was not achieved
	black-olive processing	COD: 7000 (B) COD: 966 (C)	I*: 8.15	B: 86 C: 90 BC: 99	[105]	BOD$_5$: 4500 (raw WW)
	cork processing[b]	COD: 1930 (B)	I*: 1.24	B: 27 C: 68 BC: 77	[106]	BOD$_5$: 1150 (raw WW); COD discharge level to WWTPs (500 mg l^{-1}) met in BC-effluent

Process	Substrate	I*/A*	η(M)	Ref.	Remarks
O₃ + Bio (type 1) cork processing[b]	COD: 1900 (C)	I*: 1.84	C: 54 B: 24 **CB: 65**	[106]	
pulp-mill effluent	TOC: 700 (C) (COD: 1590)	I*: 1.0	C: 16 B: 48 **CB: 56**	[107]	
Kraft mill bleaching (CEH-)[b] effluent	COD: 1145 (C)	I*: 1.07	C: 37 B: 55 **CB: 62**	[108]	
distillery–thermally pretreated	COD: 10800 (C) (TOC: 7500 approx.)	I*: 0.16 (I*: 0.24) (TOC)	C: 13 B: 45 **CB: 48**	[109]	raw WW diluted by approx 1:8

Remarks: The values of I*, A* and η(M) always refer to the indicated type of substrate concentration (COD or TOC); substrate concentrations as well as I*- and A*-values in brackets are only given for reasons of comparison; the degree of removal η(M) in the individual stages (B or C), is always calculated as: (initial concentration – residual concentration) / initial concentration ×100, where "initial" means the concentration at the beginning of the treatment in the stage referred to.

a Synthetic WW; I*: not available.
b Ozonation in a continuous process.
c CEH: chlorination–extraction–hypochlorite.

to decide on the most effective treatment combination, it is necessary to compare the combinations with the same water.

In an attempt to optimize overall removal and minimize ozone consumption in a CBP-type 1 system to remove DOC from textile dyebath waste water, Libra and Sosath [91] tested various combinations and operating conditions in a continuous sequential CBP system. The results showed that a biological pretreatment did not reduce ozone requirements, although it reduced the color. Furthermore, it was found that in order to achieve a treatment goal of >80% DOC removal, high ozone doses were required. When the ozone feed rate $F(O_3)$ was decreased from 0.6 to $0.1\,gO_3\,m^{-3}\,s^{-1}$, the degree of DOC removal in the chemical stage decreased from 82 to 38%. Since the biological post-treatment consistently contributed only 20–30% to the overall removal, this resulted in a decrease in the overall DOC removal from 95 to 62%.

The CBP-type 1 combination can be advantageously used to remove organic compounds that are inherently toxic to microorganisms. Pharmaceutical substances are a group of waste-water contaminants that have gained considerable attention throughout the last years. These substances can exhibit high toxicity even in the $ng\,l^{-1}$ to $\mu g\,l^{-1}$ range. Detailed investigations on ozonation, AOP-application ("peroxone": $O_3 + H_2O_2$) as well as combined chemical-biological treatment of various effluents from penicillin formulation have been conducted by the group of Arslan-Alaton [110–113]. Chemical-biological batch treatment (CBP-type 1) of penicillin formulation effluent resulted in overall COD removal efficiencies of 84% and 79% for ozonation ($I^* = 5.6\,gO_3\,g^{-1}\,DOC_o$, pH = 12) and perozonation (O_3 at the same specific ozone dose and an initial concentration of $2\,mM\,H_2O_2$) and stand-alone treatment of segregated streams was recommended for industrial application [110]. Despite the comparatively high COD removals the authors pointed out that biorefractory or toxic compounds might still have been contained in the CBP-effluent, although they did not measure individual substances. However, various biodegradation and toxicitiy tests have been applied in a separate study [113].

In conclusion, the treatment efficiency in the sequential chemical-biological process (CBP-type 1) is rather substrate and system specific. Up to now, a general prediction of the required treatment effort is not possible.

9.3.2.2.3 Application of CBP-type 2 and CBP-type 3

Starting in the early 1990s, reports on investigations of the more sophisticated types of CBPs in Figure 9.3, the multisequential (CBP-type 2) as well as the integrated treatment (CBP-type 3) schemes have shown up. In contrast to the (single) sequential treatment (O_3 + Bio; CBP-type 1), chemical and biological treatment is repeated for two or three times in the multisequential system (CBP-type 2): each stage treats the effluent from the preceding stage with no recycle [114–116]. Examples of the treatment of textile and tannery waste water show that a high degree of substrate removal (i.e., >90% TOC or COD) was achieved with specific ozone absorptions A^* of roughly $5.0\,gO_3\,g^{-1}\,DOC_o$ (Table 9.8). The multisequential process can be regarded as an approximation of the integrated process (CBP-type 3) in which the water is continuously recycled between the two process steps several times.

9.3 Chemical-Biological Processes (CBP) | 311

Table 9.8 Sequential chemical-biological treatment of waste waters employing CBP-type 2 or CBP-type 3.

Type of CBP	Origin of waste water	Initial substrate conc. $c(M)$ in stage (B) or stage (C) (mg l^{-1})	Specific O$_3$ dose I^* or O$_3$ absorption A^* (g g^{-1})	Degree of substrate removal $\eta(M)$ in each stage (C, B) and overall ($n \times$ CB or C $\leftrightarrow$ B) (%)	Reference	Remarks
$n \times$ (O$_3$ + Bio) multisequential (type 2)	textile (dyehouse liquor) effluent[a]	TOC: 2000 $n = 3$ [1)]	A^*: 5.0	3 $\times$ (CB): 90	[103]	[1)] n = no. of repetitions of O$_3$ + Bio
	tannery	COD: 3140 $n = 3$ [1)]	$A^{*,2)}$: 2.05	3 $\times$ (CB): 93	[114]	[2)] based on DOC$_0$: 1205 → A^*: 5.34
(O$_3 \leftrightarrow$ Bio) integrated (type 3)	pulp bleaching	TOC: 380 ±9 n_R = n.a.	A^*: 0.86 A^*: 2.46	(C $\leftrightarrow$ B): 49 (C $\leftrightarrow$ B): 59	[119]	(C $\leftrightarrow$ **B**) = (O$_3$ $\leftrightarrow$ Bio) integrated
	synthetic (11 chloro-, nitro- and chloronitrobenzenes)	TOC: 128 ... 191 n_R = 1 ... 2	A^*: 4.0 A^*: 6.0	(C $\leftrightarrow$ B): 75 (C $\leftrightarrow$ B): 95	[121]	[3)] ozonation in tube reactor
	textile (dyehouse liquor) effluent[a]	TOC: 2000 n_R = 10 [3)]	A^*: 2.2	(C $\leftrightarrow$ B): 90	[103]	
	EMP[b]	TOC: 192 n_R = 10 ... 15	I^*: 4.17	(C $\leftrightarrow$ B) 80	[120]	
	landfill leachate[c]	COD: 1100 ± 100 (B) n_R = 4	I^*: 2.8	(C $\leftrightarrow$ B) 85	[122]	

Remarks: The values of I^*, A^* and $\eta(M)$ always refer to the indicated type of substrate concentration (COD or TOC); substrate concentrations as well as I^* and A^*-values in brackets are only given for reasons of comparison; the degree of removal $\eta(M)$ in the individual stages (B or C), is always calculated as: (initial concentration–residual concentration) / initial concentration ×100, where "initial" means the concentration at the beginning of the treatment in the stage referred to; n.a. = not available.

a Segregated stream of 1.1 m^3 d^{-1}.
b 5-ethyl-2-methyl-pyridine.
c Biologically pretreated, example BS.

In several studies the continuous-flow integrated process (CBP-type 3) has proven to be superior to the single sequential process (CBP-type 1), in terms of much lower ozone consumption per DOC removed [117–120]. Treating chloro- and nitro-substituted benzenes with both treatment schemes, Stockinger [121] found the degree of overall DOC removal to increase from approximately 50% to between 75 and 95% at specific ozone doses of $I^* = 3.5–6.0\,gO_3\,g^{-1}\,DOC_0$ in the integrated process. Similar results were found for several landfill leachates, for which 75% COD removal was achieved with almost 20% less ozone, applying specific ozone doses of $I^* = 1.6–3.0\,gO_3\,g^{-1}\,COD_0$ in the integrated process [122]. The recycle ratios n_R were between 1–2 [121] and 3–4 [122], respectively. Higher recycle ratios did not bring any additional benefit. However, in a batch-integrated process on textile waste-water treatment Hemmi et al. [103] found that a recycle ratio of $n_R = 10$ resulted in 90% DOC removal at a very low specific ozone absorption of $A^* \sim 2.2\,gO_3\,g^{-1}\,DOC_0$ compared with $A^* \sim 5.0\,gO_3\,g^{-1}\,DOC_0$ in a threefold multisequential process with the same waste water (Otto et al. (1998) cited in [103]; see also Table 9.8). While Wiesmann et al. [123] also obtained an improvement when applying a CBP-type 3 compared to CBP-type 1 to the treatment of real or synthetic textile waste waters containing the dye Reactive Black 5, for example 87% of overall DOC removal with a specific ozone absorption of $A^* \approx 3.6\,gO_3\,g^{-1}\,DOC_0$ versus more than $6\,gO_3\,g^{-1}\,DOC_0$, the values are still higher than in the study of Hemmi et al. [103]. Since a plug-flow tube reactor was used for ozonation in [103], Libra and Sosath [91] suggested that not only the integration of the two stages but also the type of the ozone reactor may be responsible for the observed reduction in ozone consumption.

In conclusion, the integrated process (CBP-type 3) is also substrate and system specific. The specific reasons for the encouraging results are still not fully understood and the question of how to achieve the optimal reaction conditions still cannot be answered properly. It seems that much work can be done in this field to clarify how the reaction mechanisms and kinetics in the chemical stage affect the biodegradability of various substances. Such knowledge is the basis for optimization in designing appropriate chemical reactors.

Moreover, improvements in the biological stage can also reduce the ozone consumption. A very favorable effect was observed in some cases in the biological stage of CBP-type 3. Extended operation of such systems allowed the biomass to adapt to the problematic compounds in the waste-water influent, which were originally classified as biorefractory. Thus, the amount of ozone necessary for their near-to-complete elimination decreased considerably, or even fully, as in the case of 3-methylpyridine [118]. However, this effect depends very much on the substrate and the reaction conditions, e.g., no adaptation occurred with 5-ethyl-2-methylpyridine [120]. Also no adaptation occurred during CBP-type 1 treatment of 4-nitroaniline [124, 125] or 2,4-dinitrotoluene [126]) although these substances are meanwhile known to be biodegradable by adapted microorganisms immobilized in fixed-bed bioreactors [127–130].

Finally, it is important to note that full-scale industrial application of CBPs is still rare in waste-water treatment. The above-mentioned operational uncertainties may be one major reason. This may also explain why, despite the considerable ozone savings observed, full-scale application of the integrated process (CBP-type 3) has not yet gained much attention in waste-water treatment. Another reason holding back application of integrated processes may be due to the comparatively complicated operation of such a process. For example, the stream to the oxidation stage should be free of biomass; if a suspended biomass is used in the bioreactor, separation of biomass before it is recycled increases the capital costs and consumption of energy [103].

Nevertheless, examples are found in the treatment of landfill leachates or waste waters from the textile, pulp bleaching or chemical industry (compare Section 3.4). All types of process schemes are in operation, and most often the CBPs are applied to waste waters that were biologically pretreated.

9.3.2.2.4 Consideration of Treatment Costs

In general, industrial CBP applications have to be cost efficient in terms of operation as well as investment costs. Operating costs can generally be reduced by removing the oxidation products using the less-expensive biological process instead of extended chemical oxidation. A progressive adaptation of the microorganisms to the oxidation products can lead to a considerable reduction in the amount of ozone being necessary to achieve the overall treatment goals. This has been observed in a lab-scale integrated process (CBP-type 3; [117]). However, complex treatment schemes with many stages might negatively affect those adaption processes as well as add to investment costs. Therefore, careful consideration of various treatment options has to be conducted [91].

Although cost efficiency of chemical-biological processes was sometimes mentioned as important for industrial applications, a systematic approach to the optimization of ozonation with respect to a minimized (specific) ozone input along with cost minimization was, for example, reported in only one of the studies listed in Table 9.7 [103] and is seldom discussed in the literature.

Since treatment is very substrate and system specific, the comparison between treatment combinations with similar waste water is of interest in this area. This was possible for lab studies of CBP for segregated streams of textile waste waters [91, 103]. Libra and Sosath [91] compared their own experiments with CBP-type 1 combinations, in which an overall removal of approximately 80% DOC was achieved by $A^* \sim 6 g O_3 \, g^{-1} \, DOC_0$, with the results of Hemmi et al. [103], who achieved 90% DOC removal with a very low $A^* \sim 2.2 g O_3 \, g^{-1} \, DOC_0$ in a CBP-type 3 process. Possible reasons for this difference are discussed in Section 9.3.3. In spite of the much lower A^*, the treatment costs calculated for the CBP-type 3 process was twice as high. This highlights how different reactor configurations can lead to very different costs. The biological pretreatment in the continuous-flow sequential CBP in Libra and Sosath's investigation used immobilized bacteria. In the study of Hemmi et al. [103] sequencing batch reactors were coupled with a tube reactor for ozonation as well as a crossflow filtration unit to prevent suspended biomass from the SBR from entering the ozone reactor. This unit was, however, a major factor in investment as well as operating costs, the latter being due to a tenfold internal recycle between the ozonation and the biological stages.

All in all, not only the unsolved technological questions but also the need for thorough cost considerations leaves a broad field for future research work.

9.3.3
Experimental Design

This section deals mainly with what aspects should be considered when experimenting with biological stages, supplementing the general aspects of experimental design discussed in Chapter 4. Table 9.9 summarizes the main aspects of experimental design in chemical-biological processes (CBPs). Some of the most important aspects of experimenting with combined processes are discussed in detail afterward. Further methodological suggestions can be found in the review [87].

Table 9.9 Main aspects of experimental design in chemical-biological processes in overview.

Design step	Main aspects or main tasks	Remarks
Define goals	• complete elimination of biorefractory or toxic substances $\eta(M) \approx 100\%$	relevant for DW & WW treatment
	• high degree of biodegradable DOC (BDOC) formation by chemical oxidation	relevant for DW & WW treatment
	• high overall degree of mineralization, $\eta(DOC) > 85\%$	relevant for WW: check effluent requirements
	• cost-efficient process → optimize or minimize oxidant consumption	WW: treat segregated streams, initial biodegradation (CBP-type 0); use CBP-type 3 to reduce oxidant consumption
Define system	• type of water or waste water (synthetic, real, mixtures)	DW: pH, TIC, buffer WW: pH, buffer, raw WW dilution ratio, select segregated streams, biological pretreatment (CBP-type 0)
	• choice of biomass (mixed cultures)	adapted, nonadapted, suspended, immobilized
	• type of oxidant	ozone, AOPs, Fenton's, WAO, etc.
	• choice of chemical (C) and biological (B) stages	C: e.g., bubble column or STR → mixing conditions, mass-transfer rate B: e.g., STR or FBR → aerobic, anaer.
	• choice of CBP-types 1–3	see Figure 9.3 and Section 9.3.1

Table 9.9 Continued

Design step	Main aspects or main tasks	Remarks
Select analytical methods	• oxidants in liquid and gas phase	see Section 5.4
	• substrate and oxidation products	individual substances (odor), COD, DOC, color
	• biodegradability	BOD_5, COD, DOC, OUR
	• toxicity (see Section 1.4)	Microtox, Daphnia, OUR
Determine experimental procedure	• operating conditions	batch (e.g., CBP-types 1 & 2) continuous-flow (esp. CBP-type 3)
	• measurement conditions	offline, online
	• assure reproducibility	repetition of experiments and quality control program
Evaluate data	• chemical stage (C) (* → compare also Figure 4.3, Section 4.1 and Table 4.2 and 4.3, Section 4.2)	$\eta(M)$, $\eta(DOC)$ as a function of I^* or A^*, $\eta(O3)$, * → DW: O3 dose and conc. (liq. phase)
	• biological stage (B)	BOD_5, BOD_5/COD, DOC removal, OUR, toxicity (OUR, EC_{25}, EC_{50})
	• overall process (CB)	degree of individual and overall pollutant removal (esp. DOC)
Assess results	• compare results	to experimental goals; with threshold limits; with those found in the literature
	• iteratively optimize experiments	esp. reduce oxidant consumption

9.3.3.1 Define System

9.3.3.1.1 Type of Water or Waste Water

- Check the biodegradability of the original solution.

  If there are biodegradable compounds in the WW, the experimental plan should include an evaluation to decide whether the use of an initial biological treatment will decrease ozone consumption in the chemical stage and/or increase overall removal efficiency of the biorefractory and/or toxic constituents in all stages. A comparison with and without biopretreatment should be made.

- As far as possible try to theoretically analyze which oxidation products may be formed from the (main) compounds contained in the (waste-) water, determine methods for their measurement and estimate their toxicity and/or biodegradability.

9.3.3.1.2 Choice of Biomass

- aerobic or anaerobic process;
- mixed or pure culture of microorganisms;
- biomass acclimated or nonacclimated.

Most often, aerobic processes employing nonacclimated mixed bacterial cultures are used for the biodegradation stages. On the contrary, fungi or algal treatment of ozonated waste waters has been reported very seldom. Aerobic processes are advantageous for bio-stages following oxidation, since the water is normally rich in oxygen from the oxygen/ozone mixture transferred in the chemical-treatment step. However, for waste waters with a high load of organics, an anaerobic process may be effective, since a considerable amount of the oxidation products may be low molecular weight organic acids that are also biodegradable in an oxygen-free environment. This may reduce operating costs; though the sensitivity of methanogenic bacteria to the by-products (and their variability) must be carefully evaluated.

For example, the ozonation of a highly concentrated olive oil mill effluent (OME, $COD_0 = 50–250\,g\,l^{-1}$) created by-products that inhibited methanogenic bacteria, especially *para*-hydroxybenzoic acid. Since the ozonated mixture was not inhibitory to acidogenic bacteria the authors proposed the applicability of a two-step acidogenic and methanogenic process but did not present results themselves [131].

The use of pure bacterial cultures may be an option for specific situations, though they are not of general interest in waste-water applications.

Choose an inoculum as appropriate as possible. If the ozonated solution is to be fed to an existing process or if there is a preadapted culture able to consume (some of) the main oxidation products [132] (if known), use this specific inoculum. Also, the opposite is quite often successful: use a highly diverse mixed culture, preferably from a WWTP treating a wide variety of compounds, and develop a culture capable of degrading the oxidation products. Such a culture can be used to test the "general biodegradability" of the oxidation products (e.g., [133]).

Acclimation of the biomass to the oxidation products is basically possible. In their review study Mantzavinos and Psillakis [87] found that acclimation is a difficult question and no general answer is possible. Since it is not clear how long the process has to be operated until an adaptation occurs – or if ever – the laboratory effort required may be unsatisfactorily high. Changes in reaction-system configuration and operating conditions can change the oxidation products. In addition, microorganisms acclimated to the original substrate may not degrade the oxidations products or *vice versa*.

Nevertheless, in several studies adaptation was reported to cause a dramatic reduction in the amount of expensive ozone needed for the desired degree of removal, or with the same amount of ozone much higher degrees of DOC removal were achieved [117, 118]. Both batch as well as continuous modes of operation of the chemical-biological system were applied. In the study of Balçioglu *et al.* [108] on the ozonation of waste water from the forest industry (Kraft pulp mill CEH stage bleaching effluent) the most successful treatment occurred with an adapted algal treatment in a sequencing batch reactor.

9.3.3.1.3 Choice of Chemical-Biological Reaction System

- sequential (CBP-types 1 or 2) or integrated (CBP-type 3) process;
- batch or continuous-flow processes;
- suspended biomass (flocs) or biofilm immobilized on support material.

Most laboratory studies on chemical-biological treatment have been conducted as a single sequential process (CBP-type 1) in batch tests, mainly due to the ease of operation. The multisequential (CBP-type 2) as well as the integrated process (CBP-type 3) require increased laboratory effort but also promise higher degrees of substrate removal with lower ozone consumption (see Table 9.8). This is due to the higher reaction rates that can be achieved in continuous-flow multistage systems that approach the conditions in a plug-flow reactor with no longitudinal mixing [134]. (The benefits of PFR vs. CFSTR were discussed in detail in Chapter 4.) In spite of this benefit, multistage continuous-flow ozonation reactors are not often used in the laboratory, mainly due to their higher constructive and analytical effort. However, the inverse is true for many full-scale applications – continuous flow is often chosen for large liquid flow rates. In conclusion, the variety of multistage reactors (sequential or integrated, continuous-flow or batch) combined with their ability to reduce operating costs offers large research potential.

In designing the type and size of the individual reactors, the principles of chemical reaction engineering, for example, the influence of reactor type and mode of operation on reaction kinetics and mass transfer as well as difficulties in scaling up discussed in [134] and Chapters 4 and 5, should be considered. Generally, any combination of batch or continuous-flow processes for the individual stages can be employed. For example, the effluent from a continuously operated chemical stage can be collected in a storage tank and from this be fed to a batch biological stage or *vice versa*. Storage makes any combination easy to handle, however, checks should be made for unwanted modification of the substrate in the storage tank.

For the chemical reactor, semibatch bubble columns or stirred-tank reactors are often used. However, the benefits and difficulties in achieving plug flow in gassed ozonation reactors warrant more research in this area. For example, Hemmi et al. [103] used a tube reactor equipped with a semi-permeable membrane for ozone transfer to achieve high DOC removal with low ozone consumption treating a real textile waste water within an integrated system (CBP-type 3). The question of which type of process should be operated is further discussed below in the section on the experimental procedure.

Both batch and continuous-flow biological systems can be easily operated in lab-scale experiments. Batch systems using suspended biomass are commonly used to determine the biodegradable DOC fraction, with volumes ranging from ml to liters. Here, the same initial concentration of biomass can be easily added to each batch when comparing treatments and biomass growth may also easily be measured [114, 116]. Continuous-flow systems with immobilized biofilms in fixed-bed bioreactors may be advantageous if the oxidation products are only slowly biodegradable. Slow-growing microorganisms can be retained efficiently in biofilm systems, whereas in continuously operated systems with suspended biomass, microorganisms can be washed out when the hydraulic retention time is less than or equal to the population doubling time (i.e., the reciprocal of the specific growth rate). In several studies materials such as polyurethane foams [114, 135] or quartz sand particles have been used for immobilization [117, 118, 120, 127]. No material has shown a decisive advantage in performance.

The volumes of fixed-bed bioreactors can vary from miniaturized versions with as little as 25 ml liquid volume, to large ones with over 50 l. The charm of smaller systems is that smaller volumes of chemically oxidized substrate are needed. In general, high concentrations of biomass can be achieved in biofilm systems, resulting in high biological mineralization rates [130]. The drawback is that biomass concentrations are not easy to measure and high concentrations can lead to plugging and flow-distribution problems.

Miniaturized continuous-flow bioreactors have been successfully applied for the aerobic biological mineralization of several nitroaromatic substances, which have long been considered scarcely biodegradable, for example, 2,4-dinitrotoluene (DNT) and 2,4-dinitrophenol [136], 3- and 4-nitronaniline (NA) [128] and 4,6-dinitro-*ortho*-cresol (DNOC) [128, 137]. In all these experiments the mixed-culture bacterial biomass had been preadapted in batch systems before being transferred to the continuous system.

9.3.3.2 Select Analytical Methods

The measurement of *oxidants, substrates* and *their oxidation products* is generally based on appropriate physical and chemical analyses. This has already been discussed in Sections 4.2 and 5.4. In CBP investigations it is also important to develop routine measurements for the biological parameters *biodegradability and toxicity*, which will be discussed here. When planning the quality control program, the variability inherent in biological systems must be considered.

9.3.3.2.1 Measurement of Biodegradability

The biodegradability of a substance, the ability of bacteria to degrade a substance, depends on many parameters (type of bacteria, pH, T, nutrient availability, etc.) and the time allowed for it to occur. The discussion here is restricted to a short overview of practical methods to measure it. For a very good overview of the principles of biodegradation, its definition and measurement, the reader is referred to Grady [138] as well as to Page [90], ISO [139], and Pagga [140] for overviews of standardized methods.

Various methods can be used to measure biodegradability. The disappearance of individual compounds can be monitored directly or changes in lumped parameters such as DOC, COD or optical density can be measured, as well as the cumulative consumption of oxygen or evolution of CO_2. In ozonated waters biodegradation is mainly measured by the lumped parameters and their principles were described already in Section 9.3.1. These methods differ not only in the parameter measured (oxygen demand−BOD, organic carbon−BDOC or optical density−AOC) but also in the length of time required for the analysis. This time plays an important role for the classification−fast or easily degradable compounds can be measured within hours, inherently degradable in 28 d [141].

A method to differentiate between "fast", "slow" and "nonbiodegradable" TOC based on the kinetics of its degradation is meanwhile well established – especially in drinking-water treatment. The assessment is made by measuring TOC concentration as a function of the empty-bed contact time (EBCT) in the continuous-flow biological system (see e.g., Yavich et al. [92]). In this method, the discrimination between "fast" and "slow" biodegradable TOC is done graphically (or "operationally") by determining the intersection of the two (extreme) tangent lines of the curve that relate to the very beginning and to the "practical end" of the decrease in TOC concentrations versus EBCT. The latter also separates the "slow" and "nonbiodegradable" TOC. In an operating CBP system the hydraulic retention time is the determining time period for evaluating whether compounds are biodegradable. In evaluating the impact of the remaining compounds in the discharged effluent the inherent biodegradability may be important.

Practically, two methods are frequently used to evaluate the biodegradability of CBP waste waters [87]. In both cases bioassays with nonadapted bacterial biomass are employed to assess:

- The oxygen consumed over a certain time period x (BOD_x), for example, mostly run for five days (BOD_5), but an extended test duration may also be applied.

- The mineralization by measuring the organic carbon removal (dissolved or total) over a certain period of time (BDOC), for example, 5 days (similar to BOD_5) or 28 days (similar to the inherent biodegradability test).

The concurrent measurement of CO_2 evolution while measuring BOD offers an elegant way to estimate the extent of mineralization [142]. AOC measurements are more prevalent in the treatment of drinking water to evaluate the bacterial-regrowth potential.

The use of DOC measurements to quantify biological mineralization makes a direct comparison to the mineralization in the chemical stage possible and is therefore superior to BOD_5, which cannot be easily correlated with DOC removal in the chemical stage [87, 113]. Nevertheless, BOD_5 measurement was not only used in the early work of Gilbert [101], but is still frequently employed (e.g., [107, 108, 113, 116, 143–145]).

Unfortunately, all measures of biodegradability depend heavily on the test organisms and the test duration chosen. For example, by extending the BOD test from 5 to 21 days, its value was doubled [146]. This highlights the importance of the standardization of the tests used, at least within the investigation.

9.3.3.2.2 Measurement of Toxicity Toxicity tests (also called bioassays) are generally used to measure a response induced by test substances under controlled conditions in the laboratory, generally using cultured organisms in the tests. A short overview of toxicity and ecotoxicity testing is given in Chapter 1. Single substances or whole effluents can be tested. Standardized bioassays have been developed and optimized over the last decades to quantify effects on bacteria, daphnia and fish, for example, [147]. In addition to the direct toxicity assessments, *indirect methods* can be employed such as oxygen uptake rate (OUR) inhibition tests with mixed bacterial cultures. With the introduction of simple and inexpensive toxicity testing kits, toxicity testing in waste-water investigations has become more widespread. Most often, the acute toxicity is measured using tests of the inhibition of bioluminescence in marine microorganisms.

Various studies on chemical-biological processes have measured the occurrence and elimination of acute toxicity during the treatment (e.g., [113, 114, 135, 145, 146, 148–150]). An increase in toxicity was frequently observed during the early stages of chemical pretreatment but it was often reduced by further oxidation [87]. This increase may be due to typical oxidation products, such as organic peroxides, hydrogen peroxide, low molecular weight alcohols, carboxylic acids and aldehydes. Measurements of acute toxicity, however, do not necessarily indicate that biodegradation in a following biological stage will be affected. Subsequent biotreatment has often been found to reduce the acute toxicity; sometimes an adaptation phase of 4 weeks may be necessary [135].

This highlights that a systematic approach such as carrying out a "toxicity balance" around the whole process is important for evaluating combined chemical-biological treatment. For example, such a study was carried out by Moerman et al. [135]. In addition, the question which toxicity test(s) would be necessary or most suitable is difficult to answer. While acute toxicity tests with Daphnia and bioluminescence may provide useful information if the effluent is to be discharged to an aquatic environment, they do not assess the effects on the biological process in CBPs [87]. The oxygen uptake rate (OUR) of a mixed bacterial culture can reveal inhibitory effects more applicable to biological treatment stages. This was shown for example in the treatment of a waste water from pharmaceutical production (procain penicillin G) by applying the activated sludge inhibition test (ISO method 8192) [113] or by making use of the activated-sludge model no. 3 (ASM3) [111]. Tests with unadapted bacteria may give an indication of possible effects due to an indirect discharge to an off-site WWTP, while adapted bacteria should be used to predict effects in a coupled bioreactor.

9.3.3.3 Determine Experimental Procedure

Both the chemical stage and the biological stage require attention when developing the experimental procedure, not only individually, but also how they fit together. As is to be expected, the design and operating conditions in one stage affect the decisions for the design and operation of the coupled stage(s). This is actually self-evident, but often forgotten. A rather simple aspect is, for example, the pH-value, which often decreases during ozonation to rather acidic conditions, whereas a neutral pH is normally required in aerobic biological systems. pH adjustment is then required. This leads to an optimization question to be answered experimentally – in which stage should the pH be adjusted?

Some recommendations to help avoid common pitfalls in coupled systems are given in this section; most apply especially to waste-water treatment. Special attention is given to the biological process since many recommendations concerning the chemical stage are already contained in the checklist in Section 4.4.

Batch operation of the biological stage requires the consideration of the following aspects:

- Check that the biomass is active and has not suffered from inadequate storage conditions (e.g., at high temperature or due to starving and/or lysis) before the experiments are started.

- If a series of batch ozonations is to be compared, for example, varying the specific ozone dose I^* or the specific ozone absorption A^*, make sure to use the same inoculum in each biodegradation batch tests.

- Take care that the concentration of DOC that is introduced into the batch (bottle) with the inoculum is far lower than the DOC of the ozonated solution. Otherwise, it will be difficult to differentiate between the origins of the DOC and to track the mineralization of the oxidation products.

- The experimental conditions and the duration of the biotreatment should be constant throughout the whole set of experiments.

- Substrate concentrations over time should be monitored (at least in developing the procedure) since they can rise again when lysis of the biomass occurs.

- Last, but not least, work in a way so that the results are comparable to others from the literature.

Batch tests are recommended before starting any kind of continuous-flow experiments, especially if the application of an integrated system (CBP-type 3) is intended. For example, Karrer *et al.* [116] proposed a batch "applicability test" – simply measuring the COD removal in a multisequential chemical-biological treatment (CBP-type 2). Such testing is quick and easy to perform, reliable due to the underlying standardized methods of measurement, and can also be used to roughly estimate the costs for a combined process. Similar approaches, which take into account the amount of COD that is partially oxidized in the chemical process step and then biodegraded, were proposed by other authors [114, 115, 151].

In the *continuous-flow* operation mode of a chemical-biological process, it is important to prevent elevated concentrations of ozone (gaseous as well as liquid) from entering the biological stage. Regardless of whether operating a sequential or an integrated process, ozone will at least partly kill (oxidize) the biomass. This in turn will cause the biological process to slow down or cease completely. In the integrated system (CBP-type 3) it will also cause additional and completely unnecessary ozone consumption when additional organic carbon (DOC) from destroyed biomass is transported into the chemical system [116, 117]. Thus, it is important to construct the system in a way that effectively prevents such problems. A pressure equalizer or gas trap to prevent gaseous as well as dissolved ozone entering the biological stage can be installed between reactors [125]. Another approach is to automatically control the dissolved-ozone concentration at near zero in the integrated process (CBP-type 3) [121].

Although application of a sequential chemical-biological treatment (CBP-type 1) in industrial scale often employs a direct coupling between the two stages as well

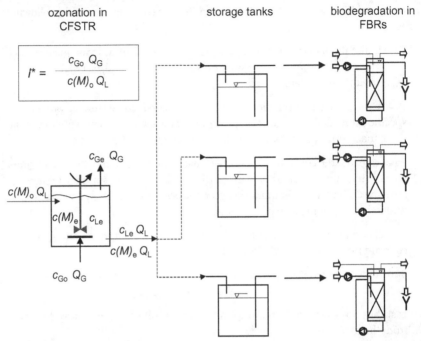

Figure 9.4 Proposed experimental setup for a sequential chemical-biological continuous-flow waste-water-treatment process (CBP-type 1) with intermediate storage of ozonated substrate.

as continuous-flow mode, in laboratory work such a system lacks flexibility. This is due to the fact that changes in the operating conditions of the chemical stage can be performed rather quickly (e.g., in a few hours or days), but to establish a new equilibrium in the biological stage normally more time is needed. Therefore, in order to increase flexibility in experiments and/or to shorten the total experimentation time, a single continuous-flow (or batch) ozone reactor may be used together with one or more bioreactors, for example, miniaturized fixed-bed reactors (FBRs), using a storage tank in between (Figure 9.4). This procedure allows the ozone reactor to be operated and optimized independently from the biological treatment, reducing the total experimentation time required compared to a direct coupling of the two stages.

Care, however, must be taken that unwanted reactions (biodegradation and/or polymerization) do not occur. An additional concern is that the biomass in the different bioreactors might develop differently with extended treatment time, even if the original inoculums were the same in each system. This must not be a disadvantage, since acclimation of the biomass to the specific oxidation products might occur.

Nevertheless, for either *continuous-flow* or *batch* ozonation, such a chemical-biological system could be operated as follows:

- In the *continuous-flow* mode:

  - run the ozonation stage at a number of desired operating conditions, for example, at previously calculated I^* (or A^*; this, however, is only possible if the ozone-transfer efficiency is already known, which would mean previous experimentation at the same working conditions);
  - collect enough effluent for the subsequent biodegradation experiments;
  - store effluent in tanks (cool, dark);
  - feed it to the appropriate number of continuous-flow miniaturized fixed-bed bioreactors as long as necessary to reach best biodegradation. The biological treatment might need a number of days or even weeks.

- *Batch* ozonation can be applied as well: in this case the liquid volume of one batch might not be enough to operate the biological system for a longer time. However, batch ozonation can easily be repeated as often as necessary.

In both cases care must be taken that the oxidation products are not changed during storage due to chemical, physical or undetected biological processes.

And be aware that even at the same level of A^* the degree of substrate removal in the chemical stage will often be very different from each other in the case of continuous-flow or batch ozonation. This is, on the one hand, due to the different reactional behavior of continuous-flow and batch systems (see Section 4.1) and, on the other hand, different kinetic regimes of ozonation may be present. Such different results were, for example, shown in the work of Saupe [125].

A stepwise increase in the bioreactor loading rate by reducing the hydraulic retention time can be realized with the goal to realize high DOC removal or biodegradation rates. In this way the best conditions for the combined chemical-biological process can be assessed, with least ozone consumption and highest degree of overall DOC removal.

9.3.3.4 Evaluate Data

In general, not only the total, but also the individual degrees of removal in each stage are of interest. Here, care has to be taken to indicate exactly the reference value for the calculation of the degree of removal in the individual stages. This can either be the influent concentration to the whole system $c(M)_o$ (Equation 4.9) or to the individual stage. If the values are calculated relative to $c(M)_o$ they are additive and the relative contribution of each stage to the overall removal is clearly seen, while the values relative to the influent concentration in each stage allow better analysis and comparison between similar reactors.

In addition, it is highly recommended to evaluate all experimental results of substrate removal as a function of the specific ozone dose I^* and related parameters such as the specific ozone absorption A^* and consumption $D(O_3)^*$, especially in waste water. Although these parameters are system-specific the advantage is that they include all parameters of influence (e.g., mass-transfer rate in the chemical stage including mass-transfer enhancement, operating mode and waste-water composition), and thus facilitate comparison between the results of reactor configurations within the investigation and from other researchers on a more standardized level. Sometimes, the ozone yield coefficient $Y_{O3/M}$, which is the ratio between the amount of ozone consumed and the amount of substrate (M) or DOC removed, is also of interest.

9.3.3.5 Assess Results

This step is important in all types of experiments, of course, but when working with process combinations, the researcher usually has to go through more iterations of the experimental design process than is required with single stages. The assessment of whether the goals have been achieved and renewed planning are frequent activities, so it is important to develop good evaluation routines early on in the experimental cycle. Some specific aspects for CBP are covered here; tips on how to plan CBP to improve their performance are also given below.

9.3.3.5.1 Assess Goal Achievement
Check that removal requirements have been attained cost efficiently across the whole combination. An (almost) complete and consistent elimination of toxic or (bio-) refractory substances ($\eta(M) \approx 100\%$) is usually required in most applications. Furthermore, a high degree of total DOC removal (e.g., ≥85%) is often a goal in waste-water treatment, while in drinking-water treatment the removal focus with concern of DOC is on the biodegradable organics.

When evaluating results from process combinations, especially from CBP-types 2 or 3, it becomes hard to retain an overview of results. Suitable combinations of parameters can help to analyze the values. Figure 9.5 shows an example of how laboratory experiments on three subsequent cycles of combined treatment

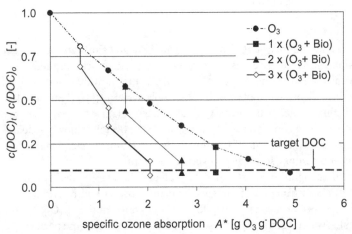

Figure 9.5 Typical degradation profile ($c(DOC)_t / c(DOC)_o$) of a multisequential chemical-biological batch treatment process (CBP-type 2) (after [114]).

(O_3 + Bio) can be evaluated by plotting the remaining concentration of the DOC as a function of the specific ozone absorption A^* (as $gO_3\ g^{-1}$ DOC).

As can be seen, each O_3 + Bio cycle reduced the amount of ozone necessary to reach the targeted DOC. A^* was reduced by 60% with 3 O_3 + Bio cycles compared to ozonation alone. In such a plot, though, it is harder to see the removal achieved in each stage. In Figure 9.5, the chemical stage contributed quite heavily to the overall DOC removal.

To visualize the contribution of each stage to overall removal, a plot similar to Figure 9.2 can be helpful, where η(DOC) is plotted against specific ozone absorption A^* and differentiated for each stage, or to gain a quick overview of where the majority of mineralization is taking place, a plot of the degree of overall removal versus chemical removal can be made. In Figure 9.6 it is easy to see that in order to achieve the targeted DOC removal, more removal due to biodegradation is achieved at the operating conditions in case 2.

9.3.3.5.2 Plan Further Experiments
Based on the result assessment, the next iteration in the experimental cycle can be planned. Usually, the first iteration is used to assess whether ozonation in combination with a biological step is a treatment option. The next steps are to improve the process and minimize the ozone consumption, usually adding complexity. The two main steps are to: (1) modify the chemical reactor design and operation, and (2) improve the integration of the chemical and biological stages. These steps are explained in detail below.

(1) Optimization focused on the chemical reactor is a starting point for all CBP types. Here, modifications based on reaction engineering concepts can be

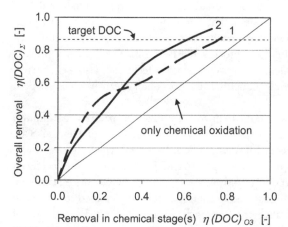

Figure 9.6 Comparison of the DOC removal due to chemical oxidation versus overall removal for a multisequential chemical-biological batch treatment process at two different operating conditions (after [152]).

evaluated to decrease ozone-generation costs. Changes in the reactor hydrodynamics, i.e., the mixing status of the liquid and the gas phase in the reactor, and the reaction rates can be made.

Constructive changes in the chemical reactor can improve overall performance. For continuous-flow applications, increases in ozone-transfer efficiency $\eta(O_3)$ and reaction rates can be reached by changing the mixing regime from completely mixed to plug flow (tube reactor or stages in series (n-CFSTR)). This carries with it that the application of the same specific ozone dose I^* often results not only in higher ozone-transfer efficiencies $\eta(O_3)$ (and thus in different values of the specific ozone absorption A^*) but also in different oxidation products or degrees of substrate removal.

In contrast, for a change from batch to plug flow or *vice versa* less change in oxidation products to be treated in the biological reactor can be expected. However, although reactions in a plug-flow reactor theoretically correspond to those in a batch reactor, this is valid only when the reactants are present at the same concentration in both systems. Ozone concentrations are not necessarily the same in the semibatch reactors with continuous gassing and the plug-flow reactors with single dosing stations. Therefore, all changes in the chemical reactor entail evaluation of the resulting changes in the bioreactor.

(2) Optimization focused on the combination of chemical and biological reactor usually involves trying to maximize biomineralization by reducing the extent of oxidation carried out in one step. For example, the ozone feed rate can be reduced so that less of the ozonation by-products are oxidized during the product oxidation. This results in high ozone-transfer efficiencies, but also in a rather low degree of DOC removal in the chemical stage. And depending on the biodegradability of the oxidation products, this may also result in a comparatively low degree of overall DOC removal during one pass through a chemical-biological system.

Therefore, multiple repetitions of ozonation and subsequent biodegradation may be necessary. The ideal system that combines an efficient chemical oxidation with as much biological mineralization as possible is the continuous-flow integrated chemical-biological process (CBP-type 3). Since such a process is rather difficult to operate in this system, its effects can also be approximated by employing a multisequential chemical-biological batch treatment (CBP-type 2).

Combining the two steps should help to minimize the ozone consumption and maximize the degree of overall mineralization. The study of Hemmi *et al.* [103] is an example of a successful optimization. In a CBP-type 3 process using a tube reactor for ozonation, as much as 90% of DOC removal was achieved with a specific ozone absorption of only $A^* \sim 2.2 \, gO_3 \, g^{-1} DOC_0$. Since such very positive effects were seldom observed and are not fully understood today, more work is necessary to generalize the approach, especially for the treatment of highly loaded (waste) waters that require a large amount of gaseous ozone to be continuously dosed.

9.4
Applications in the Semiconductor Industry

In the last 15 years interest in ozone application in the semiconductor industry has increased as an alternative to the standard cleaning methods in order to lower chemical consumption and costs as well as to improve process performance [153]. Its use in the cleaning process is well established. Looking at the 0.13-μm technology around 500 process steps are included, about a third is used for cleaning and again 30% involving wet cleaning steps. About 20% of wet cleaning applications in the FEOL (front end of line: all process steps until the interconnects are established) area are using ozone in production for cleaning steps. Here, its ability to oxidize organic and metallic contaminants in the aqueous phase is utilized. New applications exploiting its ability to oxidize organics as well as inorganics in the solid phase are now well established, for example, photoresist removal (step 4) and in fast oxidation of silicon (step 1 below). Another use is the disinfection of deionized water to keep the water system free from microbial contaminants.

For a better understanding of the use of ozone in the semiconductor industry, the general process used for the production of chips from wafers is briefly explained in this section. The main principles and goals of how ozone is used is addressed in Section 9.4.1. Since ozone is mainly used as a cleaning agent to remove various contaminants, existing cleaning processes, the conventional ones as well as those employing ozone, are outlined in Section 9.4.2. Section 9.4.3 gives valuable advice for setting up experiments to design and/or improve such cleaning processes.

The general process used for the production of chips from wafers has five basic steps. These simplified steps are shown in Figure 9.7. Further information on the whole process in the semiconductor industry can be found in Gise and Blanchard [153] and Kern [154].

9.4.1
Production Sequence

A wafer is a thin slice of a crystal, grown from pure silicon and used to produce electronic components like integrated circuits (IC). Silicon itself does not conduct electricity; additional ions must be introduced into its matrix to make it conductive (ion implantation). These changes in the crystal structure are made in intricate geometrical patterns to achieve the desired conductive properties. The principal production steps are briefly explained below (Figure 9.7). Ozone alone or in combination with other oxidants (e.g., hydrogen peroxide) can be used for initial wafer oxidation (step 1) but primarily as a cleaning agent during post-treatment (step 5).

1. **Oxidation:** An oxide layer is produced on the surface of the silicon wafer (SiO_2). This layer of silicon dioxide is an isolating layer on the surface. It is usually grown in an atmosphere containing oxygen, water vapor or other oxidants, for example, ozone or hydrogen peroxide.

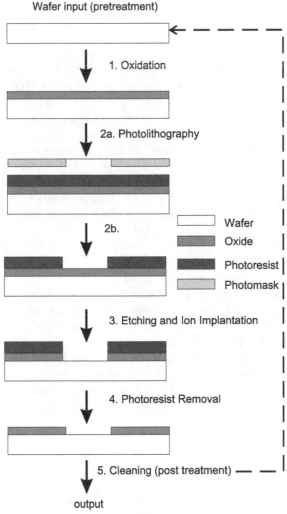

Figure 9.7 Simplified chip-production process sequence, ozonation and/or hydrogen peroxide is employed during initial oxidation (step 1) and final cleaning (post-treatment; step 5).

2. **Photolithography:** In the photolithographic process, the geometrical pattern that produces the desired electrical behavior is transferred to the surface of the wafer.

    a) The wafer is coated with photoresist, which is applied as a thin film and acts like the photographic film in a camera. An image is developed in the photoresist by using a mask. Through a photomask the wafer is exposed using UV-radiation. The radiation alters the chemical bonding of the photoresist, making it more soluble where it has been exposed (positive photoresist).

b) After the development of the photoresist and its removal, the positive image stays on the wafer in the resist.

3. **Etching and Ion Implantation:** In this process step an etchant (gas or liquid) removes the SiO_2 where it is not protected by the photoresist. Ions are implanted into the unprotected silicon. With the implantation of ions the structure of the surface will be changed.

4. **Photoresist Stripping:** The photoresist is removed.

5. **Cleaning (Post-Treatment):** Because the geometrical patterns are so intricate, multiple passes through this sequence may be necessary to obtain the desired structure. Every wafer processing step is a potential source of contamination. Consequently, cleaning of the wafers must take place after each processing step and so is the most frequently repeated step during manufacturing. Here, ozone is one important cleaning agent. The following section summarizes the main principles and goals of the cleaning process.

9.4.2
Principles and Goals

In the semiconductor industry cleanliness is an absolute requirement. Even small traces of contaminants can cause modification of the structure of the wafer surface area. Cleaning processes employing ozone in the chip production have been used since the late 1980s. Interest continues to grow as modifications and new methods are developed. Meanwhile, ozone use is well established in the chip-production process.

The general requirement for an efficient cleaning process is the removal of all contaminants that would affect the functionality or reliability of elements. The contaminants to be removed can be classified as follows:

- organic impurities: heavy (photoresist) and light organics (solvents, impurities from humans: skin, hair, clothes);
- particles: mainly from the ambient environment and from humans, but solvents and moving parts can also act as a particle source;
- metallic / ionic contamination: elemental metal films from solvents or machines, ions from humans and solvents.

Every wafer-processing step is a potential source of contamination, each step with its specific type of contaminant. This means that an efficient cleaning process consists of several cleaning steps in order to remove all contamination from the crystal. Thus, the following section mainly deals with the cleaning processes and how ozone has found its place therein and substituted conventional ones.

Since at the end of a cleaning process (post treatment), an optional oxidation step can be used to create a hydrophilic surface, the use of ozone for oxide growth is also discussed in the following section.

Table 9.10 Dissolved-ozone concentration ranges for different applications in the semiconductor industry.

Application	Range of dissolved-ozone concentration in mg l^{-1}
Light organic clean	10–30
Heavy organic clean	60–120
Particle/metal removal	10–30
Oxide growth	1–20

The requirements for all these different applications with respect to ozone concentration and flow rate depend on the application (Table 9.10).

Compared to the used concentration in drinking-water and waste-water treatment the range of dissolved-ozone concentration is relatively high, especially for heavy organic cleaning. This requires very efficient reactor systems that fulfill the semiconductor specification. In this area all contact material needs to be extremely clean. Metals and particles are undesirable due to the electric damage they can cause. Commercial systems are available [156].

9.4.3
Existing Processes for Cleaning and Oxidation

The existing cleaning methods can be divided into wet and dry cleaning. The wet cleaning process uses a combination of solvents, acids, surfactants and deionized (DI) water to spray and dissolve contaminants from the surface area. The DI water is used to rinse after each chemical use. The oxidation of the wafer surface is sometimes integrated into the cleaning steps. Dry cleaning, also called gas-phase cleaning, is based on excitation energy such as plasma, radiation or thermal excitation. This section will concentrate on the wet processes, the area where ozone is of interest.

To better illustrate the purpose behind the cleaning steps, the conventional RCA (Radio Company of America) cleaning process will first be examined in detail. It was developed for wafer cleaning in 1965, published in 1970 [157] and is still by far the predominant technique. The following Table 9.11 shows the steps of the preliminary cleaning and the conventional RCA cleaning, as well as the purpose of each step.

The purpose of the preliminary cleaning step is to remove heavy organic contamination, for example, photoresist.

During the following RCA cleaning process organic, particular and metallic contaminants are removed and a final oxide layer is created and can, optionally, be removed.

The RCA cleaning was developed at a time when the semiconductor industry was much smaller and the environmental restrictions were not as strict as today. Since then, the goal in the development of new processes has been to reduce the

Table 9.11 Preliminary and RCA cleaning [157, 158].

Process	Procedure	Goals
Preliminary cleaning		
H_2SO_4/H_2O_2 (4:1), 120–150 °C	SPM (Sulfuric acid, hydrogen Peroxide, DI water Mixture), often called Piranha	Removal of organic carbon, Photoresist removal
DI water	UPW (Ultra Pure Water)[a]	Rinse
HF (0.5%)	DHF (Diluted Hydrofluoric acid)	Removal of oxide
DI water	UPW[a]	Rinse
RCA		
Standard Clean 1 (SC1)		
$NH_4OH/H_2O_2/H_2O$ (1:1:5)[b], 70–90 °C	APM (Ammonium hydroxide, hydrogen Peroxide, DI water Mixture)	Removal of particles, organics, some metals
DI water	UPW[a]	Rinse
Standard Clean 2 (SC2)		
$HCl/H_2O_2/H_2O$ (1:1:6)[a], 70–90 °C	HPM (Hydrochloric acid, hydrogen Peroxide, DI water Mixture)	Removal of metals
DI water	UPW[a]	Rinse
Oxide growth (possible after SC1 or SC2)		
(HF (0.5%)[a]	DHF	Removal of oxide)

a Ozonated UPW is often used today (see below).
b Various concentrations and mixtures are used.

number of necessary cleaning steps, chemical consumption and waste disposal. Improvements in wet cleaning have been very successful in further reducing costs, chemical and water usage. Many advances are based on the use of ozonated ultrapure water (UPW) as a replacement for hydrogen peroxide or even sulfuric-based mixtures [159].

The so-called IMEC clean (from the Interuniversity Microelectronic Center, set up in 1984 by the Belgian Government) is one possible improvement. It is a simple two step process with an optional third step [160].

1. **SOM:** sulfuric acid/ ozone mixture: to remove organic contamination and to grow a thin chemical oxide layer. Under optimized conditions ozonated ultra pure water (UPW) can replace the SOM. The SOM step or ozonated UPW step replaces the SPM (sulfuric acid, hydrogen peroxide, DI water mixture) step.
2. **HF/HCl:** step to remove particles and metals, and the built-up oxides from the previous step, so that the surface becomes hydrophobic.

optional

3. **HCl/O$_3$**: or other ozone or hydrogen peroxide mixtures: regrowth of a thin oxide layer, hydrophilic passivation.

Comparison of the IMEC process with the RCA cleaning shows a drastic reduction in the number of necessary steps. The RCA with preliminary cleaning includes nine steps, this is reduced to two or three steps with the IMEC process. The number of chemicals reduces from six (RCA) to four (IMEC).

While ozone is primarily used to eliminate organics including photoresist, it can be found in all four application areas listed in Table 9.10. Process differentiations depend on the primary purpose of the cleaning steps. As can be seen above, especially for particle and metal removal, combinations of ozone with other oxidative chemicals have to be applied.

Removal of Light and Heavy Organics Dissolved ozone has been found to be very effective in the removal of trace organic contamination (light organic) from wafer surfaces. The fundamental chemistry of ozone involves both direct and indirect reactions [161]. Its removal efficiency depends on the type of organic, the ozone concentration and reaction regime.

Higher removal rates of photoresist (heavy organic) can be achieved by increasing the dissolved-ozone concentration as well as the process temperature [162]. However, the instability of ozone and its reduced solubility at elevated temperatures (up to 95 °C) lead to a process that must be carefully optimized to achieve the maximum photoresist removal rate.

Limitations of the ozone process are observed in the case of ion implantation with high density (10^{15}–10^{16} cm^2) and high energy (100 keV). The surface modifies to a hardened layer often referred to as a "crust" [163]. This crust cannot be attacked by ozone.

Removal of Particles and Metals Ozonated water can effectively remove particulate and metal contamination when combined with acids (e.g., HF and/or HCl), sequentially or mixed.

If ozone is used alone, the oxides/hydroxides could bind to the wafer surface. The acids added to the cleaning water act as ion exchangers to solve the metal species, keeping them in solution.

The metals and particles can be incorporated into the formed oxide layer. This oxide layer together with the contaminants will then be removed by adding HF. Repeating steps of creating an oxide layer by ozone that is subsequently etched by HF may be necessary to achieve an acceptable result [164]. Step by step more and more contaminants are removed.

Oxide Growth In the chip manufacturing process silicon will be oxidized and form silicon dioxide, so an oxide layer is grown on the wafer surface. The layer is either used as a protective layer, as a mask, as an isolator or an intermediate layer.

In the first three cases the silicon is oxidized until saturated, in the last a defined thin oxide layer (~4 nm) is desired.

Silicon surfaces that are treated with ozone exhibit self-limiting oxide growth and a hydrophilic state in the resultant surface. Silicon dioxide growth with ozone at room temperature self-limits at 10 to 11 Å, a relatively thick layer compared to that formed with other chemical oxidants. For example, oxide growth with hydrogen peroxide self-limits at a thickness of about 7 Å. The addition of platinum (Pt) to H_2O_2 enhances the oxidation rate by the generation of radical species, increasing the thickness to 10 to 11 Å. The maximum observed thickness of SC1 is around 8 Å [165]. The main parameters influencing oxide growth rate are dissolved-ozone concentration, process time, pH, temperature and the presence and type of additives coupled with equipment configuration [165].

The effective control of the oxide growth in the process by exposure time and ozone concentration is desirable since the full self-limited oxide is not always advantageous in the design of a chip. For example, establishing high-quality electrical characteristics of the interface between a high-k oxide film (higher dielectric constant than silicon oxide) and a silicon substrate requires the presence of a well-defined, extremely thin and uniform layer of silicon dioxide. This can be produced by saturating the surface with oxygen and subsequently performing an etch back using HF, or by a more easily controlled oxide growth process, that, for example, simply adjusts the process time to yield the required thickness.

9.4.4
Process and / or Experimental Design

General considerations for the design of ozonation processes, or for experimental work on developing new applications or improving existing methods are contained in the following section.

9.4.4.1 Define System
The variability in water found in other applications is greatly reduced here since ultrapure water (>18 MΩ) is used in all applications sometimes in combination with defined chemicals. The industry requirement of "absolutely clean" also extends to clean equipment (ozone generator, see Section 5.2, contact system), which means no particle generation, metal, ion or organic contamination. A whole industry has been developed to supply devices that fulfill these requirements.

9.4.4.2 Select Analytical Methods
In order to ensure reproducibility process control is required. In particular, the liquid ozone concentration must be measured (see Section 5.5).

9.4.4.3 Determine Procedure
The basic reactions are the same as in drinking-water and waste-water treatment. Therefore, knowledge about necessary equipment (Chapter 5), ozone mass

transfer (Chapter 6) and reaction kinetics (Chapter 7) including influencing parameters are very helpful for the development of new cleaning methods or recipes.

The unit used to transfer ozone into the water and the wafer processes (cleaning, photoresist removal, etc.) are often separated by a distance of up to 40 m. Therefore, the decay rate of ozone may significantly affect the ozone concentration. The liquid-ozone concentration close to the point of use should be measured to insure reproducibility of the process.

9.4.4.4 Evaluate Data and Assess Results
See Chapter 4.

References

1 Glaze, W.H., Kang, J-W and Chapin, D.H. (1987) The chemistry of water-treatment processes involving ozone, hydrogen peroxide and ultraviolet radiation. *Ozone: Science & Engineering*, 9, 335–352.

2 Lunak, S. and Sedlak, P. (1992) Photoinitiated reactions of hydrogen peroxide in the liquid phase. *Journal of Photochemistry and Photobiology A: Chemistry*, 68, 1–33.

3 Taube, H. (1956) Photochemical reactions of ozone in solutions. *Transactions of the Faraday Society*, 53, 656–665.

4 Oppenländer, T. (2003) *Photochemical Purification of Water and Air, Advanced Oxidation Processes (AOPs): Principles, Reaction Mechanisms, Reactor Concept*, Wiley-VCH Verlag GmbH, Weinheim.

5 Fenton, H.J.H. (1894) Oxidation of tartaric acid in the presence of iron. *Journal of the Chemical Society*, 65, 899–910.

6 Peyton, G.R. (1990) Oxidative treatment methods for removal of organic compounds from drinking water supplies, in *Significance and Treatment of Volatile Organic Compounds in Water Supplies* (eds N.M. Ram, R.F. Christman and K.P. Cantor), Lewis Publisher, Michigan, pp. 330–361.

7 Camel, V. and Bermond, A. (1998) The use of ozone and associated oxidation processes in drinking-water treatment. *Water Research*, 32, 3208–3222.

8 Staehelin, J. and Hoigné, J. (1982) Decomposition of ozone in water: rate of initiation by hydroxide ion and hydrogen peroxide. *Environmental Science & Technology*, 16, 676–681.

9 Brunet, R., Bourbigot, M.M. and Doré, M. (1984) Oxidation of organic compounds through the combination ozone-hydrogen peroxide. *Ozone Science & Engineering*, 6, 163–183.

10 Duguet, J.P., Brodard, E., Dussert, B. and Malleville, J. (1985) Improvement of effectiveness of ozonation in drinking water through the use of hydrogen peroxide. *Ozone Science & Engineering*, 7, 241–258.

11 Glaze, W.H. and Kang, J.W. (1988) Advanced oxidation processes for treating groundwater contaminated with TCE and PCE: laboratory studies. *Journal of the American Water Works Association*, 80, 57–63.

12 Baus, C., Sacher, F. and Brauch, H-J. (2005) Efficiency of ozonation and AOP for Methyl-tert-Butylether (MTBE) removal in waterworks. *Ozone: Science & Engineering*, 1, 27–35.

13 Prengle, H.W., Hewes, C.G. and Mauk, C.E. (1975) Oxidation of refractory organic materials by ozone and ultraviolet light. Proceedings, 2nd International Symposium for Water and Waste

Treatment, Montreal, Canada, May, pp. 224–252.
14 Paillard, H., Brunet, R. and Dore, M. (1988) Optimum conditions for application of the ozone-hydrogen peroxide oxidizing system. *Water Research*, **22**, 91–103.
15 Glaze, W.H., Peyton, G.R., Lin, S., Huang, F.Y. and Burleson, J.L. (1982) Destruction of pollutants in water with ozone in combination with ultraviolet radiation. 2. Natural trihalomethane precursors. *Environmental Science & Technology*, **16**, 454–458.
16 Peyton, G.R., Huang, F.Y., Burleson, J.L. and Glaze, W.H. (1982) Destruction of pollutants in water with ozone in combination with ultraviolet radiation. 1. General principles and oxidation of tetrachloroethylene. *Environmental Science & Technology*, **16**, 448–453.
17 Beltran, F.J., Gonzalez, M.J., Rivas, J. and Acedo, B. (2000) Determination of kinetic parameters of ozone during oxidations of alachlor in water. *Water Environment Research*, **72**, 689–697.
18 Beltran, F.J., Acedo, B., Rivas, J. and Alvarez, P. (2000) Comparison of different treatments for Alachlor removal from water. *Bulletin Environmental Contamination Toxicology*, **65**, 668–674.
19 Pettinger, K-H (1992) Entwicklung und Untersuchung eines Verfahrens zum Atrazinabbau in Trinkwasser mittels UV-aktiviertem Wasserstoffperoxid. Thesis. Fakultät für Chemie, Biologie und Geowissenschaften, Technische Universität München.
20 Wimmer, B. (1993) Metabolismus chlorierter 1,3,5-Triazine bei der Trinkwasseraufbereitung mit UV-aktiviertem Wasserstoffperoxid. Thesis. Fakultät für Chemie, Fakultät für Chemie, Biologie und Geowissenschaften, Technische Universität München.
21 Mohey El-Dein, A., Libra, J. and Wiesmann, U. (2006) Cost analysis for the degradation of highly concentrated textile dye wastewater with chemical oxidation H2O2/UV and biological treatment. *Journal of Chemical Technology and Biotechnology*, **81**, 1239–1245.
22 Prados, M., Roche, P. and Allemane, H. (1995) State-of-the-art in pesticide oxidation field. Proceedings of the 12th Ozone World Congress, Lille, France, pp. 99–113.
23 Andreozzi, R., Caprio, V., Insola, A. and Marotta, R. (1999) Advanced oxidation processes (AOP) for water purification and recovery. *Catalysis Today*, **53**, 51–59.
24 Pines, D.S. and Reckhow, D.A. (2003) Solid phase catalytic ozonation process for the destruction of a model pollutant. *Ozone: Science & Engineering*, **25**, 25–39.
25 Kasprzyk-Hordern, B., Ziolek, M. and Nawrocki, J. (2003) Review: catalytic ozonation and methods of enhancing molecular ozone reactions in water treatment. *Applied Catalysis B: Environmental*, **46**, 639–669.
26 Beltrán, F.J. (2004) *Ozone Reaction Kinetics for Water and Wastewater Systems*, Lewis Publisher, Boca Raton, London, New York, Washington D.C., pp. 227–276.
27 Kaptijn, J.P. (1997) The Ecoclear® Process. Results from full-scale installations. *Ozone: Science & Engineering*, **19**, 297–305.
28 Bahorsky, M., Billing, E.M., Deemer, D.D., Johns, J.D. and Lompe, D. (1999) Technologien zur Entfärbung von Textilabwässern mit dem Schwerpunkt katalytische Ozonierung, Preprints of 4. VDI-GVC-congress. Verfahrenstechnik der Abwasser- und Schlammbehandlung, Bremen, 6.-8.9.1999.
29 Cooper, C. and Burch, R. (1999) An investigation of catalytic ozonation for the oxidation of halocarbons in drinking-water preparation. *Water Research*, **33**, 3695–3700.
30 Ma, J., Sui, M-H., Chen, Z-L. and Wang, L-N. (2004) Degradation of refractory organic pollutants by catalytic ozonation–activated carbon and Mn-loaded activated carbon catalysts. *Ozone: Science and Engineering*, **26**, 3–10.
31 Trapido, M., Veressinina, Y., Munter, R. and Kallas, J. (2005) Catalytic ozonation of m-Dinitrobenzene. *Ozone: Science and Engineering*, **27**, 359–363.

32 Valdés, H. and Zaror, C.A. (2006) Heterogeneous and homogeneous catalytic ozonation of benzothiazole promoted by activated carbon: Kinetic approach. *Chemosphere*, **65**, 1131–1136.

33 Qu, X., Zheng, J. and Zhang, Y. (2007) Catalytic ozonation of phenolic wastewater with activated-carbon fiber in a fluid bed reactor. *Journal of Colloid and Interface Science*, **309**, 429–434.

34 Paillard, H., Doré, H. and Bourbigot, M.M. (1991) Les perspectives d'application de l'ozone catalytique en traitement des eaux à potabiliser. 10th Ozone World Congress, IOA, Monaco, 19–21 March.

35 Pi, Y., Ernst, M. and Schrotter, J-C (2003) Effect of phosphate buffer upon CuO/Al2O3 and Cu(II) catalyzed ozonation of oxalic acid solution. *Ozone: Science & Engineering*, **25**, 393–397.

36 Beltrán, F.J., Rivas, F.J. and Montero-de-Espinosa, R. (2005) Iron type catalysts for the ozonation of oxalic acid in water. *Water Research*, **39**, 3553–3564.

37 Beltrán, F.J., Rivas, F.J. and Gimeno, O. (2005) Pyruvic acid removal from water by the simultaneous action of ozone and activated carbon. *Ozone: Science & Engineering*, **27**, 159–169.

38 Carbajo, M., Rivas, F.J., Beltrán, F.J., Alvarez, P. and Medina, F. (2006) Effects of different catalysts on the ozonation of pyruvic acid in water. *Ozone: Science & Engineering*, **28**, 229–235.

39 Abd El-Raady, A.A., Nakajima, T. and Kinchhayarasy, P. (2005) Catalytic ozonation of citric acid by metallic ions in aqueous solution. *Ozone: Science & Engineering*, **27**, 495–498.

40 Fontanier, V., Farines, V., Albet, J., Baig, S. and Molinier, J. (2005) Oxidation of organic pollutants of water to mineralization by catalytic ozonation. *Ozone: Science and Engineering*, **27**, 115–128.

41 Volk, C., Roche, P., Joret, J-C. and Paillard, H (1997) Comparison of the effect of ozone, ozone-hydrogen peroxide system and catalytic ozone on the biodegradable organic matter of a fulvic acid solution. *Water Research*, **31**, 650–656.

42 Karnik, B.S., Davies, S.H., Baumann, M.J. and Masten, S.J. (2007) Removal of Escherichia coli after treatment using ozonation-ultrafiltration with iron oxide-coated membranes. *Ozone: Science & Engineering*, **29**, 75–84.

43 Lompe, D. (2007) Test reports of catalytic ozonation and data from full-scale installations by Eco Purification Systems B.V., Unpublished.

44 Sánchez-Polo, M., von Gunten, U. and Rivera-Utrilla, J. (2005) Efficiency of activated carbon to transform ozone into · OH radicals. Influence of operational parameters. *Water Research*, **39**, 3189–3198.

45 Beltrán, F.J., Rivas, J., Alvarez, P. and Montero-de-Espinosa, R. (2002) Kinetics of heterogeneous catalytic ozone decomposition in water on an activated carbon. *Ozone: Science & Engineering*, **24**, 227–237.

46 Lompe, D. and Plugge, M. (2000) Selektiver Abbau gefährlicher Inhaltsstoffe in Abwässern der pharmazeutischen Industrie durch katalytische Ozonierung, Preprints of VDI-GVC-congress. Abwässer der pharmazeutischen Industrie und Krankenhäuser, Bremen, 6.-7.9.2000.

47 Bailey, P.S. (1958) The reaction of ozone with organic compounds. *Chemical Review*, **58**, 925–1010.

48 Criegee, R. (1975) Die Ozonolyse. *Chemie in unserer Zeit*, **7**, 75–81.

49 Kornmüller, A. and Wiesmann, U. (2003) Ozonation of polycyclic aromatic hydrocarbons in oil/water-emulsions: mass transfer and reaction kinetic. *Water Research*, **37** (5), 1023–1032.

50 Hayduk, W. and Laudie, H. (1974) Prediction of diffusion coefficients for nonelectrolytes in dilute aqueous solutions. *AIChE Journal*, **20** (3), 611–615.

51 Hassan, I.T.M. and Robinson, C.W. (1977) Oxygen transfer in mechanically agitated aqueous systems containing dispersed hydrocarbon. *Biotechnology and Bioengineering*, **XIX**, 661–682.

52 Stich, F.A. and Bhattacharyya, D. (1987) Ozonolysis of organic compounds in a

53. Bhattacharyya, D., Van Dierdonck, T.F., West, S.D. and Freshour, A.R. (1995) Two-phase ozonation of chlorinated organics. *Journal of Hazardous Materials*, 41, 73–93.
54. Freshour, A.R., Mawhinney, S. and Bhattacharyya, D. (1996) Two-phase ozonation of hazardous organics in single and multicomponent sytems. *Water Research*, 30, 1949–1958.
55. Gromadzka, K. and Nawrocki, J. (2006) Degradation of diclofenac and clofibric acid using ozone-loaded perfluorinated solvent. *Ozone: Science & Engineering*, 28, 85–94.
56. Guha, A.K., Shanbhag, P.V., Sirkar, K.K., Vaccari, D.A. and Trivedi, D.H. (1995) Multiphase ozonolysis of organics in wastewater by a novel membrane reactor. *American Institute of Chemical Engineers Journal*, 41, 1998–2012.
57. Shanbhag, P.V., Guha, A.K. and Sirkar, K.K. (1996) Membrane-based integrated absorption-oxidation reactor for destroying VOCs in air. *Environmental Science & Technology*, 30, 3435–3440.
58. Kornmüller, A., Cuno, M. and Wiesmann, U. (1997) Selective ozonation of polycyclic aromatic hydrocarbons in oil/water-emulsions. *Water Science & Technology*, 35, 57–64.
59. Kornmüller, A. and Wiesmann, U. (1999) Continuous ozonation of polycyclic aromatic hydrocarbons in oil/water-emulsions and biodegradation of oxidation products. *Water Science & Technology*, 40, 107–114.
60. Ward, D.B., Tizaoui, C. and Slater, M.J. (2004) Extraction and Destruction of organics in wastewater using ozone loaded solvent. *Science & Engineering*, 26, 475–486.
61. Leeuwen, J. (1992) A review of the potential application of non-specific activated sludge bulking control. *Water SA*, 18, 101–105. van
62. Collignon, A., Martin, G., Martin, N. and Laplanche, A. (1994) Bulking reduced with the use of ozone – study of the mechanism of action versus bacteria. *Ozone: Science & Engineering*, 16, 385–402.
63. Saayman, G.B., Schutte, C.F. and van Leeuwen, J. (1996) The effect of chemical bulking control on biological nutrient removal in a full scale activated sludge plant. *Water Science & Technology*, 34, 275–282.
64. Yasui, H., Nakamura, K., Sakuma, S., Iwasaki, M. and Sakai, Y. (1996) A full-scale operation of a novel activated-sludge process without excess sludge production. *Water Science & Technology*, 34, 395–404.
65. Sakai, Y., Fukase, T., Yasui, H. and Shibata, M. (1997) An activated-sludge process without excess sludge production. *Water Science & Technology*, 36, 163–170.
66. Kamiya, T. and Hirotsuji, J. (1998) New combined system of biological process and intermittent ozonation for advanced wastewater treatment. *Water Science & Technology*, 38, 145–153.
67. Scheminski, A., Krull, R. and Hempel, D.C. (1999) Oxidative treatment of digested sewage sludge with ozone. Proceedings of the Conference "Disposal and Utilisation of Sewage Sludge: Treatment Methods and Application Modalities", IAWQ, 13–15 October, Athens, pp. 241–248.
68. Park, K.Y., Ahn, K-H, Maeng, S.K., Hwang, J.H. and Kwon, J.H. (2003) Feasibility of sludge ozonation for stabilization and conditioning. *Ozone: Science & Engineering*, 25, 73–80.
69. Stensel, H.D. and Strand, S.E. (2004) Evaluation of feasibility of methods to minimize biomass production from biotreatment. WERF Report 00-CTS-10T, IWA Publishing.
70. Lee, J.W., Cha, H.Y., Park, K.Y., Song, K-G. and Ahn, K-H. (2005) Operational strategies for an activated-sludge process in conjunction with ozone oxidation for zero excess sludge production during winter seasons. *Water Research*, 39, 1199–1204.
71. Dytczak, M.A., Londry, K.L., Siegrist, H. and Oleszkiewicz, J.A. (2007) Ozonation reduces sludge production and improves denitrification. *Water Research*, 41, 543–550.

72 Chu, L-B., Yan, S-T., Xing, X-H., Sun, X-L. and Jurcik, B (2009) Progress and perspectives of sludge ozonation as a powerful pretreatment method for minimization of excess sludge production. *Water Research (2009)*, doi:10.1016/j.watres.2009.02.012.

73 Eberius, M., Berns, A. and Schuphan, I. (1997) Ozonation of pyrene and benzo(a)pyrene in silica and soil–14C-mass balances and chemical analysis of oxidation products as a first step to ecotoxicological evaluation. *Fresenius Journal Analytical Chemistry*, **359**, 274–279.

74 Choi, H., Lim, H-N, Kim, J., Hwang, T-M. and Kang, J-W. (2002) Transport characteristic of gas phase ozone in unsaturated porous media for in-situ chemical oxidation. *Journal of Contaminant Hydrology*, **57**, 81–98.

75 Fujiwara, K., Kadoya, M., Hayashi, Y. and Kurata, K. (2006) Research note–effects of ozonated water application on the population density of Fusarium oxysporum f. sp. lycopersici in soil columns. *Ozone: Science & Engineering*, **28**, 125–127.

76 Ohlenbusch, G., Hesse, S. and Frimmel, F.H. (1998) Effects of ozone treatment on the soil organic matter on contaminated sites. *Chemosphere*, **37**, 1557–1569.

77 Eichenmüller, B. (1997) Entfernung polyzyklischer aromatischer Kohlenwasserstoffe aus Abwässern: Selektive Adsorption und Regeneration der Adsorbentien. Dissertation. TU Berlin, Germany.

78 Mourand, J.T., Crittenden, J.C., Hand, D.W., Perram, D.L. and Notthakun, S. (1995) Regeneration of spent adsorbents using homogeneous advanced oxidation. *Water Environment Research*, **67**, 355–363.

79 Teermann, I.P. and Jekel, M.R. (1999) Adsorption of humic substances onto β-FeOOH and its chemical regeneration. Trondheim, Norway, 24–26. June 1999 (Ed. H. Ødegaard), *Water Science & Technology*, **40**, pp. 199–206.

80 Kornmüller, A., Schwaab, K., Karcher, S. and Jekel, M. (2000) Oxidative regeneration of granulated iron-hydroxide adsorbed with reactive dyes. Proceedings of the International Conference "Oxidation Technologies for Water and Wastewater Treatment", Clausthal-Zellerfeld, May 28–31.

81 Kornmüller, A., Karcher, S. and Jekel, M. (2001) Cucurbituril for water treatment. Part II: ozonation and oxidative regeneration of cucurbituril. *Water Research*, **35** (14), 3317–3324.

82 Kasprzyk, B. and Nawrocki, J. (2002) Preliminary results on ozonation enhancement by perfluorinated bonded alumina phase. *Ozone Science & Engineering*, **24**, 63–68.

83 Kasprzyk, B. and Nawrocki, J. (2005) Catalytic ozonation of gasoline compounds in model and natural water in the presence of perfluorinated alumina bonded phases. *Ozone: Science & Engineering*, **27**, 301–310.

84 Kim, J. and Choi, H. (2002) modeling *in situ* ozonation for the remediation of nonvolatile PAH-contaminated unsaturated soils. *Journal of Contaminant Hydrology*, **55**, 261–285.

85 Bader, H. and Hoigné, J. (1981) Determination of ozone in water by the indigo method. *Water Research*, **15**, 449–456.

86 Williams, M.E. and Darby, J.I. (1992) Measuring ozone by indigo method: interference of suspended material. *Journal of Environmental Engineering*, **118**, 988–993.

87 Mantzavinos, D. and Psillakis, E. (2004) Review–enhancement of biodegradability of industrial wastewaters by chemical oxidation pre-treatment. *Journal of Chemical Technology and Biotechnology*, **79**, 431–454.

88 Escobar, I.C. and Randall, A.A. (2001) Assimilable organic carbon (AOC) and biodegradable dissolved organic carbon (BDOC): complementary measurements. *Water Research*, **35** (18), 4444–4454.

89 Khan, E., Babcock, R.W., Suffet, I.H. and Stenstrom, M.K. (1998) Method development for measuring biodegradable organic carbon in

reclaimed and treated wastewaters. *Water Environment Research*, **70** (5), 1025–1032.

90 Page, D. and Dillon, P. (2007) Measurement of the biodegradable fraction of dissolved organic matter relevant to water reclamation via aquifers. Water for a Healthy Country National Research Flagship report. CSIRO, Canberra, http://www.clw.csiro.au/publications/waterforahealthycountry/2007/wfhc_Measurement BiodegradableFractionDissolved OrganicMatter.pdf (accessed December 2008)

91 Libra, J.A. and Sosath, F. (2003) Combination of biological and chemical processes for the treatment of textile wastewater containing reactive dyes. *Journal of Chemical Technology and Biotechnology*, **78**, 1149–1156.

92 Rice, R.G. and Browning, M.E. (1981) *Ozone Treatment of Industrial Wastewater*, Section 7 Biological Activated Carbon, Noyes Data Corp., Park Ridge, N.J., USA, pp. 332–371, ISBN 0-8155-0867-0.

93 Jekel, M.R. (1998) Effects and mechanisms involved in preoxidation and particle separation processes. *Water, Science & Technology*, **37**, 1–7.

94 Masschelein, W.J. (1994) Towards one century application of ozone in water treatment–scope, limitations and perspectives. Proceedings of the International Ozone Symposium "Application of Ozone in Water and Wastewater Treatment", Warsaw, Poland, May 26–27 (ed. A.K. Bín), pp. 11–36.

95 Bèle, C., Habarou, H., Ambonguilat, S., Croué, J-P, Djafer, M. and Heim, V. (2006) How to Optimize NOM Removal on Drinking Water Plants?: First Step: Making a Good Diagnosis Thanks to HPSEC. Water Practice & Technology 001:04 (2006), http://www.iwaponline.com/wpt/001/0082/0010082.pdf (accessed January 2008).

96 Yavich, A.A., Lee, K-H., Chen, K-C., Pape, L. and Masten, S.J. (2004) Evaluation of biodegradability of NOM after ozonation. *Water Research*, **38**, 2839–2846.

97 Kainulainen, T., Tuhkanen, T., Vartiainen, T. and Kalliokoski, P. (1994) Removal of residual organic matter from drinking water by ozonation and biologically activated carbon. Ozone in water and wastewater Treatment, Proceedings of the 11th Ozone World Congress Aug./Sept. 1993, Vol. 2, San Francisco CA, S-17-88-S-17-89.

98 Melin, E.S. and Ødegaard, H. (2000) The effect of biofilter loading rate on the removal of organic ozonation by-products. *Water Research*, **34**, 4464–4476.

99 Pitter, P. (1976) Determination of biological degradability of organic substances. *Water Research*, **16**, 231–235.

100 Tabak, H.H., Quave, S.A., Mashni, C.I. and Barth, E.F. (1981) Biodegradability studies with organic priority pollutant compounds. *Journal Water Pollution Control Federation*, **53**, 1503–1518.

101 Gilbert, E. (1987) Biodegradability of ozonation products as a function of COD and DOC elimination by example of substituted aromatic substances. *Water Research*, **21**, 1273–1278.

102 Scott, J.P. and Ollis, D.F. (1995) Integration of chemical and biological oxidation processes for water treatment: review and recommendations. *Environmental Progress*, **14**, 88–103.

103 Hemmi, M., Krull, R. and Hempel, D.C. (1999) Sequencing batch reactor technology for the purification of concentrated dyehouse liquors. *The Canadian Journal of Chemical Engineering*, **77**, 948–954.

104 Saroj, D.P., Kumar, A., Bose, P., Tare, V. and Dhopavkar, Y. (2005) Mineralization of some natural refractory organic compounds by biodegradation and ozonation. *Water Research*, **39**, 1921–1933.

105 Beltran-Heredia, J., Torregrosa, J., Dominguez, J.R. and Garcia, J. (2000) Treatment of Black-Olive wastewaters by ozonation and aerobic biological degradation. *Water Resaerch*, **34**, 3515–3522.

106 Benitez, F.J., Acero, J.L., Garcia, J. and Leal, A.I. (2003) Purification of cork processing wastewaters by ozone, by activated sludge, and by their two sequential applications. *Water Research*, **37**, 4081–4090.

107 Bijan, L. and Mohseni, M. (2005) Integrated ozone and biotreatment of pulp mill effluent and changes in biodegradability and molecular weight distribution of organic compounds. *Water Research*, **39**, 3763–3772.

108 Balçioglu, I.A., Saraç, C., Kivilcimdan, C. and Tarlan, E. (2006) Application of ozonation and biotreatment for forest industry wastewater, *Ozone: Science & Engineering*, **28**, 431–436.

109 Sangave, P.C., Gogate, P.R. and Pandit, A.B. (2007) Ultrasound and ozone assisted biological degradation of thermally pretreated and anaerobically pretreated distillery wastewater. *Chemosphere*, doi:10.1016/j.chemosphere.2006.12.052.

110 Arslan-Alaton, I., Dogruel, S. and Gerone, G. (2004) Combined chemical and biological oxidation of penicillin formulation effluent. *Journal of Environmental Management*, **73**, 155–163.

111 Cokgor, E.U., Arslan-Alaton, I., Karahan, O., Dogruel, S. and Orhon, D. (2004) Biological treatability of raw and ozonated penicillin formulation effluent. *Journal of Hazardous Materials B*, **116**, 159–166.

112 Arslan-Alaton, I. and Caglayan, A.E. (2005) Ozonation of procaine penicillin G formulation effluent Part I: process optimization and kinetics. *Chemosphere*, **59**, 31–39.

113 Arslan-Alaton, I. and Caglayan, A.E. (2006) Toxicity and biodegradability assessment of raw and ozonated procaine penicillin G formulation effluent. *Ecotoxicology and Environmental Safety*, **63**, 131–140.

114 Jochimsen, J.C. (1997) Einsatz von Oxidationsverfahren bei der kombinierten chemisch-oxidativen und aeroben biologischen Behandlung von Gerbereiabwässern, VDI-Fortschritt-Berichte Reihe 15 (Umwelttechnik) Nr. 190, VDI-Verlag Düsseldorf.

115 Kayser, R. (1996) Wastewater treatment by Combination of Chemical Oxidation and Biological Processes. Clausthaler Umwelt-Akademie, Oxidation of Water and Wastewater, Goslar 20–22 Mai 1996 (ed. A. Vogelpohl), CUTEC–Schriftenreihe Nr. 23, Clausthalet–Umwelttechnik–Institut GmbH, International Conference Oxidation Technology for Water and Wastewater Treatment, A. Vogelpohl (ed.).

116 Karrer, N.J., Ryhiner, G. and Heinzle, E. (1997) Applicability test for combined biological-chemical treatment of wastewaters containing biorefractory compounds. *Water Research*, **31**, 1013–1020.

117 Stern, M., Ramval, M., Kut, O.M. and Heinzle, E. (1995) Adaptation of a mixed culture during the biological treatment of p-toluenesulfonate assisted by ozone. 7th European Congress on Biotechnology Nice February, 19–23, 1995.

118 Stern, M., Heinzle, E., Kut, O.M. and Hungerbühler, K. (1996) Removal of Substituted Pyridines by Combined Ozonation / Fluidized Bed Biofilm Treatment. Clausthaler Umwelt-Akademie Oxidation of Water and Wastewater, Goslar 20–22 Mai 1996 (ed. A. Vogelpohl), CUTEC–Schriftenreihe Nr. 23, Clausthaler–Umwelttechnik–Institut GmbH, International Conference Oxidation Technology for Water and Wastewater Treatment, A. Vogelpohl (ed.).

119 Heinzle, E., Geiger, F., Fahmy, M. and Kut, O.M. (1992) Integrated ozone-biotreatment of pulp bleaching effluents containing chlorinated phenolic compounds. *Biotechnology Progress*, **8**, 67–77.

120 Heinzle, E., Stockinger, H., Stern, M., Fahmy, M. and Kut, O.M. (1995) Combined biological-chemical (ozone) treatment of wastewaters containing chloroguaiacols. *Journal of Chemical and Technical Biotechnology*, **62**, 241–252.

121 Stockinger, H. (1995) Removal of Biorefractory Pollutants in Wastewater by Combined Biotreatment-Ozonation. Dissertation ETH No. 11063, Zurich.

122 Steensen, M. (1996) Chemical oxidation for the treatment of leachate–process comparison and results from fullscale plants. Clausthaler Umwelt-Akademie: Oxidation of Water and Wastewater, Goslar 20–22 Mai 1996 (ed. A. Vogelpohl), CUTEC–Schriftenreihe Nr. 23, Clausthaler–Umwelttechnik– Institut GmbH, International Conference Oxidation Technology for Water and Wastewater Treatment, A. Vogelpohl (ed.).

123 Wiesmann, U., Sosath, F., Borchert, M., Riedel, G., Breithaupt, T., Mohey El-Dein, A. and Libra, J.A. (2002) Versuche zur Entfärbung und Mineralisierung des Azo-Farbstoffs C.I. Reactive Black 5; gwf-Wasser. Abwasser, **143**, 329–336.

124 Langlais, B., Cucurou, Y.A., Capdeville, B. and Roques, H. (1989) Improvement of a biological treatment by prior ozonation. Ozone Science & Engineering, **11**, 155–168.

125 Saupe, A. (1997) Sequentielle chemisch-biologische Behandlung von Modellabwässern mit 2,4-Dinitrotoluol, 4-Nitroanilin und 2,6-Dimethylphenol unter Einsatz von Ozon, VDI-Fortschritt-Berichte Reihe 15 (Umwelttechnik) Nr. 189, VDI-Verlag Düsseldorf.

126 Saupe, A. and Wiesmann, U. (1996) Abbau von nitroaromatischen Xenobiotika durch Ozonisierung und biologische Nachbehandlung. Acta hydrochimica et hydrobiologica, **24**, 118–126.

127 Saupe, A. and Wiesmann, U. (1998) Ozonization of 2,4-dinitrotoluene and 4-nitroaniline as well as improved dissolved organic carbon removal by sequential ozonization-biodegradation. Water Environment Research, **70**, 145–154.

128 Saupe, A. (1999) High-rate biodegradation of 3- and 4-nitroaniline. Chemosphere, **37**, 2325–2346.

129 Heinze, L., Brosius, M. and Wiesmann, U. (1995) Biologischer Abbau von 2,4-Dinitrotoluol in einer kontinuierlich betriebenen Versuchsanlage und kinetische Untersuchungen. Acta Hydrochimica et Hydrobiologica. **23**, 254–263.

130 Wiesmann, U., In Su, C. and Dombrowski, E-M (2007) Fundamentals of Biological Wastewater Treatment, Wiley-VCH Verlag GmbH & Co. KGaA, Weinheim, ISBN: 978-3-527-31219-1.

131 Andreozzi, R., Longo, G., Majone, M. and Modesti, G. (1998) Integrated treatment of olive oil mill effluents (OME): study of ozonation coupled with anaerobic digestion. Water Research, **32**, 2357–2364.

132 Jones, B.M., Saakaji, R.H. and Daughton, C.G. (1985) Effects of ozonation and ultraviolet irradiation on biodegradability of oil shale wastewater organic solutes. Water Research, **19**, 1421–1428.

133 Contreras, S., Rodriguez, M., Al Momani, F., Sans, A. and Esplugas, S. (2003) Contribution of the ozonation pre-treatment to the biodegradation of aqueous solutions of 2,4-dichlorophenol. Water Research, **32**, 2357–2364.

134 Levenspiel, O. (1999) Chemical Reaction Engineering, 3rd edn, John Wiley & Sons Inc., New York.

135 Moerman, W.H., Bamelis, D.R., Vergote, P.M. and Van Holle, P.M. et al. (1994) Ozonation of activated sludge treated carbonization wastewater. Water Research, **28**, 1791–1798.

136 Heinze, L. (1997) Mikrobiologischer Abbau von 4-Nitrophenol, 2,4-Dinitrophenol und 2,4-Dinitrotoluol in synthetischen Abwässern. Fortschr.-Ber., VDI Reihe 15 Nr. 167, VDI-Verlag, Düsseldorf 1997, ISBN 3-18-3-16715-8.

137 Gisi, D., Stucki, G. and Hanselmann, K.W. (1997) Biodegradation of the pesticide 4,6-dinitro-ortho-cresol by microorganisms in batch cultures and in fixed-bed column reactors. Applied Microbiology & Biotechnology, **48**, 441–448.

138 Grady C.P.L. Jr. (1985) Biodegradation: Its Measurement and Microbiological Basis. Biotechnology & Bioengineering, **27**, 660–674.

139 International Organization for Standardization (1997) ISO 15462. Water Quality. Selection of Tests for Biodegradability (Technical Report), International Organization for Standardization, Geneva, Switzerland.

140 Pagga, U. (1997) Testing biodegradability with standardized methods. *Chemosphere*, **35**, 2953–2972.
141 ISO (1993) DIN EN 29 888. *Verfahren zur Bestimmung der inhärenten biologischen Abbaubarkeit von Abwasserinhaltsstoffen und Chemikalien (Zahn-Wellens-Test)*.
142 Strotmann, U., Reuschenbach, P., Schwarz, H. and Pagga, U. (2004) Development and evaluation of an online CO_2 evolution test and a multicomponent biodegradation test system. *Applied and Environmental Microbiology*, **70** (8), 4621–4628.
143 Shiyun, Z., Xuesong, Z. and Doatang, L. (2002) Ozonation of naphthalene sulfonic acids in aqueous solutions, Part I: elimination of COD, TOC and increase of their biodegradability. *Water Research*, **36**, 1237–1243.
144 Yung-Chien, H., Hsiang-Cheng, Y. and Jyh-Herng, C. (2004) The enhancement of the biodegradability of phenolic solution using preozonation based on high ozone utilization. *Chemosphere*, **56**, 149–158.
145 Goi, A., Trapido, M. and Tuhkanen, T. (2004) A study of toxicity, biodegradability, and some by-products of ozonised nitrophenols. *Advances in Environmental Research*, **8**, 303–311.
146 Diehl, K., Hagendorf, U. and Hahn, J. (1995) Biotests zur Beurteilung der Reinigungsleistung von Deponiesickerwasserbehandlungsverfahren. *Entsorgungs Praxis*, 3/95, 47–50.
147 US Environmental Protection Agency (1999) Toxicity Reduction Evaluation Guidance for Municipal Wastewater Treatment Plants, EPA-833B-99-002, Office of Water, Washington, D.C.
148 Sosath, F. (1999) Biologisch-chemische Behandlung von Abwässern der Textilfärberei. Dissertation am Fachbereich Verfahrenstechnik. Umwelttechnik, Werkstoffwissenschaften der Technischen Universität Berlin, Berlin.
149 Zenaitis, M.G., Sandhu, H. and Duff Sh, J.B. (2002) Combined biological and ozone treatment of log yard run-off. *Water Research*, **36**, 2053–2061.
150 Kaludjerski, M. and Gurol, M.D. (2004) Assessment of enhancement in biodegradation of dichlorodiethyl ether (DCDE) by pre-oxidation. *Water Research*, **38**, 1595–1603.
151 Jochimsen, J.C. and Jekel, M. (1996) Partial oxidation effects during the combined oxidative and biological treatment of separated streams of tannery wastewater. CUTEC–Schriftenreihe Nr. 23, Clausthaler–Umwelttechnik–Institut GmbH, International Conference Oxidation Technology for Water and Wastewater Treatment, A. Vogelpohl (ed.).
152 Krull, R. (2003) *Produktionsintegrierte Behandlung industrieller Abwässer zur Schließung von Stoffkreisläufen*, Band 15, FIT-Verlag, Paderborn, ISSN: 1431-7230, ISBN 3-932252-20-9.
153 Gise, P. and Blanchard, R. (1998) *Modern Semiconductor Fabrication*, Englewood Cliffs, New Jersey.
154 Kern, W. (1993) *Handbook of Semiconductor Wafer Cleaning Technology*, Noyes Publication, New Jersey.
155 Reinhardt, K.A., and Ken W. (2007) Silicon Wafer Cleaning Technology, William Andrew, Norwich.
156 Kern, W. and Puotinen, D.A. (1970) Cleaning solutions based on hydrogen peroxide for use in silicon semiconductor Technology. *RCA Review*, **31**, 187–206.
157 Ohmi, T. (1998) General introduction to ultra clean processing, UCPSS, Tutorials, 20 September, Oostende Belgium.
158 Kern, W. (1999) Silicon Wafer Cleaning: A Basic Review, SCP Global Technologies, 6th International Symposium, May 11, Boise, Idaho.
159 Heyns, M.M., Anderson, N., Cornelissen, I., Crossley, A., Daniels, M., Depas, M., De Gendt, S., Gräf, D., Fyen, W., Hurd, T., Knotter, M., Lubbers, A., McGeary, M.J., Mertens, P.W., Meuris, M., Mouche, L., Nigam, T., Schaekers, M., Schmidt, H., Snee, P., Sofield, C.J., Sprey, H., Teerlink, I., Van Hellemont, J., Van Hoeymissen, J.A.B., Vermeire, B., Vos, R., Wilhelm, R., Wolke, K. and Zahka, J. (1997) New Process

developments for improved ultra-thin gate reliability and reduced ESH-impact, 3rd Annual Microelectronics and the Environment Forum, Semicon Europe, April 15.

160 Abe, H., Iwamoto, H., Toshima, T., Iino, T. and Gale, G.W. (2003) Novel Photoresist Stripping Technology Using Ozonated Water/Vaporized Water Mixture, *IEEE Transactions on Semiconductor Manufacturing*, **16**, 401–415.

161 Truscello, A. (2003) DIO$_3$ post ash cleaning implementation for significant cost savings. 3rd annual FSI Surface Conditioning Symposium.

162 Mertens, P., Vereecke, G. and Voss, R. (2006) Post-etch residue and photoresist removal challenges for the 45 nm technology node and beyond. *Semiconductor Fabtech*, **31**, 86–94.

163 Hattori, T. (2003) Implementation a single-wafer cleaning technology suitable for minifab operations. *Micro*, **21**, 49–59.

164 Claes, M., Röhr, E., Conrad, T., De Gent, S., Storm, W., Bauer, T., Mertens, P. and Heyns, M.M. (2001) Surface charakterisation after different wet chemical cleans. *Solid State Phenomena*, **76**, 67–70.

165 Onsia, B., Caymax, M., Conrad, T., De Gendt, S., Delabie, A. and Green, M. (2004) On the Application of a thin ozone-based wet chemical oxide as an interface for ALD high-k deposition. Proceedings in the 7th International Symposium Ultra Clean Processing of Silicon Surfaces, pp. 37–38.

Glossary of Terms

Symbols (*variables* and constants)   units[1]

a	specific (volumetric) area	m^{-1} ($m^2 m^{-3}$)
A	absorption	$mg\,l^{-1}$
A	total bubble surface area	m^2
A_E	surface area of electrode	m^2
A^*	specific ozone absorption	$g\,O_3\,g^{-1}\,DOC$
A'	frequency factor	–
$c(A)$	concentration of the compound A	$mg\,l^{-1}$
c^*	saturation concentration	$mg\,l^{-1}$
c_G	gas concentration (in reactor)	$mg\,l^{-1}$
c_{Go}	influent-gas concentration	$mg\,l^{-1}$
c_{Ge}	effluent-gas concentration	$mg\,l^{-1}$
c_L	liquid concentration (in reactor)	$mg\,l^{-1}$
c_{Lo}	influent-liquid concentration	$mg\,l^{-1}$
c_{Le}	effluent-liquid concentration	$mg\,l^{-1}$
d	diameter	m
D	depletion factor	–
D	diffusion coefficient	$m^2\,s^{-1}$
D_{appl}	applied ozone dose	$mg\,l^{-1}$
$D(O_3)$	consumed ozone dose or ozone consumption	–
$D(O_3)^*$	specific ozone consumption	–
d_B	bubble diameter	mm
d_P	pore diameter, pore size	µm
d_R	reactor diameter	m
E	mass-transfer enhancement factor	–
E_i	instantaneous mass-transfer enhancement factor	–
E_o	electrochemical potential under STP	V
E_A	activation energy	$J\,mol^{-1}$

1) Mass based units may as well be replaced by molar units.

Ozonation of Water and Waste Water. 2nd Ed. Ch. Gottschalk, J.A. Libra, and A. Saupe
Copyright © 2010 WILEY-VCH Verlag GmbH & Co. KGaA, Weinheim
ISBN: 978-3-527-31962-6

Symbols (*variables* and constants) units[1)]

f	frequency of applied voltage	s^{-1} [Hz]
F	dose rate or feed rate	$mg\,l^{-1}\,s^{-1}$
F^*	specific dose or feed rate	$mg\,l^{-1}\,s^{-1}$
$F(H_2O_2)/F(O_3)$	hydrogen peroxide/ozone dose ratio	$mg\,mg^{-1}$
g	gravitational constant	$m\,s^{-2}$
h	height	m
h_d	thickness of dielectric	m
Ha	Hatta number	–
H/D	ratio height/diameter	–
H	Henry's Law constant, with dimension	$atm\,l\,mol^{-1}$ [= atm mole fraction)$^{-1}$]
H_C	Henry's Law constant, dimensionless	–
I	current	A
I	ozone dose or ozone input	$mg\,l^{-1}$
I^*	specific ozone dose	$g\,O_3\,g^{-1}$ DOC
I_o	intensity before absorption cell	–
I_l	intensity after absorption cell	–
k	film mass-transfer coefficient	$m\,s^{-1}$
k'	reaction rate coefficient, pseudo-first order	s^{-1}
k	reaction rate constant, first order	s^{-1}
k	reaction rate constant, second order	$l\,mol^{-1}\,s^{-1}$
k_C	pseudo-1st-order ozone-decomposition rate constant	s^{-1}
k_d	2nd-order ozone-decomposition rate constant	$l\,mol^{-1}\,s^{-1}$
k_D	reaction rate constant for direct reaction of ozone	$l\,mol^{-1}\,s^{-1}$
k_G	gas film mass-transfer coefficient	$m\,s^{-1}$
$k_G\,a$	gas-phase volumetric mass-transfer coefficient	s^{-1}
k_L	liquid-film mass-transfer coefficient	$m\,s^{-1}$
$k_L\,a$	liquid-phase volumetric mass-transfer coefficient	s^{-1}
$K_L\,a$	overall mass-transfer coefficient	s^{-1}
k_R	reaction rate constant for hydroxyl radicals	$l\,mol^{-1}\,s^{-1}$
l	internal width of the absorption cell	m
l_R	reactor length	m
m	specific mass-transfer rate or mass flow rate	$mg\,l^{-1}\,s^{-1}$
$m_{PR}(O_3)$	production capacity of the ozone generator	$g\,h^{-1}$
$MW(O_3)$	molecular weight (48 for ozone)	$g\,mol^{-1}$
n	reaction order	–
n	molar flow rate	$mol\,l^{-1}\,s^{-1}$

Symbols (*variables* and constants) units[1]

Symbol	Description	Units
n_{STR}	stirrer speed	s^{-1}
N	mass-transfer flux	$mg\,m^{-2}\,s^{-1}$
$N_{L,abs}$	mass-transfer flux absorbed from gas into liquid	$mg\,m^{-2}\,s^{-1}$
$N_{L,bulk}$	mass-transfer flux from liquid film to bulk liquid	$mg\,m^{-2}\,s^{-1}$
$N_{L,film}$	mass-transfer flux reacted in film	$mg\,m^{-2}\,s^{-1}$
p	partial pressure	Pa
P	power	$kW, N\,m\,s^{-1}$
P_{abs}	pressure, absolute	Pa
P_G	pressure in gas	Pa
P_{gauge}	pressure, gauge	Pa
pK_a	dissociation constant	–
P_L	pressure in liquid	Pa
Q_G	gas-flow rate	$l\,s^{-1}$
Q_L	liquid-flow rate	$l\,s^{-1}$
Q_{LC}	cooling-water flow rate	$l\,s^{-1}$
r	reaction rate	$mg\,l^{-1}\,s^{-1}$
r_G	ozone-consumption rate in gas phase	$mg\,l^{-1}\,s^{-1}$
r_L	ozone-consumption rate in liquid phase	$mg\,l^{-1}\,s^{-1}$
$r(O_3)$	ozone-consumption rate in liquid phase	$mg\,l^{-1}\,s^{-1}$
$r_A(O_3)$	ozone-absorption rate in liquid phase	$mg\,l^{-1}\,s^{-1}$
$r(M)$	compound removal rate	$mg\,l^{-1}\,s^{-1}$
R	reaction factor	–
R_{CT}	constant defined by Hoigné / von Gunten	–
R_G	gas-phase resistance	s
R_L	liquid-phase resistance	s
R_T	total resistance	s
$\mathfrak{R}$	ideal gas law constant (8.314)	$J\,mol^{-1}\,K^{-1}$
s	solubility ratio	–
t	time	s
t_H	hydraulic retention time	s
t_R	reaction time	s
$t_\Sigma t_T$	total time	s
T	temperature	°C or K
T_L	temperature of or in liquid	°C or K
T_{LC}	cooling-water temperature	°C or K
U	voltage (across the discharge gap, peak volts)	V
V_B	bubble volume	m^3
V_G	gas volume	m^3
V_L	liquid volume	m^3
V_n	molar volume	$l\,mol^{-1}$

Symbols (*variables* and constants)

		units[1]
w	width (of gap in ozone generator)	m
v_S	superficial gas velocity	m s^{-1}
y	mole fraction in gas phase	–
$y(O_3)$	yield of ozone (in ozone generation)	gO$_3$ kWh$_{el.}^{-1}$
$Y(O_3/M)$	ozone yield coefficient	gO$_3$ g^{-1}M or DOC?
z	stoichiometric coefficient	–

Greek Alphabetic

		unit
α	*alpha* factor	–
β	hydroxyl-radical initiating rate	–
δ	width of film	m
ε	extinction coefficient	l mol^{-1} cm^{-1}
ε	dielectric constant	?
ε	porosity	–
ε_L	liquid hold-up	–
$\eta(M)$	degree of pollutant removal	–; %
$\eta(O_3)$	ozone-transfer efficiency	–; %
λ	wavelength	nm
μ	ionic strength	mol l^{-1}
ν	kinematic viscosity	kg m^{-1} s^{-1}
θ	temperature correction factor	–
ρ	density	kg m^{-3}
σ	surface tension	N m^{-1}
σ	conductivity	S m^{-1}
τ	half-life of the reaction (sometimes $t_{1/2}$)	s
X	tortuosity	–
Ω	oxidation competition value	–

Dimensionless Numbers

Bo	Bodenstein number
Re	Reynolds number
Sc	Schmidt number
Si*	coalescence number
σ*	dimensionless surface tension

Abbreviations

AC	activated carbon
ACGIH	American of Conference of Governmental Industrials Hygienists
ADM	axial dispersion model
Alk	alkalinity
AOC	assimilable organic carbon

Glossary of Terms

AOP	advanced oxidation processes
APM	ammonium hydroxide, hydrogen peroxide, DI water mixture
B	biological stage, bioreactor
BAC	biological-activated carbon
BC	bubble column
BC	biological-chemical (process)
BDOC	biodegradable dissolved organic carbon (DOC)
BOD	biochemical oxygen demand
BOM	biodegradable organic matter
C	chemical stage
CA	catechol
CB	chemical-biological (process)
CBP	chemical-biological process
CFD	computational fluid dynamics
CFSTR	continuous-flow stirred-tank reactor
CIP	clean in place
CL	chemiluminescence
CMF	completely mixed flow
COD	chemical oxygen demand
CFSTR	continuous flow stirred-tank reactor
DBD	dielectric barrier discharge
DBDOG	dielectric barrier discharge ozone generator
DBP	disinfection by-products
DBPFP	disinfection by-product formation potential
DC	direct current
DCDE	dichlorodiethyl ether
DCP	dichlorophenol
DHF	diluted hydrofluoric acid
DI	deionized water
DOC	dissolved organic carbon
DPD	N,N-diethyl-1,4 phenyldiammonium
DW	drinking water
DWTP	drinking-water treatment plant
EBCT	empty-bed contact time
EC	effective concentration
ED	electrical discharge
ED	effective dose
EDC	endocrine-disrupting chemicals
EfOM	effluent organic matter
EL	eletrolysis, electrolytic
EDOG	electrical-discharge ozone generator
ELOG	electrolytic ozone generator
EPA	Environmental Protection Agency
ETBE	ethyl-*tert*-butylether
FA	formic acid

FAD	formaldehyde
FDA	Food and Drug Administration
FEOL	front end of line
FIA	flow injection analysis
GA	glyoxylic acid
HPM	hydrochloric acid, hydrogen peroxide, DI water mixture
HPYR	2-hydroxypyridine
HY	hydrochinone
I	initiator, intermediate
IC	integrated circuit
IMEC	Interuniversity Microelectronic Center
IOD	instantaneous ozone demand
LC	lethal concentration
LD	lethal dose
LOX	liquid oxygen
M	micropollutant, compound, substrate
M	molar (mol l^{-1})
MA	maleic acid
MAK	maximal allowable working concentration
MCL	maximum contaminant level (of bromate $10\,\mu g\,l^{-1}$, USEPA, 1998)
MUA	muconic acid
MEP	methyl-pyridine
MLR	multiple linear regression
MTBE	methyl-*tert*-butylether
MWWTP	municipal waste-water treatment plant
NIOSH	National Institute for Occupational Safety and Health
NOM	natural organic matter
NTP	normal temperature and pressure ($T = 273.15\,\text{K}$, $P = 10^5\,\text{Pa}$)
NTU	number of turbidity
OA	oxalic acid
ODS	octadecyl silica gel
OEBAC	ozone-enhanced biological-activated carbon
OME	oil-mill effluent
OUR	oxygen uptake rate
OSHA	Occupational Safety and Health Administration
OTE	ozone-transfer efficiency
P	promoter, product
PAH	polyaromatic hydrocarbon
pCBA	*para*-chlorobenzoic acid
PCE	tetrachloroethylene
PEL	permissible exposure limit
PER	tetrachloroethene
PCP	pentachlorophenol
PFA	perfluoralkoxy
PFR	plug-flow reactor

Ph/PH	phenol
PMMA	polymethylmethacrylate
PSA	pressure swing adsorption
POTW	publicly owned treatment work
PTFE	polytetrafluoroethylene
PVA	polyvinylalkoxy
PVC	polyvinylchloride
PVDF	polyvinylidenfluoride
RCA	Radio Company of America
S	scavenger
SAC	spectral absorption coefficient
SBR	sequencing batch reactor
SC	standard clean
SOM	sulfuric acid, ozone mixture
SPE	solid-polymer electrolyte membrane
S_{PER}	selectivity
SPM	sulfuric acid, hydrogen peroxide, DI water mixture
ss	steady state
STP	standard temperature and pressure (T = 273.15 K; P = 1.013 × 10^5 Pa)
STPR	stirred photochemical reactor
STR	stirred-tank reactor
SUVA	specific ultraviolet absorbance
TBA	*tert*-butanol
TCE	trichloroethylene
TDS	total dissolved solids
THM	trihalomethane
TIC	total inorganic carbon
TOC	total organic carbon
TP	tap water
UPW	ultrapure water
UV	ultraviolet
VPSA	vacuum pressure swing adsorption
WW	waste water
WWTP	waste-water treatment plant

Index

a

Acid-base equilibrium (pKa) 14–15, 20, 217, 246
Actinometry 273
Activated carbon 38, 273–275, 294, 303
Adsorption 277–278, 301–303
Advanced oxidation processes (AOPs) 13, 20–24, 43–44, 45, 205, 212, 215–219, 268–279, 297, 270–298, 299, 316
– comparison 21–23, 271–272
– definition 269–270
– chemical AOPs 269–273
– – experimental design 272–273
– – existing processes 269–272
– – principles and goals 269
– catalytic ozonation 273–281
– – heterogeneous 268–269, 274–275
– – homogeneous 268–269, 274
– – experimental design 280–281
– – existing processes and current research 275–280
– – principles and goals 273, 275
– principles and goals 269
Agitated cell 136, 137, 138, 197
Alkalinity 9
Alpha factor 183–185, 187, 286
Ambient air ozone monitor 8, 114, 153, 155
Amperometric method 156, 296
Analysis of error and sensitivity 197–199
Analytical methods 147–154, 156
– ambient air ozone monitor 8 154, 156
– amperometric method (*see* Electrode, ozone)
– biodegradability 320, 321
– chemilumenescence 150–151
– detection limit 153
– DPD method 150, 153
– indigo method 149, 153, 156, 297

– interferences 153
– iodometric method 147, 153
– ozone electrode 151, 153, 156, 297
– practical aspects 152
– in three-phase systems 297–298
– UV-absorption 147–148, 153
– visible light absorption 149
Applications 8–9
– advanced oxidation processes 20–24, 267–279
– agriculture 32, 34
– coagulation 28
– chemical-biological processes 297–325
– – type 0 300, 306–314
– – type 1 300, 301, 304–307, 310, 312–313, 321
– – type 2 300, 301, 310–313
– – type 3 300, 301, 310–313
– combined processes 106, 218, 267–334
– differences between 71–73, 229–237
– excess sludge treatment 55–56, 290
– food processing 30, 34
– full-scale drinking water treatment 37, 40–45
– full-scale waste water treatment 114, 123–126, 134–135, 250, 302–304
– in countries
– – China 56
– – Europe (EU) 28–30, 47–48, 56
– – Germany 28, 52–53, 54, 57
– – USA 29, 30, 47–48, 56, 57
– in the gas phase 32–33
– in the liquid phase 33–35
– industrial processing 34, 35, 51, 56–57
– medical 32, 34
– overview 28–30, 31–35
– ozonation and biodegradation 28, 299–328
– particle separation 34, 44–46

- pulp bleaching processes 35, 51, 52, 185, 309, 313, 318
- filtration 28, 37, 42, 44, 281, 286, 315
- sedimentation 28, 55, 286
- semiconductor industry 116–117, 119, 124, 143, 329–336
- swimming pools 10, 34, 40
- synthesis of organic chemicals 31
- three-phase systems 155, 278–279, 289–296

Aquatector® 139, 143
Arrhenius' law 217
Assessment of results 89–93
Assimilable organic carbon 38, 75, 298

b

Bacterial regrowth 10, 42–43, 275, 279–280, 292, 300–304, 319
Batch (see also Semibatch) 71–73, 81–106, 131, 144–146, 165, 179, 184–187, 190–197, 208, 213–214, 270–271, 290, 305–326
Bicarbonate 17, 213, 218–219, 279–289
- radical 218, 279–289
Bioassay (see also Toxicity testing) 6, 319
Biodegradability 53–55, 75, 292, 305–306, 318–319
Biodegradable organics 38, 42–45, 48–55
Biodegradable organic carbon (BDOC) 42–43, 304
Biological process (see Ozonation and biodegradation)
Biological oxygen demand (BOD) 300, 310, 316–317, 318–319
Biomass adaptation 314, 316
Bodenstein number 132, 139
Bromate 9–11, 18–19, 37, 39, 43, 48, 243
Bromide 9–10, 18, 37, 43, 48, 208, 243
Bromine 18, 206
Bubble coalescence 142, 271, 283–285
Bubble column 47, 132–133, 135–136, 142–143, 246
- equilibrium concentration in 167–168
- mass transfer coefficient 135–136, 163–167, 173–197, 285
- material 141
Buffer solution 16, 135, 171–172
- phosphate 209, 213, 220, 279
Bunsen coefficient 168–169
By-products
- disinfection 9–10, 28–34, 36–39, 47–48, 267
- ozonation 7, 9–11, 48, 231–233, 300–301, 316
- reduction of 10–11, 38–39

c

Carbonate 17, 75, 142, 171, 218–219, 246–250
- radical 246–250
Clogging
- - calcium carbonate 52, 278
- - calcium oxalate 52, 142, 277–278, 280
Catalysts 275–283
Chain reaction 14–17, 20–22, 208, 215, 217, 220
Chemical-biological processes 297–326
Chemical kinetics (see also Reaction kinetics) 181–182, 215–221
Chemical model 227, 231–242, 251–252, 258–259
Chemical oxygen demand 75–77, 298, 270, 298, 306, 318–319
Chemilumenescence 150–151
Cleaning processes 327–334
- conventional (RCA) 330–332
- improvements (IMEC UCT cleaning) 331–332
- wet and dry 330
Collision partner 121–123, 125
Combined processes (see also Multi-stage treatment; Applications) 106, 218, 267–334
Competition kinetics 213–215
Computational fluid dynamics (CFD) 31, 39, 134, 250
Continuous-flow 82–85, 95–99, 131, 144–146, 187–195, 312–313, 317–323, 326
- determination of mass transfer coefficient 187–195, 317–323
- equations 82–85, 95–99
- stirred tank reactor (see also Stirred tank reactor) 116, 132, 136–147, 184, 188–189, 317
Cooling system 30, 124–125, 129
Corona discharge (see Ozone generator) 121–124
Corrosion resistance 114–117
Criegee mechanism 17
Costs 292, 302–304, 313, 314
- capital 29–30, 57
- efficiency 32, 35–36, 56–59
- operating 30, 49, 52, 54–55, 58
ct-value (see Disinfection)
Cyanide (see also oxidation of) 48–49

d

Data evaluation 72, 81–83
- methods 81–85, 86–87

Decomposition of ozone 16–17, 21, 152, 169, 180, 187, 191, 196, 218–219, 230–235, 237–249
Destruction of ozone 31–34, 42–46
- in off-gas 116
- in samples 156
Depletion factor 178–182, 256–257
Dielectric Barrier Discharge (*see also* Ozone generator) 116–120
Diffusers 46–47, 133–134, 142–143
Diffusion coefficients 187–188, 286
- in solvents 289
- in water (for O_2 and/ or O_3) 168–170
Discharge standards 29, 47, 53
Disinfection 47–49
- by-products 9–10, 29, 36–37, 39, 42–43, 47–48, 304–305
- *ct-value* 31, 34, 36–39, 40
- in drinking water treatment 9–10, 28, 37–39
- in waste water treatment 47–48
Dissolved Organic Carbon 209, 219–220, 261, 398, 304–315, 324–325
- overall degree of removal 299, 302, 324, 325
Doping (of feed gas) 116, 123, 125
DPD method 150, 153
Drinking water 208, 212, 214–216, 219–220, 239–251
- disinfection 28–29
- humic acid 220–221
- kinetic regime 180–182, 212
- mass transfer 208, 212–213, 215–217, 219–220
- micropollutants 214, 218, 220, 239–251
- legal regulations 29
- treatment 29, 36–46, 220, 239–251, 297, 301–304
- treatment plants (full-scale) 36–46, 115, 123–126, 134–135, 138–143
- in three-phase systems 284–289
- in water 163–167, 173

e

Economical aspects 56–59, 302
Ecoclear® 50, 277–279
Ecotoxicity (*see* Toxicity)
Ecotoxicology (*see* Toxicology)
Electrical discharge ozone generator (*see* Ozone generator, electrical discharge)
Electrode
- in Electrical Discharge Ozone Generators 120–125

- oxygen 192
- ozone 148–154, 156, 280
Electrolytical ozone generator (*see* Ozone generator, electrolytic)
Empirical correction factor 182–184
- *alpha factor* 156, 183–185, 187, 254, 286, 295
- *theta factor* 183
Empty bed contact time 304, 319
Endocrine disrupting chemicals 44, 52, 57–58
Energy consumption 29–30
- specific 119, 124, 126, 129, 141
Energy efficiency (*see* Energy consumption, specific)
Enhancement factor 158, 175–182, 232–234, 251–263
Equilibrium concentration 167–175, 189–191
- Bunsen coefficient 168
- Henry's 167–175, 189–191, 287–288
- partition coefficient 168, 284–289
- solubility ratio 168–170
Equipment 113–145
- ancillary 78–81, 101–103
Etching 329
Experimental design 69–111
- checklists for 105–108
- in advanced oxidation processes
- – chemical AOPs 268–271
- – catalytic ozonation 280–282
- in ozone and biodegradation 316–328
- in semiconductor industry 126, 143, 333–334
- in three-phase systems 282–284, 296
- process of 69–93
Experimental error 197, 199
Experimental goals (*see* Goals)
Experimental procedure 70, 72, 85, 145–146
- batch experiments (*see* Batch)
- CBP-processes 320–323, 325, 326
- continuous-flow experiments 85, 146
- determination of direct reaction rate 180–182, 197–199
- determination of mass transfer coefficient 192, 194–199
- determination of reaction order 208
- process combinations 146–147
Experimental set-up 113–162
- analytical methods 147–154
- design 73–80
- material 113–118
- mode of operation 78, 94–95, 144

– ozone generator 77–79, 113–152
– reactor 94–104, 130–145
– safety aspects 152–155
Explosive gas mixtures 154–155
Extinction coefficient ε
– hydrogen peroxide 270–271
– ozone 147, 273

f

Feed gas
– doping 116, 123
– preparation 125–127
– type of 30, 125
Fick's first law of diffusion 164
Film mass transfer coefficient 164–167
Film theory 163–173, 285
Film
– laminar gas 165
– laminar liquid 164, 175
Full-scale applications 115–117, 142–143, 174, 182, 252
– drinking water 28–30, 36–46, 134, 142, 284, 302–304
– material 116
– ozone generator 28, 30, 119–120, 123–125
– waste water 35, 46–56, 277–280, 290, 304–305, 312–313

g

Gas 32–33
– contactor (see diffuser)
– diffuser 135–136, 141–143
– explosive mixtures 154–155
Gas hold-up 182
Gas/liquid interface 173–176, 197, 287
Generator (see Ozone generator)
Goals 34, 73–74
Groundwater 29, 33, 245, 270–271

h

Henry's Law constant 109, 167–175, 189–191, 287–288
Heterogeneous systems 130, 135–138, 181, 196, 214, 268–269, 274–275
History 27–31
Homogeneous systems 130, 138–141, 181, 213, 243, 268–269, 274
Humic acid 16, 220–221, 292
Hydrazine 194–195
Hydraulic retention time 81, 85, 134, 193, 195, 318–319
Hydrodynamics 31, 39, 82, 85, 130–141, 182–186, 226–228, 250–264

Hydrogen peroxide (see also Advanced oxidation processes) 20–24, 216, 218–219, 239–251
– and ozone 20–24, 216, 219, 246–251, 268–273
– and UV-radiation 22–24, 212, 216
– dose ratio 216, 250, 270
– extinction coefficient (ε_{254} nm) 22
– feed rate 216, 248
– in-situ formation 216
– photolysis 21–24, 271
Hydroxyl radical (see Radical hydroxyl)

i

Ideal gas law 217
Indigo method 149–150, 295
Influence on reaction rate 16–24, 205–217
– inorganic carbon 75, 218–220
– pH 18, 20–21
– organic carbon 220–221
– oxidants 216–217
Inhalation of ozone 7–8
Inhibitor (see Scavenger)
Initial reaction rate method 210
Initiator 13, 16, 20, 218, 227, 231, 234, 238–250
Inorganic carbon (see Total inorganic carbon)
Interface (see Gas/liquid interface)
Interfacial surface area 136, 167, 176, 181, 185–186
Intermediate ozonation 29, 37, 40
Iodometric method 147, 150
Ion implantation 327–332
Ionic strength 75, 157, 169–172, 193

k

$k_L a$ (see Mass transfer coefficient)
Kinetic regime 145, 158, 175–182, 194, 196–198, 212, 226, 229–240, 250–261
– effect on determination of mass transfer coefficients 180–182
– fast reaction 151, 156, 175, 176–180, 197, 208, 210, 213, 217, 230–231, 235
– instantaneous reaction 175, 176–180, 181–182, 196–199, 214, 231, 235
– moderate reaction 175, 177–181, 233, 235, 251
– slow reaction 175, 176, 178, 180–181, 195, 208, 210, 230–231, 235, 239
Kinetics of reaction (see Reaction kinetics)

l

Lab-scale application 113–118, 137
– material 113–118
– ozone generator (*see also* Experimental design) 119–120, 126–127, 130
Lambert-Beer's law 147–151
Landfill leachates 49–53, 277–279
Large installations 28–29, 35, 57
Legislative regulations (*see* Regulations)
Liquid oxygen 31, 123, 125
Logarithmic concentration difference 189–190

m

Mass balance 82–85, 97, 188–189, 191–196, 225, 227, 232–263
Mass transfer 72, 117, 130–134, 141–143, 156–158, 163–200, 226–264, 286–289
– and electric discharge ozone generator 127
– bubble coalescence 142, 158, 171, 174, 183–187
– coefficient 91, 133, 137, 156, 158, 163–167, 172–173, 182–200, 252–263
– – bubble column 133, 135, 182, 189–190, 262
– effect of kinetic regime 180–182
– – determination 136, 156–158, 180–182, 186–200
– error and sensitivity analyses 197–198
– direct or indirect 187–188
– general experimental considerations 187–190
– inherent problems 199–200
– methods with mass transfer enhancement 196–199
– nonsteady state methods without mass transfer enhancement 191–194, 199
– steady state methods without mass transfer enhancement 194–196, 199–200
– membrane reactor 133, 139–141
– packed tower 133
– prediction 182–184
– plate tower 133
– stirred tank reactor 133, 137–138, 173–175, 182
– depletion factor 178–182
– driving force 164–166, 173
– empirical correction factor (*see also* Alpha factor) 156, 182–187
– enhancement 87, 157–158, 173, 175–182, 187, 191, 196–199, 226, 230, 232–237, 250–262
– film mass transfer coefficient 163–166, 176, 178–180
– film theory 164–166, 255–260, 285
– flux 166, 176, 178–180
– interdependencies with chemical reactions 178–182
– interface 163–167
– interfacial surface area 136, 167, 181–182, 185–186, 193, 197
– limitation 72, 93, 128, 158, 181, 208, 212, 228, 230, 237
– one phase 163–166
– over-all 166–167, 172–173, 180–182
– parameters that influence 173–175
– penetration theory 164–166
– rate 87, 135, 141–143, 176, 181–182, 208, 212–213, 215–216, 219, 230, 232–2491, 253
– – control 172–173, 180–182
– reaction plane 177, 181–182, 196
– resistance 163–167, 172–175, 284–287
– specific mass transfer rate 167, 176
– surface renewal 164–166
– surface tension 173–175, 182–186
– surfactants 158, 185–186
– three-phase 284–287
– tube reactor 133
– two film theory 166–167, 172–173
– two phase 166–167
– with chemical reaction 157–158, 175–182, 231–236, 250–262
Material 113–116, 128–129, 135–143, 155
– concrete 115–116
– ceramics 115–116, 121, 124
– dielectric 121–124
– fused alumina 115–116
– glass (quartz) 115, 117
– Kalrez® 115
– Kynar® 115
– PFA 115
– piping 114–116, 118
– Plexiglass 117
– PTFE 115, 118
– PVA 115
– PVC 115, 117, 138–141
– PVDF 114, 118
– Viton 115
– stainless steel 115–117
Material balance (*see* Mass balance)
Matrix (*see* water matrix)
Membrane
– solid phase (in electrolytic ozone generator) 128–129
Membrane reactor 115, 117, 139–141
– flat-sheet 117

– hollow-fiber 118, 139–140
– hydrophobic 117, 134
Measurement (*see* Analytical methods)
Mechanism (*see* Reaction mechanism)
Method (*see* Semi-empirical method based on)
Micropollutants 14, 34, 43–45, 52, 214, 218, 220, 235, 237–249
Mineralization 42–43, 52, 221, 251–260, 275–279
Mixing 95–97
Model (*see also* Modeling) 225–263
– chemical based on 228, 231–234, 241–249
– – hydroxyl radical initiating rate (ß) 246, 248–249
– – empirical selectivity for scavengers (SPER) 247
– – number of pollutants 251–262
– – observable parameters rate equation and experimental data 247–248
– – R_{CT}-concept (indirect measurement) 243–245
– chemical and physical 228
– development 228
– empirical 260
– mathematical 228, 234–240, 251–260
– onion 78–82
– semi-empirical 225
Modeling of ozonation processes (*see also* Model) 31, 225–263
– drinking water 7–11, 27–57
– general description 228–237
– interdependent chemical and physical processes 222, 251–263
– software 225, 236
– three-phase systems 284–287
– water type 226–237, 315
– waste water 234–241, 251–263
Mode of operation 94–95, 144–145, 184–189
– batch (*see* batch)
– continuous-flow 144–146, 186–187, 192–196
– semibatch (*see* batch)
Multi-stage treatment
– chemical-biological systems 28, 53, 146, 297–326
– combining continuous-flow/batch 320–323
– in drinking water treatment 36–37
– reduced ozone consumption in 310–315
– sequential or integrated processes 146, 301–315, 324, 325–326

n

Natural organic matter (*see also* Humic acid) 10, 35, 38, 42–43, 75, 89, 292, 302, 302–304
Natural water 19, 27, 39, 54, 71, 76, 221, 240–243, 245, 268
Numerical methods 236, 254

o

Odor
– control (deodorization) 27, 33, 42, 45, 48
– compounds 32–33, 45
– improvement 28, 33, 42, 48, 51
– of electricity (ozone) 27
Off gas treatment 30, 33
Oil emulsions 34, 283–289, 294–296
Ordinary differential equations 235
Organic carbon (*see* Dissolved organic carbon)
Organic compounds (*see* Oxidation of)
Oxalic acid 43, 252, 255, 262, 270, 274, 276, 278–279
Oxidation competition value 243
Oxidation of 13, 17–24, 225–266
– aliphatics 17, 270
– aromatics 17, 45, 214, 262, 298
– biosolids 55–56
– cyanide 48–49, 270
– halogenated aromatics 270
– inorganics 18, 32–34, 36, 40–42, 48–49, 218–219
– landfill leachate 49–53, 277–278, 311–313
– metals 327, 329–331
– micropollutants 14, 35, 43–44, 46, 49, 52, 57, 233–235, 245–246, 249
– nitroaromatics 54, 258, 262, 278, 288–289, 311–312
– organics 18, 31–32, 33–34, 42–45, 49–55, 258–260, 264, 268–270
– – dissociated 18
– – nondissociated 18
– particles 32–34, 44–47
– pesticides 44–45, 270–271, 302–303
– pharmaceutical compounds 290, 302, 310
– polyaromatic hydrocarbons (PAH) 45, 289, 291
– sludge 55–56, 290–292
– soil 270, 292–293
– solids 31–32, 41, 46, 48, 290–293
– solvents 33, 287–289
Oxidative regeneration 282, 284, 292–294, 296
Oxidation (ozonation)
– by-products (*see also* By-products) 44, 230, 243

– catalytic   53, 273–281
– efficiency improved   275–278, 291
– intermediates (*see* products)
– phases I–IV (in semi-batch)   229–230, 235, 258
– products (*see also* By-products)   229–234, 251–262
Oxide layer   327–328
Oxygen uptake rate (OUR)   315, 320
Ozonation and biodegradation   297–326
– biomass   316
– – adaptation   312–313, 316
– existing processes   301–314
– experimental design   314–326
– mineralization   298–303
– principles and goals   297–301
– reduction in operating costs   301–304, 306, 313–314
Ozone   13–24, 205–209, 212–213, 215–219, 226–263
– absorption specific   53, 87–89, 92–93, 145–146, 302, 305–312, 321–326
– and hydrogen peroxide   20–21, 216, 242–251, 270–275
– and UV-radiation   21–22, 212, 216, 218, 247, 249, 268–273
– consumption, specific   42–43, 57, 84–87, 89, 92–93, 143–144
– discovery of   27
– exposure (R_{CT})   242
– dose, specific   86–87, 91–93, 144–145
– installations   29, 30, 47–56
– molar weight   109
– physical properties   109–110
– solubility   109, 167–172
– – in water   170, 286
– – in solvents   286, 298
– consumption specific   84, 87, 92, 99–100, 102, 103–104, 108, 288, 290
– input specific   86–88, 92
– transfer efficiency   86–89, 92, 323, 326
– yield coefficient   50–51, 87, 89, 92, 298, 324
Ozone concentration   208–209, 213, 215–217, 225–263
– conversions   108–111
– equilibrium (*see* Equilibrium concentration)
– in gas   30, 120–123, 126, 143, 147–153, 229–230
– in liquid   130, 143, 147–153, 154, 156, 209, 213, 215–217, 228–237, 288, 333
Ozone consumption   84, 86, 99–100, 102–103, 290, 306–312, 315, 317, 321, 323, 325–326

– rate   87, 89, 92, 99, 108, 233–234, 237–261
– specific   42–43, 57, 84–87, 89, 92–93, 143–144
Ozone data sheet   108–110
Ozone decomposition   16–17, 21, 117–118, 120–123, 141, 156, 169, 171, 180–181, 186–187, 189, 191–192, 195–196, 208–209, 213, 215, 218–219, 230–236, 239–240, 256, 258
– rate   16–17, 156, 208, 213, 240
Ozone destructor   77–78, 80–81, 91, 102–103, 114, 116, 130, 152–155
– activated carbon   152, 295
Ozone dose   49, 53–55, 58, 86–87
– rate   86, 216, 237
– ratio   87, 216–217, 250
– specific   87, 92–93, 144–145, 299, 302, 305–312, 321–326
– transferred   38, 86
Ozone enhanced biological activated carbon (OEBAC)   48, 52, 302–304
Ozone extinction coefficient (ε254 nm)   22, 109, 148, 270
Ozone feed rate   86, 91, 101, 103, 248, 310, 326
Ozone formation (*see* Ozone generator)
Ozone generator   29–31, 35, 56, 118–130
– cooling   30, 120, 124
– dielectric barrier discharge (*see* electrical discharge)
– electrical discharge   27–28, 118–127
– – chemistry   121–123
– – engineering and operation   123–125
– electrolytic   117–130, 138
– – material   128–129
– energy consumption   30
– feed gas   30–31, 118–120, 121–127
– silent discharge (*see* electrical discharge)
Ozone measurement (*see also* Analytical methods)   28, 147–152
Ozone modeling (*see* Modeling)
Ozone production
– by lightening   27
– by radiation chemistry   28, 119
– capacity   119–120, 126–127, 129
– photochemical   119
– thermal (light arc ionization)   119

p

Packed column   133
Particle removal   32, 44–46, 55–56
Partition coefficients (*see also* Henry's Law constant)   109, 167–175, 189–191

Pathway of reaction   225–263
- direct   17–20, 205, 207, 212–220, 228
- indirect (radical)   13–17, 205, 207–208, 212–216, 220, 228, 231, 234–235, 237
Penetration theory   164–166
Pathogenes   32, 34, 36, 291–293
Peroxone   42, 268, 310
pH   18, 20–21, 145, 157, 171–172, 187, 191, 217–219, 225–263
Phosphate   16, 171, 209, 213, 220
Photolithography   328–329
Photolysis   21–24, 247, 270–271
Photomask   328
Photoresist   328–329
Plate tower   131
Plug-flow   80, 94, 96, 99, 131–134, 140, 138, 250, 317, 326
Pre-ozonation   28–29, 37, 40, 42, 45–46, 284, 300, 303, 306
Probe lag effects   192–193
Post-ozonation   28–29, 37–38, 40, 54, 300, 306–307
Process control   30–31, 102, 104
Promoter   15–16, 20, 209, 219–221
Pseudo-first order   236

r

Radical   13–14, 207–209, 213–221, 225–263
Radical chain   13, 15, 22, 208, 215, 219–220, 225–263
initiator   13, 16, 218, 237–250
- inhibitor   15, 214, 237–250
- promoter   15–16, 20, 209, 219–221, 237–250
- reactions   13, 15, 22, 231–237
-- nonselective   13, 16, 20, 268, 303
- termination   13–15, 22, 237–250
Radical hydroxyl (OH°)   13–24, 205–208, 214–221, 225–263, 274
- concentration   24, 236–250
- exposure (OH-ct)   242–245
- formation by UV-radiation   21
- initiating rate (ß)   245–247, 249
- nonselective   268, 303
- reaction rate constants   13–24, 205–221, 237–250
- stoichiometric yield   24
Radical   225–263
- hydroperoxyl (HO°)   14, 244
- organic (R°)   14–15, 220
- ozonide anion ($O_3^{°-}$)   14, 22, 221, 244
- peroxy (ROO°)   14–15
- superoxide anion ($O_2^{°-}$)   14, 22, 221, 244
R_{CT}-concept   241–243

Ratio of exposures (*see* R_{CT}-concept)   241–243
Reaction   205–208, 213–215, 220–221, 225–263
- direct   17–20, 205–208, 213–215, 220–221, 228–232, 234, 252–261, 302
- indirect   13–17, 205–208, 213–215, 220–221, 228–232, 234, 237–250, 302
- liquid film   208, 230, 258
- nonelementary   206–207
- nonselective   13, 16, 20, 214, 268, 303
Reaction kinetics   205–221
- competition method   213–215
- influence of oxidants   216–217
Reaction mechanism   13–20, 231
- direct   17–20, 226–228, 234–236
- indirect   13–17, 226–228, 234–236
Reaction order   205–211
- determination of   205–211
-- halflife method   210
-- initial reaction rate method   210
-- trial and error method   210–211
- first   214
- nth   207, 260
- pseudo-first   181, 207–208, 214, 236, 260
- second   208, 234–235, 239–240, 251, 257, 259
- total   206
Reaction pathway   13–20
- direct   17–20
- indirect (radical)   13–17, 226, 231, 234
Reaction rate   225–263
- equations   97, 206, 234–235, 238, 254, 256
- parameters that influence   215–221, 234–250
Reaction rate constant   212–215, 225–263
- direct reactions   234–237, 252–253, 255–256
-- determination of   156, 180–182, 197–199, 213–215
- indirect reactions   234–237
-- determination of   214
Reaction rate coefficient   238, 260
Reaction regime (*see* Kinetic regime)
Reaction system (*see* Reactor)
Reactor   130–147
- agitated cell   136–138
- batch (*see* Batch; *also* Semibatch)   94–95, 97–99, 144–146, 184–185
- bubble column   47, 132–133, 135–136, 139, 142, 182, 189–190, 227, 258
- completely mixed   82, 91, 96 132–133
- continuous-flow stirred tank (*see* Continuous flow stirred tank reactor)

- design   100–104
- directly gassed   130, 135–138
- fixed bed   276–278, 291, 301, 318, 322–323
- for drinking water treatment   132–134
- gas diffuser   135–137, 141–143
- geometry   174
- heterogeneous system   94, 130–131, 135–138, 181, 196, 214
- homogeneous system   94, 130–131, 138–141, 181, 213
- ideally-mixed   82, 95–97, 131–133, 136, 138, 188–189
- indirectly gassed   130, 138–139
- material   114–118, 135–141
- membrane   130, 133, 139–141
- non-gassed   130
- plug-flow (see Tube reactor; also Plug-flow)
- –packed tower   132–133
- plate tower   133
- semibatch (see Semibatch)
- sequencing batch   95, 307, 313, 316
- static mixer   133, 139, 142–143
- stirred photochemical tank   272
- stirred tank (see Stirred tank reactor)
- tube (see Tube reactor; also Plug-flow)
- volume   97–101

Regulations
- air   8
- drinking water   29, 40
- food processing   29
- waste water   47, 52–54

Residence time (see Retention time)
Retention time   85, 94, 102–103

S
Safety aspects   152–154
- ambient air ozone monitor   8, 77, 114, 155
- ozone gas destruction   103, 114, 152–155
Sampling   152, 296
Scale-up   174, 182, 184, 231, 249–252
Scavenger   14–17, 20, 156, 213, 217–221, 227, 231, 234, 237–250
Schönbein   27–28
Selectivity term (S_{PER})   247–248
Semibatch   144–146, 227–230, 233, 250–253, 255–263
- determination of mass transfer coefficient   192, 195–197
- equations   85–86, 97–98
- ozonation Phases I–IV   229–231
Semiconductor application   114, 116–117, 119, 123–126, 143, 327–334

- cleaning processes   327, 329–333
- contaminants   327, 329–330, 332
- feed-gas doping   116, 123
- material   116–117, 124, 141
- principles and goals   329–330
- process design   333–334
- reactor   119
- ultra-pure water   116, 119, 139
- wafer production   327–334

Semi-empirical methods (see also Model semiempirical)   225, 237–250
- hydroxyl radical initiating rate (ß)   246–247, 249
- empirical selectivity for scavengers (SPER)   247–249

Siemens   28
Silent discharge (see also Ozone generator electrical discharge)   28
Silicon   327, 333
Sludge   55–56, 290–292
Sodium sulfite   193, 199
Soil   270, 291–292
Solubility (see Ozone solubility)
Solvents   283–284, 287–289, 296
Specific energy consumption   30, 119–120, 124, 126–129, 141

Spectral absorption coefficient
- at 436 nm   148
- at 254 nm   22, 154, 270

Static mixer   133, 139, 142–143
Steady-state   83–85, 97
- concentration of ozone   235–236, 239–240, 244–245
- concentration of hydroxyl radicals   239, 242–249
- experiments   258, 260

Stirred tank reactor (see also Continuous-flow stirred tank reactor)   82, 89, 95–99, 117, 131–134, 136–139, 142, 182, 227, 259, 262
- equilibrium concentration in   169, 171, 173, 175
- mass transfer in   133, 136, 171, 173–175, 182
- material   137

Stoichiometry   24, 157, 206–207, 216–217, 234–241, 252–253, 255, 257
Stoichiometric coefficient   14, 22, 205–206, 234, 239, 253, 257
Superoxide anion radical ($O_2^{\cdot-}$)   237, 244, 274
Surface renewal   164–166
Surface tension   173–175, 182–186
Surface water   27–29, 37, 44–45, 47, 55

Surfactants 51, 54, 145, 158, 185–186
Swimming pools 10, 27, 34–35, 37, 39–40

t

Temperature 262
– influence on ozone solubility 169–170
– in waste water experiments 169–172, 262
Tertiary butyl alcohol (TBA) 16, 213
Terminator (see Radical chain inhibition or termination) 156, 158, 184–185
Taste 28, 45, 271
Textile waste waters 52–54
Theta factor 183
Three-phase systems 155, 281–297
– definition 155, 281–282
– existing processes and current research 289–296
– – gas/water/solid 290
– – gas/water/solvent 288–290
– – change in the solids 290–292
– – change in compounds adsorbed on the solids 292
– – soil ozonation 292–293
– – regeneration of adsorbents 293–294
– experimental design 294–297
– mass transfer in 284–287
– principles and goals 282–284
– – gas/water/solid 284
– – gas/water/solvent 283–284
Total inorganic carbon 17, 20, 152, 154, 209, 237–250, 301, 303
Total organic carbon 20, 261
Toxicity 5–11, 76–77, 305–307, 310, 315, 318–320
– acute 5, 7–8, 320
– balance 320
– byproducts 5
– chronic 5, 7
– ecotoxicity 6, 319–320
– human 5, 7
– regulations 8
– testing 6, 319–320
Toxicology (see also Toxicity) 5–11
Treatment goals (see Goals)

Treatment train
– drinking water 28–29, 36–37, 40
– multistage 28, 36
– waste water 47, 55–56
Trial and error method 210–211
Tube reactor (see also Plug-flow) 96, 99
– material 131–135, 137–139, 248, 311–313, 326
– mass transfer coefficient 115, 117–118
Two-film theory 164, 171, 172, 285

u

Ultra pure water (see Water quality)
UV-radiation 21–24, 147–148, 153, 212, 216–218
– and hydrogen peroxide 22–24, 218, 268–273
– and ozone 21–22, 212, 216, 218, 247, 249, 268–273
– lamps 21–24
– photolysis 247, 270–273, 297

w

Wafer production 327–334
Waste water 226–237, 250–263
– alpha factor 156, 254
– mass transfer 156–158, 177–185, 208, 212
– – kinetic regime 145, 158, 177–182
– mode of reactor operation 180–182
– surfactant 145, 158, 280, 330
– treatment (full-scale) 34, 36, 46–59, 144–146, 158, 179, 182–184, 212, 215, 220–221, 250–261
Water
– constituents 75, 79
– define system 74–76
– effect of matrix 71–73, 227–240, 245, 304–314
Water reuse 34–36, 46, 48, 51, 57
Water quality
– high purity 119, 127, 129
– superpure 116, 128
– ultra pure 139, 331, 333